Y0-CVJ-346

STATISTICAL TABLES

By the Late Ronald A. Fisher

STATISTICAL METHODS FOR RESEARCH WORKERS. 1925-1958. Oliver and Boyd Ltd., Edinburgh.

THE GENETICAL THEORY OF NATURAL SELECTION. 1930. Oxford University Press. 1958. Dover Publications, New York.

THE DESIGN OF EXPERIMENTS. 1935-1960. Oliver and Boyd Ltd., Edinburgh.

THE THEORY OF INBREEDING. 1949. Oliver and Boyd Ltd., Edinburgh.

STATISTICAL METHODS AND SCIENTIFIC INFERENCE. 1956-1959. Oliver and Boyd Ltd., Edinburgh.

By Frank Yates

SAMPLING METHODS FOR CENSUSES AND SURVEYS. 1949-1960. Charles Griffin & Co. Ltd., London.

STATISTICAL TABLES

FOR BIOLOGICAL, AGRICULTURAL AND MEDICAL RESEARCH

BY THE LATE

SIR RONALD A. FISHER, Sc.D., F.R.S.

FORMERLY ARTHUR BALFOUR PROFESSOR OF GENETICS, UNIVERSITY OF CAMBRIDGE, AND GALTON PROFESSOR, UNIVERSITY OF LONDON. HONORARY MEMBER, AMERICAN STATISTICAL ASSOCIATION AND AMERICAN ACADEMY OF ARTS AND SCIENCES. FOREIGN MEMBER OF THE AMERICAN PHILOSOPHICAL SOCIETY. FOREIGN ASSOCIATE OF THE NATIONAL ACADEMY OF SCIENCES OF THE UNITED STATES OF AMERICA. FOREIGN MEMBER OF THE ROYAL SWEDISH ACADEMY OF SCIENCES AND THE ROYAL DANISH ACADEMY OF SCIENCES AND LETTERS. MEMBER OF THE PONTIFICAL ACADEMY OF SCIENCES

AND

FRANK YATES, C.B.E., Sc.D., F.R.S.

HEAD OF STATISTICAL DEPARTMENT, ROTHAMSTED EXPERIMENTAL STATION, AND OF THE RESEARCH STATISTICAL SERVICE OF THE AGRICULTURAL RESEARCH COUNCIL AND THE MINISTRY OF AGRICULTURE

SIXTH EDITION
REVISED AND ENLARGED

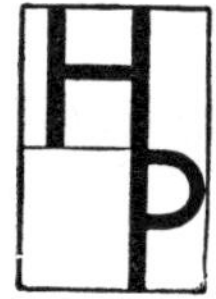

HAFNER PUBLISHING COMPANY INC.

NEW YORK

First Published	1938
Second Edition	1943
Third Edition	1948
Reprinted	1949
Fourth Edition	1953
Fifth Edition	1957
Sixth Edition	1963
Reprinted	1967

PRINTED AND PUBLISHED IN GREAT BRITAIN BY
OLIVER AND BOYD LTD., EDINBURGH

PREFACE TO FIRST EDITION

THE problems with which an active statistical department may have to deal require, if their solutions are to be made widely accessible, a great variety of special tables. A number of those in the present book are familiar to statisticians, and are already widely used. In presenting them in a convenient form, the opportunity has been taken to supplement them with a selection of others, chosen as likely also to be of value, and not accessible elsewhere. The volume is completed with a number of tables of standard functions of general utility. The experience of the authors of the problems arising in practical research, is the basis of the selection, from among those tables which from time to time have been computed for special purposes.

To perform their compilation, the main work has been to complete the tables in a form suitable for the general user, and, what is by no means easy, to ensure their entire accuracy. In addition we have been principally concerned with the design of printing and arrangement, which shall be suitable both for constant and for occasional users. Comments from both classes of reader will therefore be especially valued.

We are indebted to all workers at the Galton Laboratory, and at the Statistical Laboratory, Rothamsted, and especially to W. L. Stevens, for assistance in the completion of our task.

R. A. F.
F. Y.

August 1938

PREFACE TO FOURTH EDITION

THE second edition contained as new material Tables V1 and V2 for the test of significance of the differences between two means obtained by different methods; Table VIII1, which supplies solutions of the type of problem illustrated in Examples 1, 2 and 4, more direct than those previously available; and Table VIII2, which gives densities of organisms estimated by the dilution method. Additions to the third edition consisted of Table XI1 (due to Dr D. J. Finney) which gives the modified probit weighting coefficients necessary for dosage mortality tests involving deaths among controls; the inclusion of the 10 per cent. points of z and e^{2z} (based on the tables of the incomplete beta function of Miss Catherine M. Thompson, for which we are indebted to Prof. E. S. Pearson and Dr V. G. Panse) in Table V; Table XIV1, which gives a method of scoring linked data from intercrosses of double heterozygotes; the extension of Table XXIII of orthogonal polynomials for values of n' from 53 to 75 (due to Mr V. Satakopan); and the extension of Table XXVI of natural logarithms to cover the range ·100 to 1·00.

The combinatorial solutions of Table XVII were considerably extended and simplified in presentation in the second edition by the more extensive use of solutions of the cyclic type, and the corresponding section of the introduction was expanded to give an account of Youden's Squares, and of the method of utilising information, formerly discarded, from comparisons between blocks. Further new solutions were added in the third edition, and some further modifications have been made in the fourth edition.

Table VIII was improved in the third edition so as to allow for interpolation for values of the observed event above 10; Table XI, of weighting coefficients and probit values for adjustments of special accuracy, was recast in order to simplify its use; Table XXXIV of constants, weights and measures, etc., was slightly revised; and a small positive bias was removed from the 1 per cent. values of Table V1 for the significance of the difference between two means.

The following new material has been included in the fourth edition: Table VIII3, which provides a test for the existence of a periodic component; Table XIV2, which gives a set of functions termed "the segmental functions" which serve to specify the frequencies of non-recombinant and recombinant gametes in terms of the metrical positions of the centromere, a series of markers and the terminus of the chromosome arm; and Tables XXXIII1 and XXXIII2, which contain random permutations of 10 and 20 numbers respectively and are of use in the construction of experimental arrangements.

In addition a further table of the normal integral, Table VIII4, has been added. In this Table the deviation from the mean of the distribution is taken as the argument. The section of the introduction on dosage mortality tests involving a natural death rate has also been re-written. (We are indebted to Dr D. J. Finney for this.)

We have from time to time received a number of requests for the inclusion of simple tables, useful for special fields of work, or giving approximate tests or estimates, which can easily be derived from tables either given in this book or available elsewhere. Examples of such tables are: the factors required to obtain an estimate of the standard deviation from the mean range in samples from a normal distribution (derivable from Table XX—a note on this use of the table is included in the introduction); the multiple infection transformation $y = 100(1-e^{-x/100})$ (derivable from Table XXVI).

The inclusion of tables of this kind considerably increases the size of the book. Those giving approximate tests or estimates might also unduly encourage the use of such tests or estimates in fields in which they were not really appropriate. We have therefore considered it best to leave the compilation of such tables to the workers who require them. This has the advantage that they can then be arranged in the form which is most suitable for the particular purpose for which they are required.

For the convenience of owners of other editions, a list of errata has been included showing all the errors of any importance not previously discovered. Our thanks are due to those who have drawn our attention to these errors. Our especial thanks are also due to members of the Department of Statistics at Rothamsted, and particularly to Dr P. M. Grundy, for their pains in preparing the fourth edition for the press.

R. A. F.
F. Y.

16th June 1952

PREFACE TO FIFTH EDITION

THE increasing use of transformations in the analysis of frequency data and the recognition that all should be treated in the same manner as probits, has led us to include tables for the logit and complementary loglog transformations in the fifth edition (Tables XI and XII), together with tables for final adjustments (XI1 and XII1), in the same form as those already included for the probit and angular transformations. The tables for the angular transformation have been enlarged and an example illustrating their use has been given.

In view of the logical and mathematical interest of Behrens' test for the difference between two means, we have added a further table (VI1), based on exact formulæ, which gives values of d for a number of significance levels when n_1 and n_2 are both small and odd. We believe the later test published by Pearson and Hartley (*Biometrika Tables for Statisticians*, Table 11) to be erroneous.

A new table (XIII1) is provided for the application of the product-ratio method in linkage investigations with intercross data. The table of segmental functions (Table XIV) has also been amplified and extended.

A further two pages, primarily covering the range ·0100—·1000, have been added to the table of natural logarithms (Table XXVI), and values for even n from 30 to 70 have been included in the table of χ^2 (Table IV); an asymptotic formula for χ^2 for large n is also given in the introduction.

A few changes have been made in the numbering of the tables. These are indicated in the Contents. Of the new material, Table XII1 is based on one by D. J. Finney in "Statistical Method in Biological Assay" (Griffin, 1952). The revised values of Atomic Weights and certain physical constants, Table XXXIV, are based on Kaye and Laby (Longmans, 1956). Our thanks are due to the authors and publishers for permission to use this material. The remainder of the new tables have been computed *ab initio*, for the most part on the electronic computer at Rothamsted.

No errors in the tables of the fourth edition have been reported to us.

R. A. F.
F. Y.

November 1956

PREFACE TO SIXTH EDITION

SIR RONALD FISHER's sudden death in July 1962 delayed the completion of this edition, but most of the new material was agreed between us, and the new edition is, I believe, substantially as he would have wished to see it.

The main additions are a table (V1) giving the fiducial limits for a variance component (due to M. J. R. Healy), and solutions for balanced incomplete blocks with 11 to 15 replications (Tables XVII1 and XVIII1, due to C. R. Rao). In response to various requests, the table of the angular transformation (Table X) has been expanded to intervals of 0·1 per cent.

The discovery of Graeco-Latin squares of side $4s+2$ is a milestone in this branch of combinatorial analysis; an example of a 10×10 Graeco-Latin square is given in the Introduction, together with two sets of five 12×12 orthogonal squares. A note has been added on the derivation, by means of Table XXII, of cumulants of the binomial distribution. Examples 12 and 12·1 give an improved method of forming random permutations. Examples 3.1 and 7.2.1 are also new.

F. Y.

March 1963

CONTENTS

The Table numbers of the earlier editions, when different from those of the fifth and sixth editions, are given in brackets.

CONTENTS

INTRODUCTION

TABLES I TO V and VII. TESTS OF SIGNIFICANCE BASED ON THE NORMAL DISTRIBUTION

Tables I to V and VII constitute a group of tables, based on the normal distribution, and now widely used in making tests of significance. The common uses of these tables are fully illustrated with numerical examples in Fisher's *Statistical Methods for Research Workers*, where they were first published. Further statistical problems soluble by the same tables are, however, constantly being discovered.

If x is the deviation from the mean of a normal distribution having unit variance, the ordinate z, given in Table II, is obtained from the algebraic expression

$$z = \frac{1}{\sqrt{2\pi}} e^{-\frac{1}{2}x^2}.$$

The integral of this function gives the area of the curve, or the frequency of observations, between the limits of integration. The fraction of the frequency falling outside the limits $\pm x$ is, therefore,

$$P = 2\int_x^\infty z dx.$$

Table I gives the values of x corresponding with various values of this fraction. Table II1 gives the single-tail probabilities, *i.e.* $\frac{1}{2}P$, corresponding to various values of x.

The distribution of t, due originally to "Student", is used to test the significance of a deviation when its standard error is *estimated from the data*; t is the deviation divided by its estimated standard error. For n degrees of freedom the algebraic expression for the ordinate of the distribution is

$$\frac{\frac{n-1}{2}!}{\frac{n-2}{2}!\sqrt{\pi n}} \cdot \frac{1}{\left(1+\frac{t^2}{n}\right)^{\frac{1}{2}(n+1)}}.$$

Table III gives the values of t corresponding with various values of the probability of a random value falling outside the limits $\pm t$.

The quantity χ^2, used to test goodness of fit, homogeneity, and for a variety of other purposes, may be regarded as the sum of the squares of n variables, which vary normally and independently about zero with unit variance. It is a useful single measure of the aggregate of a set of deviations from expectation. The element of frequency may be expressed as

$$df = \frac{1}{\frac{n-2}{2}!} (\tfrac{1}{2}\chi^2)^{\frac{1}{2}(n-2)} e^{-\frac{1}{2}\chi^2} d(\tfrac{1}{2}\chi^2).$$

When n is even the probability of exceeding any given value χ^2 is reducible to the partial sum of a Poisson series:

$$P = \int_{\chi^2}^\infty df = e^{-\frac{1}{2}\chi^2}\left\{1+(\tfrac{1}{2}\chi^2)+\frac{1}{2!}(\tfrac{1}{2}\chi^2)^2+ \dots + \frac{1}{\frac{n-2}{2}!}(\tfrac{1}{2}\chi^2)^{\frac{1}{2}(n-2)}\right\}.$$

For sufficiently large values of n the value of χ^2 corresponding to a given probability level P is expressible as a series of polynomials in a normal deviate x corresponding to this probability level (single tail) by the general method of Cornish and Fisher (69), which leads to the successive terms:

$$\chi^2 = n+x\sqrt{(2n)}+\tfrac{2}{3}(x^2-1)+\frac{1}{9\sqrt{(2n)}}(x^3-7x) - \frac{2}{405n}(3x^4+7x^2-16)+\frac{1}{4860n\sqrt{(2n)}}(9x^5+256x^3-433x)+ \dots$$

At the 1 per cent. point, for example,

$$x = 2{\cdot}326348 \qquad 3x^4+7x^4-16 = 109{\cdot}7$$
$$x^2-1 = 4{\cdot}41190 \qquad 9x^5+256x^3-433x = 2828$$
$$x^3-7x = -3{\cdot}694$$

For $n = 30$, $\sqrt{(2n)} = 7{\cdot}7459667$, and thus:

Power of x	Term	Remainder
	30	20·892,2
x	18·0198	2·872,4
x^2	2·9413	— ·068,9
x^3	— ·0530	— ·015,9
x^4	— ·0181	·002,2
x^5	·0025	— ·000,3

The true value is 50·892,2. The last column shows the errors remaining after each term.

Convergence is quicker for n exceeding 30, but slower for higher levels of significance. The method is good for the general Eulerian distribution with non-integral n.

The mean of the whole Poisson series is $\frac{1}{2}\chi^2$, the sum of the whole series is unity, and the number of terms in the partial sum is $\frac{1}{2}n$. In consequence the table of χ^2 (Table IV), which gives the values of χ^2 for various values of P, can be used to find the mean of a Poisson series of which the first $\frac{1}{2}n$ terms constitute a given fraction of the whole.

Example 1

Four serious floods have occurred in a given region during a century's experience; supposing the numbers of floods experienced in different centuries are distributed in a Poisson series, find the mean number for which only one century in 20 would have four serious floods or less, and the mean number for which only one century in 20 would have 4 floods or more.

In the first case we need the first five terms of the Poisson series, so $n = 10$; the table shows the 5 per cent. value of χ^2 is 18·307; the expectation would be 9·154 floods per century. If, on the contrary, only one century in 20 experienced 4 floods or more, the first four terms of the series would constitute 95 per cent. of the whole; for $n = 8$, the 95 per cent. χ^2 is 2·733, corresponding with a frequency of 1·366 serious floods per century. Unless our experience has fallen in the upper, or the lower 5 per cent. range of the distribution, the expectation lies between 1·366 and 9·154 floods per century. More or less stringent probabilities might, of course, have been chosen. Table VIII1 has been constructed to make this test directly at the 10 per cent., $2\frac{1}{2}$ per cent. and $\frac{1}{2}$ per cent. limits. Entering the table with $a = 4$ and $N = \infty$, for instance, the $2\frac{1}{2}$ per cent. limits of the expectation are found to be 1·09 and 10·24.

In testing goodness of fit χ^2 is simply calculated from the numbers observed and expected in each distinguishable class. If m is the frequency expected and a that observed, then

$$\chi^2 = S\frac{(a-m)^2}{m} = S\left(\frac{a^2}{m}\right) - S(a)$$

where the summation is taken over all classes. The number of degrees of freedom is the number of classes the frequencies of which can be varied independently without violating any totals, sub-totals, etc., which may have been used in calculating the expectations. It is inadvisable to use classes with expectations less than 5, so that these will usually be grouped in larger classes in applying the test. (See, however, Table VIII.)

The quantity z may be defined as the difference of one-half the natural logarithms of two different estimates of variance, one based on n_1 and the other on n_2 degrees of freedom. Table V gives values of z for different values of the three variables n_1, n_2 and P. The corresponding values of the variance ratio, e^{2z}, are given on the opposite pages. We may usually consider a z test as a test whether one estimate of a variance based on n_1 degrees of freedom significantly exceeds a second estimate based on n_2 degrees of freedom. We are indebted to C. G. Colcord and L. S. Deming (9) for the extension of the z table to $P = {\cdot}001$, to H. W. Norton and Iowa State College for $P = {\cdot}2$, and to Catherine M. Thompson (40, 41) and the Department of Statistics, University College, London, and to V. G. Panse (42) for $P = {\cdot}1$.

The distributions of t and of χ^2 are equivalent to special cases of the distribution of z. Thus when $n_1 = 1$, $t = e^z$ and t^2 for n_2 degrees of freedom is the variance ratio. Equally, when $n_2 = 1$, the variance ratio is the value of $1/t^2$ corresponding with the probability $1-P$.

When $n_2 = \infty$, the variance ratio is χ^2/n_1 for n_1 degrees of freedom; and when $n_1 = \infty$ it is the value of n_2/χ^2 for n_2 degrees of freedom, corresponding with the probability $1-P$.

From the value 6 onwards the values of n_1 tabulated are in harmonic progression; *i.e.* they are the numbers for which $24/n_1$ takes the values 4, 3, 2, 1 and 0. As the corresponding values of z run smoothly other values of n_1 may be used by interpolating for the values of $24/n_1$. Similarly, the largest n_2 tabulated are those for which $120/n_2$ takes the values 4, 3, 2, 1 and 0. The z values for any pair of numbers n_1 and n_2 can thus be found from the table. With lower accuracy the variance ratio may be similarly interpolated. For high values of n_1 and n_2 a useful approximation is given below the table for each value of P tabulated.

In the table of z, like that of χ^2, the probability given, and most frequently wanted, is that of a single tail of the distribution beyond the tabulated value. For the tables of x and t it is the sum of two equivalent tails.

In addition to its general use in the analysis of variance, the z table is related to the partial sum of a binomial series in the same way as is that of χ^2 to the Poisson series. The ratio of the sums of squares is n_1/n_2 times the variance ratio, or n_1e^{2z}/n_2. If $p : q$ is the ratio of the sums of squares, P is the sum of the first $\frac{1}{2}n_1$ terms of the expansion of $(q+p)^{\frac{1}{2}(n_1+n_2-2)}$, and $1-P$ is the sum of the remaining $\frac{1}{2}n_2$ terms. Similarly if $-p$ is the ratio of the sums of squares, P is the sum of the first $\frac{1}{2}n_1$ terms in the expansion of the negative binomial $(q+p)^{-\frac{1}{2}n_2}$.

Example 2

An event has occurred 3 times out of 14 trials. The probability of its occurring 3 times or less, is represented by the sum of the first four terms out of 15, hence $n_1 = 8$, $n_2 = 22$. If $P = 1$ per cent., the table gives a variance ratio 3·45 ; hence the ratio of the sums of squares is $3{\cdot}45 \times 8/22 = 1{\cdot}255 = p/q$. Thus $p = 1{\cdot}255/2{\cdot}255 = {\cdot}557$. On the other hand, for the probability of its occurring 3 times or more, we need $n_1 = 24$, $n_2 = 6$, when the variance ratio is 7·31, and $q/p = 29{\cdot}24$, $p = 1/30{\cdot}24 = {\cdot}0331$. Unless our experience has fallen in the upper or the lower 1 per cent. range of the distribution, the probability of occurrence lies between ·0331 and ·557.

This type of problem also may be examined at other levels of significance by the use of Table VIII1. Entering the table with $a = 3$, $p = 3/14 = 0{\cdot}214$, we find that the limits of expectation, at the probability level $P = \frac{1}{2}$ per cent., are 0·360 and 8·26, giving the corresponding limits of the probability of occurrence as ·0257 and ·590. If $P = 2\frac{1}{2}$ per cent. the limits are ·0466 and ·508.

If the event had not occurred at all in 14 trials, it is still possible to assign an upper limit to the probability. For $P = 2\frac{1}{2}$ per cent., for example, the upper limit to the probability is $1-({\cdot}025)^{\frac{1}{14}}$ or ·232, and for $P = \frac{1}{2}$ per cent. it is $1-({\cdot}005)^{\frac{1}{14}}$ or ·315, corresponding to upper limits of expectation of 3·25 and 4·41 respectively. The same result may be obtained by interpolation with respect to $1/N$ in Table VIII1.

Table VII gives the exact test of significance for inter-class correlation coefficients. This is equivalent to the t test of the regression, which reduces to

$$t = \frac{r}{\sqrt{1-r^2}}\sqrt{n'-2}$$

for $n'-2$ degrees of freedom, where n' is the number of pairs on which the correlation is based. The results often differ materially from the comparison of r with its standard error estimated from the formula,

$$\sigma_r = \frac{1-r^2}{\sqrt{n'-1}}.$$

Other comparisons involving correlation coefficients are best carried out by the z transformation provided by Table VII1.

Table V1. Fiducial Limits for a Variance Component

In an analysis of variance between and within classes, whereas it is often desirable to make probability statements about the individual classes and the various comparisons between them (Cornish, 70), it is sometimes also desirable, *e.g.* in genetical work, to interpret them as a random sample of n_1+1 items from a normal population of unknown variance. Table V1 is appropriate for finding fiducial limits for this variance. If the analysis of variance takes the form

	d.f.	M.S.	Expected M.S.
Between classes	n_1	v_1	$k\sigma_1^2+\sigma_2^2$
Within classes	n_2	v_2	σ_2^2

and $v_1/v_2 = F$, the fiducial limits for σ_1^2 are found as $\phi v_1/k$, where ϕ is the quantity tabulated as a function of n_1, n_2 and F. These limits are, of course, only exact if σ_1^2 is the variance of a normal population from which n_1+1 values have been selected at random, and if the samples drawn from each class are of equal size.

The mathematical form of the distribution was given by Fisher in 1935 (71). The numerical values of Table V1 were recently computed by Healy (81).

Tables VI, VI1 and VI2. Significance of Difference between Two Means

Tables VI-VI2 give the test of significance, due originally to Behrens, for the difference of two means for which the errors are due to different causes, so that estimates of variance cannot properly be pooled. In general we need the percentile points of a distribution compounded of two Student distributions with n_1 and n_2 degrees of freedom. The ratio of the scale-factors s_1 and s_2 is given by

$$s_1/s_2 = \tan\theta,$$

and the difference is judged to be significant if it exceeds $d\sqrt{s_1^2+s_2^2}$.

For each level of significance the value of d is required given the 3 known values n_1, n_2 and θ. Table VI by P. V. Sukhatme gives the 5 per cent. and 1 per cent. points for 15° intervals of θ, and for the lattice of values 6, 8, 12, 24 and ∞ for n_1 and n_2. Four-point interpolation is sufficient for full accuracy.

Table VI1 gives, more exactly, the values of d for small odd integral values of n_1 and n_2. The analysis has been given by Fisher and Healy (67).

Table VI2 deals more fully with the marginal case $n_1 = \infty$, where the distribution is compounded of a normal component of known variance and a component of Student's type. This occurs when we compare estimates of a physical constant, one of which is based on a large quantity of data of low precision, while the other rests on few, but rather precise, observations. When the errors from these two estimates are of comparable magnitude, both can be taken into account by finding the angle, defined by

$$\tan \theta = \sigma / s,$$

while the significant value for the observed difference is

$$d\sqrt{s^2+\sigma^2}.$$

Thus for high angles the distribution tends to normality and for low angles it tends to Student's form.

Sukhatme's values for 1 per cent. have been revised to remove the very small positive bias, about ·001 in d, detected by comparison with the asymptotic calculation used in constructing Table VI2.

Example 2.1

Let us suppose that a physical constant evaluated by a new method gives a mean of twelve determinations,

$$\bar{x} = 4{\cdot}77383,$$

and that the sum of the squares of the deviations of these values from their mean is

$$S(x-\bar{x})^2 = {\cdot}011580,$$

so that from 11 degrees of freedom the variance of the mean is estimated to be

$$s^2 = {\cdot}00008773,$$

and the estimated standard deviation of the mean is

$$s = {\cdot}009366.$$

For 11 degrees of freedom Student's t is 2·718 at the 2 per cent. point, so that at this level of significance we should regard our observed mean as differing significantly from any value outside the limits

$$\bar{x} \pm ts = 4{\cdot}77383 \pm {\cdot}02546 = 4{\cdot}74837,\ 4{\cdot}79929.$$

If, now, numerous previous determinations, using a different method, have given the value,

$$\mu = 4{\cdot}744,$$

outside these limits, it would appear that there must be a real discrepancy between the results from the two methods used.

The comparison, however, is not complete, if we realise that μ also is not a perfectly precise determination, for it might be that though the value adopted is incompatible with the new observations, yet that both methods might be simultaneously compatible with some intermediate value of the constant.

Behrens' test enables this question to be answered, when the precision of μ is known, by averaging the probability of obtaining so high a deviation, when μ is assigned a frequency distribution instead of a fixed value. If, for example, the value of μ adopted, has a standard error σ, based on a large number of degrees of freedom, with

$$\sigma = {\cdot}00382,$$

we should find,

$$\tan \theta = \sigma / s = {\cdot}4079,$$
$$\theta = 22^\circ\ 12';$$

and

$$d = \frac{\bar{x}-\mu}{\sqrt{s^2+\sigma^2}} = \frac{{\cdot}02983}{{\cdot}010115} = 2{\cdot}949$$

Table VI2 shows that at the 2 per cent. level of significance d is only 2·704 at 20° and 2·637 at 30°, for 10 degrees of freedom. For 11 degrees of freedom lower values still would be significant. The discrepancy observed is therefore a significant one, even when the errors to which μ is liable are fully taken into account.

Table VIII. Tests of Significance for 2×2 Contingency Tables

Contingency tables are normally tested for significant departures from independence by means of the χ^2 test. Provided that more than one degree of freedom is involved this test is satisfactory, even when the expectation in each class is quite small; but with only a single degree of freedom (as in 2×2 contingency tables) the number of possible sets of values satisfying the given marginal totals is small, so that the discontinuity of the resultant distribution is marked, whereas the χ^2 distribution, which is an approximation to it, is continuous. The effect of this discontinuity can be overcome by applying Yates' correction for continuity when calculating χ^2. This is done by reducing by $\frac{1}{2}$ the values which are greater than expectation and increasing by $\frac{1}{2}$ those which are less than expectation.

The correction for continuity, which should be applied when the smallest expectation is less than 500, considerably increases the effective range of the ordinary χ^2 test for a single degree of freedom; but when the smallest expectation is less than 100, the possible asymmetry of the actual distribution associated with given marginal totals becomes of importance. Such asymmetry implies that deviations of a given magnitude in one direction are less probable than those in the other.

Table VIII allows this asymmetry (and other slight deviations from the ordinary χ^2 distribution) to be taken into account. It gives what are in effect the 2·5 per cent. and 0·5 per cent. points of χ_c (the square root of χ^2 corrected for continuity) for each tail separately. Since there are three independent marginal totals in a 2×2 contingency table all possible contingency tables could only be completely covered by a three-dimensional table of χ_c for each level of significance. In general, however, the distribution of χ_c is not very different for sets of marginal totals which give the same smallest expectation, and for which the observed value in any one cell can assume the same number of values (*i.e.* tables in which the smallest marginal total is the same, this being one less than the possible number of values).

This property is utilised in Table VIII, where values of χ_c are shown for chosen values of the smallest expectation m, and of p, the smallest expectation divided by the smallest marginal total. (For a detailed account of the method of construction of this table see Yates (14).)

In the contingency table :—

x	y	a'
z	u	b'
a	b	n

where $a\leqslant b$, $a'\leqslant b'$ and $a\leqslant a'$, so that a is the smallest marginal total, and x is the number in the cell with smallest expectation, we have $m = aa'/n$, and $p = m/a = a'/n$. It will be noted that as n increases with fixed m and p, a remains fixed and $a' : b' : n$ is constant. When n is very large the distribution of x tends to that of the binomial distribution $(p+q)^a$. When n takes on its smallest possible value, which is that which gives the smallest integral value of $a'\geqslant a$, the "limiting contingency distribution" of x is attained. The values of χ_c for these two extreme distributions are shown in Table VIII, the former in ordinary type and the latter in black type. In general the intermediate contingency distributions give intermediate values of χ_c.

Furthermore, as p tends to zero with fixed m, the expectations in all cells except that containing x become large, and the Poisson distribution is approached, whatever the ratio of a to a'. Thus for $p = 0$ there is only one set of values in Table VIII, representing the virtual 2·5 and 0·5 per cent. points of the Poisson distributions. Thus the table can be used to test the significance of departures from binomial distributions with known p and Poisson distributions with known m, as well as for contingency tables.

It should be noted that when there is a zero entry the exact test of 2×2 contingency tables involves a simple product of factorials easily calculated from Table XXX. When there is a small entry the exact calculation is never difficult and is often quicker than interpolation in any existing table. Such tables are intended only to save labour in more troublesome cases, and not to obviate the necessity of understanding the exact test.

It may also be noted that for a binomial distribution with $p = 0{\cdot}5$, χ^2 corrected for continuity can be expressed in the form $\chi_c^2 = (a-b-1)^2/(a+b)$, where a and b are the numbers of successes and failures $(a>b)$. This gives a rapid preliminary test, based on the numbers of + and − signs, for quantitative data, such as the differences of paired observations, in which the probabilities of positive and negative deviations are judged to be equal. The values given in line 4 of each panel of Table VIII show that the continuity correction is particularly effective in this case.

Example 3

The number of summers (April-September) with periods of over 20 days without rainfall at Rothamsted during this century and the last half of last century, are given in Table 1.

Are the summers of the present century significantly more subject to droughts than those of the last?

The value of χ^2, corrected for continuity, is given by

$$\chi_c^2 = \frac{(13\tfrac{1}{2}\times 44\tfrac{1}{2}-4\tfrac{1}{2}\times 42\tfrac{1}{2})^2\times 105}{49\times 56\times 18\times 87} = \frac{(14\times 45-4\times 42-\tfrac{1}{2}105)^2\times 105}{49\times 56\times 18\times 87} = 4{\cdot}098.$$

Thus significance is attained at the 5 per cent. point, since the 5 per cent. value of χ^2 for one degree of freedom (Table IV) is 3·841.

Reference to Table VIII confirms that χ_c^2 gives a satisfactory approximation in this example. The smallest expectation $m = 18\times 49/105 = 8{\cdot}40$, $p = m/18 = 49/105 = 0{\cdot}47$. The observed set lies on the shorter tail of the distribution, since the number 4 with the smallest expectation is less than expectation. Interpolation in Table VIII ($P =$ ·025) gives a value of 1·92 for the corresponding binomial distribution, and 1·94 for the limiting contingency distribution, compared with 1·96 ($= \sqrt{3{\cdot}841}$) from the χ^2 distribution.

TABLE 1. SUMMERS WITH AND WITHOUT DROUGHT AT ROTHAMSTED

	1852-1900.	1901-1956.	Total.
Experiencing drought . .	4	14	18
Without drought . . .	45	42	87
Total . . .	49	56	105

In cases of doubt, or when the exact probability is required, this may be calculated by the procedure given in *Statistical Methods*, Section 21·02. In this example the exact probability is ·0201, compared with the value of ·0217 given by χ_c and ·0202 if χ_c is increased by ·03, the mean of the differences of the values given by Table VIII for $P =$ ·025 from 1·96.

Example 3.1

If the observations in a 2×2 table are distinctly out of proportion (and, indeed, in other cases) we may wish to set limits to the true product-ratio. For example, the observed table

10	3
2	15

gives a crude ratio of 25. How small could the true ratio be, in reasonable consistency with the data ?

If the expectations were

$10-x$	$3+x$
$2+x$	$15-x$

the true ratio would be

$$(10-x)(15-x)/(3+x)(2+x)$$

and χ^2 for the observations would be

$$\chi^2 = x^2\left(\frac{1}{10-x}+\frac{1}{3+x}+\frac{1}{2+x}+\frac{1}{15-x}\right).$$

So, if x were 3·0,

$$\chi^2 = 3^2(\cdot 59286) = 5{\cdot}3357$$

with one degree of freedom.

The exact probability of such a small sample of 30 giving 10 or more in the first quadrant is the partial sum of a hypergeometric series, and not easy to calculate. Using Yates' adjustment, however, we can at once find

$$\chi_c^2 = (2{\cdot}5)^2(\cdot 59286) = 3{\cdot}7054.$$

Further, taking x to be 3·1, we have also

$$\chi_c^2 = (2{\cdot}6)^2(\cdot 58897) = 3{\cdot}9814.$$

Interpolating for the tabular entry, 3·841, it appears that

$$x = 3{\cdot}0491,$$

giving a cross-product ratio

$$\xi = 2{\cdot}720.$$

It may therefore be inferred from the data that the true cross-product ratio exceeds 2·720, unless a chance of one in forty has occurred.

Table VIII cannot be applied to this example, but an indication of the precision to be expected may be obtained by calculating the complete series for $\xi = 2{\cdot}72$. The terms are proportional to

$$1, \frac{12.17}{2.1.\xi}, \frac{(12.11).(17.16)}{(2.3).(1.2).\xi^2}, \text{ etc.}$$

The first three terms, corresponding to the table observed and the two more extreme possible tables, give a probability of 2·408 per cent.; slightly higher ratios should therefore be allowed within the 2·5 per cent limit. A more exact value for the limit is 2·7534.

The upper 2·5 per cent. limit may be similarly determined. The value for x, using Yates' adjustment, is found to be 1·5849, giving $\xi = 327{\cdot}1$, and calculation of the complete series gives a true probability for this value of ξ, of 2·167 per cent. Thus, even with two very small expectations, the adjustment has given a fairly close approximation, though as Table VIII shows, substantially larger discrepancies can arise in certain cases with such small expectations.

Example 4

In the course of 500 spins of a roulette wheel, zero appears 24 times. What evidence is there that the wheel is biased in favour of the bank?

If the wheel is unbiased, the chance of zero occurring in any one spin is $\frac{1}{37}$, and zero may be expected 13·514 times on the average in 500 trials. The discrepancy is therefore 10·486, $\chi^2 = \frac{10{\cdot}486^2}{13{\cdot}514} + \frac{10{\cdot}486^2}{486{\cdot}486} = 8{\cdot}362$, and $\chi = 2{\cdot}892$. (Note that the class of non-occurrences as well as the class of occurrences must be included when calculating χ^2.) Interpolation in Table II1 for the value 2·892 of the argument indicates that the probability of obtaining 24 or more zeros was only ·00191. Correcting for continuity, the discrepancy becomes 9·986, and $\chi_c^2 = 7{\cdot}584$, $\chi_c = 2{\cdot}754$, giving a probability of ·00294. Referring to Table VIII, however, for $m = 13{\cdot}5$ and $p = {\cdot}027$ (values in ordinary type) we see that the 0·5 per cent. point of χ_c is about 2·80, compared with the value of 2·576 given by normal theory. Thus the true probability is somewhat greater than ·005. (True value obtained by summing the binomial distribution = ·00559.) The omission of the correction for continuity introduces large errors, and even the use of the ordinary χ^2 table gives a probability of one half the true value.

The problem may also be solved directly by the use of Table VIII1. Since a is greater than 10 we must use the corrections given for $a = 11$ and over. $p = 24/500 = {\cdot}048$ and therefore the conventional standard error $\sqrt{\{a(1-p)\}} = 4{\cdot}79$. Multiplying this by the normal deviate for $P = \frac{1}{2}$ per cent. (single tail) gives $4{\cdot}79 \times 2{\cdot}576 = 12{\cdot}34$. The limits of expectation calculated from the conventional standard error are therefore 11·66 and 36·34. The corrections given by Table VIII1 for these limits are +1·70 and +2·80, giving the adjusted limits of 13·36 and 39·14, corresponding to probability limits ·0267 and ·0783. Since $\frac{1}{37} = {\cdot}0270$ the probability of obtaining as many as 24 zeros in 500 spins is just over ·005, agreeing with our previous solution.

Example 5

Locate the upper and lower 2·5 and 0·5 per cent. points of a Poisson distribution with an expectation of 25.

If x is the observed value and this is greater than expectation (*i.e.* lies on the longer tail) we have $\chi_c = (x-25-\frac{1}{2})/\sqrt{25}$. For $m = 25$, $p = 0$, $\chi_c = 2{\cdot}05$ for the 2·5 per cent. point, so that $x = 35{\cdot}8$. Hence a value of 36 or over has a smaller probability than ·025. For the 0·5 per cent. point, $\chi_c = 2{\cdot}76$, so that $x = 39{\cdot}3$, giving a value of 40 or over. Similarly, the lower 2·5 and 0·5 per cent. values of x will be found to be 15·2 and 12·6, giving integral values of 15 and 12 respectively. Normal theory would give x's about $\frac{1}{2}$ a unit in error for the 2·5 per cent. points and about a unit in error for the 0·5 per cent. points.

If the actual value of the probability of obtaining a value of say 14 or less is required, we may proceed as follows. At the lower 2·5 per cent. point χ_c is 1·86 and χ (normal theory) is 1·96, and at the lower 0·5 per cent. point χ_c is 2·37 and χ is 2·58. A value of 14 gives a χ_c of 2·10. Hence by linear interpolation, using χ_c as argument, the corresponding value of χ is

$$\chi = 1{\cdot}96 + \frac{2{\cdot}10-1{\cdot}86}{2{\cdot}37-1{\cdot}86}(2{\cdot}58-1{\cdot}96) = 2{\cdot}25.$$

Reference to Table II1 gives the corresponding probability as ·0122. (True value = ·0124.)

Table VIII1. Limits for the Probability of an Event

If an event is observed to occur a times out of N, a lower limit π_1 can be assigned to the probability of this event such that if the probability were actually π_1 then an observed number of occurrences as great or greater than a out of N trials would only occur by chance with a frequency of P. Similarly an upper limit π_2 can be assigned such that if the probability were actually π_2 an observed number of occurrences as small or smaller than a would occur with a frequency P. Corresponding to these probabilities π_1 and π_2 there are limits of expectation of the number of occurrences in N trials, namely $\pi_1 N$ and $\pi_2 N$. These limits of expectation are tabulated in Table VIII1 for $P = 0{\cdot}1$, $0{\cdot}025$ and $0{\cdot}005$ against values of a from 0 to 10 (with extension to higher values) and of N from $2a$ to ∞. The values for $N = \infty$ gives the limits of expectation of the Poisson distribution. When $a > \frac{1}{2}N$, the table may be entered with $a' = N - a$. Examples of the use of the table have already been given (Examples 1, 2 and 4).

Table VIII2. Densities of Organisms Estimated by the Dilution Method

One of the methods available for estimating the number of organisms of a given type present in a medium is the dilution method. This method can be employed in cases in which the presence or absence of the organism can be determined, but in which counts cannot be made.

In the dilution method a series of suspensions of the organism, s levels in all, is prepared, each of which is a times as dilute as the preceding one. Each suspension is used to inoculate n tubes with a known volume of the suspension, and the presence or absence of the organism is determined by suitable test after a period of incubation. The most accurate estimate of the number of organisms per tube at any given level is obtained by solving the equation of maximum likelihood, and a method of doing this has been discussed by Mather (62); but it has been shown (Fisher, 30) that 87·7 per cent. of the information is contained in the total number X of fertile or Y of sterile plates, counted without regard to level. If λ is the number of organisms per tube at the highest concentration, the value of λ for which the expected average number of sterile plates is equal to the observed number is given by the equation

$$Y = n(e^{-\lambda} + e^{-\lambda/a} + e^{-\lambda/a^2} + \ldots + e^{-\lambda/a^{s-1}})$$

Table VIII2 enables the solution of this equation to be obtained expeditiously for two-fold, four-fold or ten-fold dilution series of any length. All that is necessary is to calculate what may be called the mean fertile level, $x = X/n$, and the mean sterile level, $y = s - x$, and find the tabular value K corresponding to the observed x or y for the appropriate number, s, of levels. The estimate of λ is then given by

$$\log \lambda = x \log a - K.$$

In an extended series of dilutions the value of K becomes constant for all values of x and y sufficiently distant from the ends of the distribution, except for a periodic component, which, however, is only of importance for the ten-fold series. This periodic component necessitates the special form of tabulation adopted for this series.

The average value of the variance of the mean fertile level is $\frac{1}{n}\frac{\log 2}{\log a}$ except at the ends of the series, where the value increases. The average value of the variance of $\log \lambda$ is $\frac{1}{n} \log 2 \, . \log a$, and this may be used in the two- and four-fold series in the range where the tabular entries are substantially constant. The variance of the mean fertile level is, however, more stable than the variance of $\log \lambda$. Closer approximation to the fiducial limits can therefore be found by using the average value $\frac{1}{n}\frac{\log 2}{\log a}$ of this variance, increasing and decreasing the observed mean fertile level by the appropriate multiple of the corresponding standard error. The table can then be re-entered to determine limiting values of $\log \lambda$, and thence, if required, of λ. This approximation is satisfactory except at the extreme ends of the range. For the same reason the mean fertile level should be used in preference to $\log \lambda$ when performing tests of significance, etc., on comparisons between different media. Standard errors should not be attached to λ.

The inclusion of tables for the four-fold and ten-fold series should not be taken to imply that the use of these series is ordinarily advisable. If an adequate number of tubes is available the two-fold dilution series should be used, with correspondingly fewer plates at each level, in preference to a four-fold or ten-fold series covering the same range. As is shown by the formula given above, the average variance of $\log \lambda$ in the centre of the range is the same for a given total number of plates, whatever the dilution factor, but the periodicity becomes appreciable for the more coarsely spaced series.

The number of dilution levels required depends primarily on the range of populations it is desired to cover. For any given population density the majority of the information is given by the levels at which the average number of organisms per tube is near 1. Thus in a two-fold dilution series in which one dilution level has an average of exactly one organism per tube the contributions of the various levels of an infinite series to the total amount of information are as follows:

Average number of organisms per tube	8 and more	4	2	1	$\frac{1}{2}$	$\frac{1}{4}$	$\frac{1}{8}$	$\frac{1}{16}$ and less
Percentage of information	0·9	12·6	26·4	24·5	16·2	9·3	4·9	5·2

For any given population density, therefore, most of the information is contributed by five levels of the series. A series of 12 dilutions will therefore amply cover a range of 100 fold, and a series of 15 dilutions a range of 1000 fold. If more than one organism which can be detected in the same set of tubes is under investigation, then longer series may be required, the exact length depending on the relative numbers of the different types of organism.

When a large number of tests involving series of the same length and with the same number of tubes at each level have to be made, the experimenter may find it convenient to prepare a special table giving log λ and λ directly for each possible value of the total number of fertile tubes. Such a table may be easily prepared from Table VIII2.

In practical work greater precision is often required in one part of the range than in the others. Thus the tests may have as their primary object the determination of whether the population densities are less than a given standard density, but at the same time a rough estimate of the actual population densities of samples differing widely from the standard may be of interest. In such a case the number of tubes should be increased at the levels for which the population per tube would be in the neighbourhood of 1 at the standard density, *e.g.* we might increase the number of tubes at levels having 4, 2, 1, $\frac{1}{2}$ and $\frac{1}{4}$ organisms per tube at the standard density.

For the estimation of the population densities of dilution series of this type a special table will have to be prepared. If there are n_1 tubes at each of s levels and an additional n_2 tubes at each of the levels from $r+1$ to r', the average total number of sterile tubes is given by

$$Y = n_1y_1+n_2y_2 = n_1(e^{-\lambda}+e^{-\lambda/2}+ \ . \ . \ . \ +e^{-\lambda/2^{s-1}})+n_2(e^{-\lambda/2^r}+ \ . \ . \ . +e^{-\lambda/2^{r'-1}})$$

The table can be prepared directly by calculating the values of Y for a series of values of log λ, thence determining by inverse interpolation the values of log λ corresponding to integral values of Y. Alternatively Table VIII2 may be used to determine the values of y_1 and y_2 separately for a series of values of log λ, thence performing the inverse interpolation as before. Values of y_2 corresponding to values of log λ falling outside the range of the table will in any case have to be computed directly, but this can be done rapidly, using the expansion of y_2 in powers of λ for small values of λ.

The standard errors of estimates derived from a modified dilution series of this type naturally vary in different parts of the range. At the standard density, however, the standard errors will be those corresponding to an infinite series of n_1+n_2 tubes at each level, except for a slight increase due to the shortness of the subsidiary series.

Example 5·1

Tests with potato flour containing rope spores (B. *mesentericus*) gave the following observations, using 5 tubes, each of 1 c.c., of dilutions containing 4, 2, 1, . . ., 1/128 g. per 100 c.c. (E. C. Barton-Wright's data).

g. per 100 c.c.	Number fertile.	g. per 100 c.c.	Number fertile.
4	5	1/8	3
2	5	1/16	2
1	5	1/32	2
1/2	5	1/64	0
1/4	4	1/128	0

The total number of fertile plates is 31, and the mean fertile level (x) is therefore 6·2. Hence y is 3·8. The table gives $K = {·}383$. Hence

$$\log \lambda = 6{·}2 \log 2 - {·}383 = 1{·}483$$
$$\lambda = 30{·}4.$$

The number of organisms per gram is thus estimated to be 760.

The expected variance of the mean fertile level is 0·2. The fiducial limits of the number of tubes at levels ·025 to ·975 for P are therefore $6{\cdot}2 \pm 1{\cdot}960 \times \sqrt{0{\cdot}2} = 5{\cdot}323$ and 7·077. Re-entering the table with these values, we obtain limits of 1·211 and 1·761 for log λ, or 407 and 1440 for the actual number of organisms per gram. The average standard error of log λ is $\sqrt{\cdot 0181} = \cdot 135$, giving limits of $1{\cdot}483 \pm \cdot 264 = 1{\cdot}219$ and 1·747 for log λ, or 415 and 1400 for the actual number, which may be compared with the more exact limits given by the first method.

Table VIII3. Significance of Leading Periodic Components

If u_r ($r = 1, 2, \ldots, 2n+1$) constitute a random sample from a normally distributed population, the linear functions $S(a_r u_r)$ and $S(b_r u_r)$ define a Fourier component, if

$$a_r = \sqrt{\frac{2}{2n+1}} \cos \frac{2\pi p r}{2n+1}, \qquad b_r = \sqrt{\frac{2}{2n+1}} \sin \frac{2\pi p r}{2n+1}.$$

Values of p from 1 to n give n pairs of components referable to different periods. If $A = S(a_r u_r)$ and $B = S(b_r u_r)$, then the sum of squares corresponding with the two degrees of freedom for any period is

$$X = A^2 + B^2.$$

A general test for the existence of a periodic component is therefore provided by seeing whether the largest of the values of X for the n periods is greater than might be expected from chance causes. Table VIII3 shows at the 5 per cent. and 1 per cent. levels the fraction g that the largest of the n values of X may bear to the total sum of squares, $S(u_r - \bar{u})^2$.

When the number of observations in the series is even, $2n+2$, the coefficients in A and B are

$$a_r = \frac{1}{\sqrt{n+1}} \cos \frac{\pi p r}{n+1}, \qquad b_r = \frac{1}{\sqrt{n+1}} \sin \frac{\pi p r}{n+1}.$$

Values of p from 1 to n give n values of X corresponding with n different periods. The table again gives the significance levels of the ratio g that the largest X bears to the sum of all n values of X. In this case the sum of the n values of X, which is the sum of squares for the n periods, is equal to

$$S(u_r - \bar{u})^2 - \frac{1}{2n+2}(u_1 - u_2 + u_3 - \ldots - u_{2n+2})^2.$$

The application of the table is, therefore, exceedingly simple although the subdivision of the sum of squares into its periodic components is somewhat tedious. The theoretical basis of the table was discussed by R. A. Fisher (55).

Tables IX to IX3. The Probit Transformation

Many important classes of data may be interpreted on the supposition that a normal deviate is linearly dependent on some observable concomitant measurement, and that an observable frequency is that with which this deviate is exceeded in a normal distribution. For example, the frequency with which a weight is judged to be heavier than a standard increases with the actual weight used; the frequency with which a high jumper clears a rod decreases with the height at which it is placed; the frequency with which insects survive fumigation decreases with the concentration of the fumigant. In many such cases the relation between the frequency and measurement concerned is excellently represented by considering the normal deviate corresponding to the frequency to be a linear function of the measurement, or, more frequently in toxicology, of its logarithm, the measurement being the concentration of the toxin.

Table IX gives the deviate corresponding with each thousandth of the total frequency, which is exceeded by any given frequency; by means of the difference column it is simply used to ten-thousandths. To facilitate standard methods of calculation by making all values positive, 5 has been added to the deviate. The resulting values are known as "probits." It is obvious that the probits corresponding to complementary percentages, such as 40 per cent. and 60 per cent., must add to make 10. Beyond 95 per cent. the actual differences are inserted, and beyond 98 per cent. the tabulation is by ten-thousandths. By subtracting from 10, these advantages may be made available also at the beginning of the table.

Table IX1, simple quantiles of the normal distribution, gives the deviates corresponding with proper fractions. The proper fractions, up to ½, having denominators up to 30, are here arranged in order of magnitude. When small numbers are involved this table enables the probits to be found more expeditiously than does Table IX.

Example 6

What deviate is exceeded by 4/17 of a normal curve? The deviate given is 0·7215, so that if 4 animals survive out of 17 the probit is 5·7215. The same value may be obtained by expressing 13/17 as a decimal, ·7647, and referring to Table IX.

Dosage Mortality Tests

When tests have been made on several concentrations it is usual to fit a straight line expressing the probit in terms of the logarithmic concentration. The weight assigned to each point will be the number of animals used in determining it multiplied by the weighting coefficient given in Table IX2. It is best to draw a provisional line graphically, and obtain the weighting coefficients from the probits read from the graph. Some workers prefer to draw a provisional line by calculation, using weighting coefficients based on the probits empirically observed. This leads to some ambiguity when at any concentration all die (or all survive), for such observations nominally give an infinite probit with zero weight.

However the provisional line be obtained, for refined purposes, such as tests of significance, it should be replaced by one based on working probits, obtainable from Table IX2. When all animals survive, the minimum working probit in the 2nd column corresponding to the expected probit in the 1st is used; when all die, the maximum working probit in the 4th column is used. Other cases are assigned values greater than the minimum working probit by the product of the proportion killed and the range (given in the 3rd column); the same quantity is obtained by decreasing the maximum working probit by the product of the proportion surviving and the range. The working probit is given a weight found by multiplying the total number of animals on which it is based by the weighting coefficient in the 5th column of the table. Any tolerably good provisional line will give a satisfactory result when this process is applied. An excellent account of the treatment of toxicological data has been given by C. I. Bliss (4-7), and the subject is also discussed, from the point of view of amount of information, in *The Design of Experiments* (2).

Example 7

The following survival fractions were obtained on testing sets of eight brine-shrimps, *Artemia salina*, from a single brood, in arsenical solutions having concentrations in geometric progression.

Solution	*C*	*D*	*E*	*F*	*G*	*H*	*I*
Total tested	8	8	8	8	8	8	8
Survived	8	8	6	5	5	1	0

Since, using strengths *C*, *D* and *I*, all or none have survived, the tests with solutions *E* to *H* are available for plotting probit values, in order to obtain a provisional line. A sufficiently good line may usually be drawn graphically. A second serviceable method is to fit a straight line, without weighting, to the probits available. Thus, using Table IX1, we have

Solution	Empirical Probit.
E	4·33
F	4·68
G	4·68
H	6·15
Total	19·84
Mean	4·96

The slope from four equally spaced points is

$$b = \frac{3(H-E)+(G-F)}{3^2+1^2} = \cdot 546.$$

It is worth while giving some attention to obtaining a satisfactory preliminary fit, since otherwise the fitting process may have to be repeated. It is also desirable to avoid interpolation by choosing simple values. The inclusion of *C*, *D* and *I* will certainly tend to increase the slope, and may lower the average probit. For *b*, therefore, we take the value ·6, and take the probit midway between *F* and *G* to be 4·9. The provisional expected probits are those shown in the 3rd column of Table 2. The remainder of Table 2 is then filled by using Table IX2.

When all die, as for solution *I*, the working probit has its maximal value, given in the 2nd column of Table IX2. This is 6·9394, or ·5394 above expectation. (In the calculations for fitting, we may use either the working probits themselves to obtain the actual constants of the fitted line, or their deviations from provisional expectations to obtain adjustments of the constants provisionally adopted. Table 2 is arranged to show the first method.)

TABLE 2.

Solution.	Independent Variate x.	Provisional Probit Y.	Working Probit y.	Weight w.	wy.
C	3	2·8	2·4081	·7343	1·7683
D	4	3·4	2·9060	1·9002	5·5220
E	5	4·0	4·3775	3·5090	15·3606
F	6	4·6	4·6826	4·8042	22·4961
G	7	5·2	4·6776	5·0193	23·4783
H	8	5·8	6·0998	4·0208	24·5261
I	9	6·4	6·9394	2·4159	16·7649
				22·4037	109·9163

More generally, we subtract from the maximal probit a fraction of the range (3rd column of Table IX2) equal to the proportion of survivors; so G gives

Maximal working probit	6·2759	Range 2·5573
Range × 5/8	1·5983	
Working probit	4·6776	

while for H we have, similarly, 6·0998.

For E, if all had *lived*, we should have 3·3443, but as one-quarter have died we add one-quarter of 4·1327, giving 4·3775, while for F we find, similarly, 4·6826.

In all cases the weight assigned to the evidence of each solution is the weighting coefficient, corresponding with the provisional probit, multiplied by the number of individuals tested.

From the sum of the weights, and the weighted sums of x and y, we obtain the values of $\bar{x}$ and $\bar{y}$ according to the formulæ:

$$W = S(w), \qquad W\bar{x} = S(wx), \qquad W\bar{y} = S(wy).$$

It is convenient to calculate $S(wx^2)$ at the same time as $S(wx)$; for calculating these any suitable origin such as F ($x = 6$) may be used. The reduced sum of squares of the independent variate is then

$$A = Sw(x-\bar{x})^2 = S(wx^2) - \bar{x}S(wx).$$

With most types of machine, the values of wy will be written down and summed as in Table 2; from these it is easy to obtain the reduced sum of products,

$$B = Swy(x-\bar{x}) = S(wxy) - \bar{x}S(wy).$$

These quantities are all that are needed for adjusting the regression line. For testing goodness of fit we also calculate

$$C = Sw(y-\bar{y})^2 = S(wy^2) - \bar{y}S(wy).$$

Turning to the numerical values of Table 2, we may note that the three solutions first omitted contribute nearly one quarter of the total weight, and a still higher proportion to A, or to the amount of information supplied as to the slope of the line. For the others the working probits differ not greatly, but appreciably, from the empirical probits first used.

Numerically, Table 2 gives

$S(wx)$	10·7963	$S(wx^2)$	60·5641	$S(wxy)$	91·1157
$\bar{x}$	·481898	A	55·3614	B	38·1473
$\bar{y}$	4·906167			b	·68906

The recalculated slope, $b = B/A$, is ·68906. This is considerably higher than the provisional value, ·6. Accepting this solution, the 50 per cent. death point is found when $Y = 5$, that is ·093833 above $\bar{y}$. We have, therefore,

$$6{\cdot}481898 + \frac{{\cdot}093833}{{\cdot}68906} = 6{\cdot}618$$

as the estimated 50 per cent. death point.

For goodness of fit we find

$S(wy^2)$	568·6512
C	29·3835
$\chi^2 = C - Bb$	3·0977 $n = 5$

Since two constants have been fitted there are five degrees of freedom, so that there is no indication of departure from the form of the theoretical expectation. In these circumstances the variance of the estimated 50 per cent. death point may be taken to be

$$s^2 = \frac{1}{b^2}\left\{\frac{1}{W}+\frac{(x-\bar{x})^2}{A}\right\}$$

Numerically,

$1/W$	·044635
$(·136)^2/A$	·000334
	·044969
bs	·21206

The standard sampling error of the estimate is therefore ·308.

This is a large-sample method, appropriate if $\bar{y}$ is sufficiently near to 5, and the slope b sufficiently well determined. Failing this, the use of a standard error is not justified, and we should calculate the fiducial limits of the 50 per cent. death point, at any chosen level of significance, as Bliss has proposed (5). These are

$$\bar{x}+\frac{1}{Bb-t^2}\left\{(5-\bar{y})B \pm t\sqrt{A\left\{\frac{1}{W}(Bb-t^2)+(5-\bar{y})^2\right\}}\right\},$$

where t stands for the normal deviate corresponding with the level of significance chosen. *E.g.* the limits are 5·988 and 7·294, when $t = 1·96$.

Had χ^2 been excessive, indicating causes of disturbance other than random sampling, the estimated variance would be multiplied by χ^2/n, or, in the small-sample treatment, t would be taken from Student's distribution for the number of degrees of freedom in χ^2, and multiplied by $\chi/\sqrt{n}$ before use in the formula for fiducial limits. It should be remembered also that a significant value of χ^2 may throw some doubt on the linearity of the regression.

The example has been chosen to illustrate the expeditious use of the tables. No interpolation has been employed, as it would be if the process were repeated, using the results of the first fit as a new provisional line. It is, therefore, of interest, as an indication of the precision to be expected, that a much more precise fit gives 6·609±·317 for the 50 per cent. point, showing that this value and its precision have been very satisfactorily estimated. The slope, estimated at ·6891±·1344, is not so good, since the more exact value is ·7126±·1574. Here the error exceeds a tenth of the standard error of random sampling, so that, if the slope were of importance, a second fitting would have been justified. Finally, from the more exact fit we find $\chi^2 = 3·773$ in place of 3·098, confirming the agreement with expectation, but showing also that had this been doubtful a more exact fitting might have resolved the doubt.

Dosage Mortality Tests involving a Natural Death Rate

A further complication arises when there is reason to suspect that some animals have died from natural causes operating independently of the poison. If the natural mortality rate is K, the mortality rate, P, actually due to a particular dose of poison may be estimated as $p = (p^* - K)/(1-K)$, where p^* is the *total* death rate observed. The probit regression line should then be fitted to the probits of p, not to those of p^*, using modified weighting coefficients obtained by multiplying the coefficients in Table IX2 by $P\Big/\left\{P+\frac{K}{1-K}\right\}$. Values of the modified coefficients, adapted from Finney (3a), are given in Table IX3 for values of K from ·00 to ·40.

In practice, K is generally unknown. If observations have been made on a control (untreated) group of animals, the mortality rate in this group, K_c, is an estimate of K; a better estimate will take account also of evidence from the lower doses of the poison, especially if one of these shows a total mortality rate less than K_c.

For exact work, particularly when K is not well determined by the control group, the method of maximum likelihood should be applied to estimate this parameter at the same time as the coefficients a and b of the regression line are estimated. A provisional estimate of K, say K_o, should be used to give values of p from which a provisional regression line may be constructed; K_o may be taken as equal to K_c, but, if the number of animals in the control group is small (or even zero), a suitable guess should be based upon the values of p^* for the lower doses of the poison. A table like Table 2 is prepared, with the addition of a column for the auxiliary variate $x' = Q/Z$, tabulated as a function of Y in Table XI1. A weighted linear regression equation of y on x and x' is next calculated, with the modification that the quantities $n_c(1-K_0)/K_0$ and $n_c(K_c-K_0)/K_0$ are added to the sum of squares of deviations for x' and the sum of products of deviations for x', y respectively. Here n_c is the number of subjects in the control group, and all weighting coefficients come from the modified formula. The sum of squares of deviations for y, denoted by C, must be increased by $\frac{n_c(K_c-K_0)^2}{K_0(1-K_0)}$, as a contribution from the controls. The modified notation now used will be clear from examination of Table 2.01. The two regression coefficients are estimates of b and $\delta/(1-K_0)$ respectively; $(K_0+\delta)$ is a revised estimate of K, and the revised regression equation is $Y = a+b(x-\bar{x})$, where $a = \bar{y}-\bar{x}'\delta/(1-K_0)$. The process may be repeated, using the values just calculated as a new provisional set if they differ seriously from the first set.

Example 7.1

The first three columns of Table 2.01 give the results of a test of the toxicity of an extract of *Derris elliptica* sprayed on to the sawtoothed grain beetle, *Oryzæphilus surinamensis* (Martin (43)). The last line refers to a control group; this received the spray medium only, and in it 16.3 per cent. of the insects were affected.

In this example, K_0 might well be taken as equal to ·163, but, in order to simplify interpolation in Table IX3 (and to illustrate the calculations more fully), the value ·16 has been used. Values of p have been calculated from each p^*, their empirical probits plotted against x, and a straight line drawn by eye on the diagram. From this line, provisional probits, Y, have been read. If K_0 were well determined, second approximations to a and b could be obtained by the procedure of Example 7, except that the weighting coefficients would be taken from Table IX3 by interpolation for $K = \cdot 16$. The values of w, y, wx, wy and the sums of squares and products from which A, B and C are derived are shown in Table 2.01.

TABLE 2.01

Log Conc. (x).	No. Exposed.	No. Affected.	p^*.	p ($K_0 = \cdot 16$).	Emp. Probit.	Y.	x'.	y.	w.	wx.	wx'.	wy.
2·79	125	125	1·000	1·000	—	7·3	·379	7·68	7·9	22·041	2·9941	60·672
2·66	117	115	·983	·980	7·05	7·0	·421	7·05	12·8	34·048	5·3888	90·240
2·49	127	114	·898	·879	6·17	6·5	·516	6·08	28·3	70·467	14·6028	172·064
2·17	51	40	·784	·743	5·65	5·6	·823	5·65	22·6	49·042	18·5998	127·690
1·57	132	34	·258	·117	3·81	3·8	4·557	3·81	18·5	29·045	84·3045	70·485
—	129	21	·1628	—	—	—	—	—	$W = 90·1$	204·643	125·8900	521·151

$$\bar{x} = 2\cdot 2713,\quad \bar{x}' = 1\cdot 3972,\quad \bar{y} = 5\cdot 7841$$

$$129 \times \cdot 84/16 = 677\cdot 2500,\quad 129\times(\cdot 1628-\cdot 16)/\cdot 16 = 2\cdot 258,\quad 129\times(\cdot 1628-\cdot 16)^2/\cdot 16\times\cdot 84 = \cdot 01$$

$S(wx^2)$.	$S(wxx')$.	$S(wx'^2)$.	$S(wxy)$.	$S(wx'y)$.	$S(wy^2)$.
479·5467	231·7684	410·4217	1225·501	576·060	3138·30
464·8031	285·9324	175·8967	1183·684	728·165	3014·41
$A = 14·7436$	$A' = -54·1640$	234·5250	$B = 41·817$	−152·105	123·89
		677·2500		2·258	·01
		$A'' = 911·7750$		$B' = -149·847$	$C = 123·90$

If the information on K is less complete, it also must be included in the maximum likelihood estimation. The additional columns x' and wx' are then required in Table 2.01, where $x' = Q/Z$ is taken from Table IX3. From these columns, squares and products of deviations for x' with x and y must be calculated just as were

A, B and C. The lower part of Table 2.01 summarises the calculations leading to equations for the regression coefficients on x and x'. These are

$$14{\cdot}7436b - 54{\cdot}1640\left(\frac{\delta}{1-K_o}\right) = 41{\cdot}817$$

$$-54{\cdot}1640b + 911{\cdot}7750\left(\frac{\delta}{1-K_o}\right) = -149{\cdot}847.$$

The equations are best solved by the method outlined in *Statistical Methods* (1), as the inverse matrix of coefficients thus obtained leads to the variances and covariance of b and δ. The inverse matrix is

$$V = \begin{bmatrix} {\cdot}0867605 & {\cdot}0051540 \\ {\cdot}0051540 & {\cdot}0014029 \end{bmatrix}$$

whence

$$b = 2{\cdot}8558, \quad \delta/(1-K_o) = {\cdot}00530.$$

Multiplication of the second of these quantities by ·84 gives $\delta = {\cdot}0045$, so that the revised value for K, ·1645, is little different from either the provisional value or that estimated from the controls alone. Also

$$a = 5{\cdot}7841 - 1{\cdot}3972 \times {\cdot}00530 = 5{\cdot}7767.$$

The regression equation $Y = a+b(x-\bar{x})$ is easily seen to agree well with the Y column of Table 2.01, so that no repetition of the calculations is necessary.

Goodness of fit may be tested by calculating

$$\chi^2 = C - Bb - B'\delta/(1-K_o) = 5{\cdot}27,$$

for which the degrees of freedom are here three (= total number of dose levels, including controls, minus 3). This gives no indication of heterogeneity. In cases of doubt, χ^2 should be calculated from expected frequencies after grouping of classes with small expectations, in the manner described by Finney (3a). A large value of χ^2 must be interpreted and used in the same way as for the simple probit problem (Example 7).

The 50 per cent. affected point is estimated to be

$$x_{50} = \bar{x} + (5-a)/b = 1{\cdot}999.$$

On large sample theory, the diagonal elements of V, v_{11} and v_{22}, are the variances of b and $\delta/(1-K_o)$ respectively, and v_{12} is the covariance of these. Provided that b^2 is large relative to the variance of b, the variance of x_{50} may be taken as

$$s^2 = \frac{1}{b^2}\left\{\frac{1}{W} + v_{11}(x_{50}-\bar{x})^2 - 2v_{12}\bar{x}'(x_{50}-\bar{x}) + v_{22}\bar{x}'^2\right\}$$

$$= \frac{1}{(2{\cdot}8558)^2}\left\{{\cdot}0110988 + {\cdot}0867605 \times ({\cdot}2720)^2 + 2 \times {\cdot}0051540 \times 1{\cdot}3972 \times {\cdot}2720 + {\cdot}0014029 \times (1{\cdot}3972)^2\right\}$$

$$= {\cdot}002964.$$

The standard error of the 50 per cent. point is therefore ·0544. Here $(1{\cdot}960)^2 v_{11}/b^2 = {\cdot}041$, a value small enough to justify the use of s in assigning 95 per cent. fiducial limits to x_{50}; if this quantity exceeds ·05, the limits ought to be calculated by an extension of the procedure illustrated for Example 7. The variance of δ is $v_{22}(1-K_o)^2$, which is also therefore the variance of the revised estimate of the natural response rate; hence this rate is estimated to be $\cdot 1645 \pm \cdot 0314$.

Tables X to XII1. Other Transformations

The normal or probit transformation may be looked upon as one of a series of methods of transforming percentages into more suitable variables. The choice of transformation depends on the mathematical specification (the "model") which is considered most appropriate to the data under investigation. When this specification has been decided the transformation chosen should be such that the quantities that require estimation are simple functions of the transformed variate; thus in probit work it is anticipated that the transformed variate will bear a linear relation to the dosage when the latter is expressed in suitable units (usually logarithms), and the quantities that require estimation are the constant term and the slope of this regression.

The general formulæ for the maximum likelihood values of the working deviate for given provisional values are :

Maximum	$X+Q\frac{dX}{dP}$
Minimum	$X-P\frac{dX}{dP}$
Weighting coefficient	$\frac{1}{PQ}\left(\frac{dP}{dX}\right)^2$

where P, Q stand for the probability corresponding to the provisional value X.

Tables for the angular, logit and loglog transformations are given in this volume. The angular transformation

$$p = \sin^2 \phi$$

transforms a probability varying from 0 to 1 into an angle varying from 0° to 90°, and is remarkable in that the amount of information concerning ϕ is constant. In consequence, in large samples, the sampling variation tends to normality with a variance which depends only on the number of observations on which the percentage is based. The weighting coefficient for this transformation is constant. In other respects Tables X to X2 afford the same facilities for handling percentages transformed in this way as do Tables IX to IX2 for the probit transformation. This transformation is equivalent to replacing the normal curve of the probit transformation by one of which the ordinate is $\sin^2 \phi$.

When the extremes of the probability scale have to be covered the angular transformation is often inappropriate, since expected angular values outside the range 0° to 90° do not correspond to real probabilities. A transformation which is often used in such cases when there is no clearly appropriate theoretical model, and when there is no reason to differentiate between the two ends of the probability scale, is the logit transformation

$$z = \tfrac{1}{2} \log_e(p/q)$$

Tables XI and XI1 provide the necessary facilities for the use of this transformation. As in the probit transformation 5 is sometimes added to the transformed variate to avoid negative values, but this has not been done in Tables XI and XI1. The transformation is equivalent to the r, z transformation ($r = \tanh z$) given in Table VIII1, with $r = 2p-1$.

In genetics also, Kosambi's relation between the map distance (x) and the recombination fraction (y) between two loci, may be expressed as

$$2y = \tanh(2x)$$

so the table may be entered with $2y$ for r or for $2p-1$, or with $2x$ for z.

The logit transformation has the property that for equal intervals in the logit scale the odds ($\lambda p : \lambda q$) are changed by the same factor, a difference of $\frac{1}{2}\log_e 2$ (= 0·34657) being equivalent to a doubling of the odds. The logit transformation is equivalent to replacing the normal curve of the probit transformation by one of which the ordinate is $\frac{1}{2}\,\text{sech}^2 z$.

Intermediate in character between the distributions of the probit and the logit is the deviate appropriate to gene frequencies determined by selection and diffusion, where p, q are related to the standard deviate x by the equation

$$\frac{d^2q}{dx^2} = 4pqx$$

The name " legit " has been proposed for this transformation. The necessary tables have been given by Fisher (56).

In many physical and biological phenomena the probability of occurrence of a change can be expressed in the form

$$p = 1 - e^{-\beta t}$$

In such cases the transformation

$$u = \log_e\left\{-\log_e(1-p)\right\}$$

is of value, since, for example, if two experimental treatments result in different values β_1 and β_2 of β, the differences of the transformed variate when t is the same for both treatments provide estimates of $\log_e\beta_1 - \log_e\beta_2$. Tables XII and XII1 give the necessary facilities for this transformation, which we have termed the complementary loglog transformation, to distinguish it from the direct loglog transformation

$$v = \log_e(-\log_e p)$$

which is the form usually tabulated. The tabulation of the complementary form has the advantage that the transformed variate increases with p, as in the other tabulated transformations. Since the transformation is not symmetrical about $p = \frac{1}{2}$ the form required will depend on the phenomena under investigation. When p is small the complementary loglog transformation is nearly equivalent to the log transformation $u = \log_e p$, and consequently also (apart from a factor of 2) to the logit transformation.

In addition to their toxicological and similar applications, transformations are of use in the analysis of material which has a very skew distribution, for the purpose of equalising the variance of quantities of widely different magnitudes which it is desired to include in the same analysis, and for the purpose of reducing the observations to a scale on which the effects of associated variables are more nearly constant. Thus when it is required to perform an analysis of variance on a set of percentages, *e.g.* of the deaths of seedlings obtained from plots of a replicated experiment, it is frequently advantageous to transform these percentages by means of the angular transformation. As we have seen, this has the effect, in large samples, of equalising the component of variance due to sampling, when each percentage is based on the same number of observations The square root transformation

$$y = \sqrt{x}$$

has the same property when the observations on each plot are distributed according to a Poisson distribution. This transformation is equivalent to the angular transformation at each end of the percentage scale.

The logarithmic transformation

$$y = \log x$$

is useful when dealing with material on which the effects of other variables may be expected to be proportional to the numbers involved. The influence of a given increase in temperature on the number of insects caught in a light trap, for example, is likely to be proportional to the number of active insects, in which case the increment in the logarithm of the catch will be the same whatever the number of insects, unless some of the numbers are very small. In this connection it may be noted that the transformation $y = \log_{10}(x+1)$ is sometimes convenient, being substantially equivalent to $y = \frac{1}{3}\sqrt{x}$ when x is less than 10 and to $y = \log_{10}x$ for larger numbers.

Final adjustments can be made to data distributed according to the Poisson distribution in the same manner as for data distributed according to the binomial distribution. For the square root and logarithmic transformations the working values and weighting coefficients corresponding to provisional values Y and X are

	Square root	Logarithmic (base e)
Working value	$\frac{1}{2}Y + x/2Y$	$Y - 1 + x/X$
Weighting coefficient . .	4	X

The use of such adjustments makes proper allowance for the variability of the data in small samples, and thus obviates the need, which has been felt in certain quarters, for the use of transformations such as $y = \sqrt{(x+\frac{1}{2})}$ in place of $y = \sqrt{x}$. It has also the important additional advantage, which accrues also in the transformations for percentages, of eliminating the distortion that would otherwise occur when means and other estimates obtained in terms of the transformed variate are expressed in terms of the original variate. Thus, for example, if x has the values 2, 3, 4, 6, $\bar{x} = 3{\cdot}75$, but the mean of $\sqrt{x}$ is $1{\cdot}8990$, giving on squaring the value $3{\cdot}6060$. Taking a provisional value for Y of $1{\cdot}9$ we obtain an adjusted value for $\bar{x}$ of $(0{\cdot}95+3{\cdot}75/3{\cdot}8)^2 = 3{\cdot}7513$. A second approximation would give full accuracy to 6 places of decimals.

A further use of transformations is for the purpose of facilitating interpolation by obtaining a transformed function which is much more nearly linear. One example of this, the use of the reciprocals of n_1 and n_2 in interpolation in the table of z (Table V), has already been mentioned. The logarithmic transformation and the transformation of Table XI have frequently proved of use when dealing with functions whose argument is a probability.

Example 7.1.1

In a study of recombination in the short interval between *agouti* and *undulated* in mice the data shown in Table 2.01.1 were obtained. These data clearly show a decrease in recombination fraction for both male and female heterozygous parents as the age of the heterozygous parent increases. It was desired to test the data to see if there was any difference in the rate of decrease for heterozygotes of the two sexes, and whether allowing for differences in age, there was any average difference between the sexes (Fisher, 68).

The angular transformation is suitable. The basic computations are shown in Table 2.01.2 and the analysis of variance in Table 2.01.3. A preliminary fitting of a regression line to the angles derived from the pooled data for the two sexes gave the provisional angles shown in the third column of Table 2.01.2. The corresponding minimal angles and ranges derived from Table X2 are shown in the fourth and fifth columns. (Four point

TABLE 2.01.1. RECOMBINATION DATA

Age of heterozygous parent (months)	From heterozygous females: Numbers	From heterozygous females: Recombination percentage	From heterozygous males: Numbers	From heterozygous males: Recombination percentage
0–3	20/314	6·37	16/228	7·02
3–5	19/525	3·62	26/506	5·14
5–7	29/522	5·56	20/434	4·61
7–9	12/267	4·49	18/342	5·26
9–11	3/81 }		5/222 }	
11–13	0/20 }	2·91	2/105 }	2·10
13–15	0/2 }		0/6 }	
	83/1731	4·79	87/1843	4·72

TABLE 2.01.2 RECOMBINATION DATA: COMPUTATIONS

Age	t	Prov. Working Angles: Angle	Min.	Range	Females: n_1	ϕ_1	$n_1\phi_1$	Males: n_2	ϕ_2	$n_2\phi_2$	Both sexes: n	$\bar{\phi}$	$n_1\phi_1+n_2\phi_2$
0–3	−2	14·35	7·02	119·3	314	14·62	4590·68	228	15·39	3508·92	542	14·94391	8099·60
3–5	−1	13·38	6·56	127·2	525	11·16	5859·00	506	13·10	6628·60	1031	12·11212	12487·60
5–7	0	12·41	6·11	136·5	522	13·69	7146·18	434	12·40	5381·60	956	13·10437	12527·78
7–9	1	11·44	5·64	147·4	267	12·26	3273·42	342	13·40	4582·80	609	12·90020	7856·22
9–11	2	10·47	5·18	160·3	81	11·12	900·72	222	8·79	1951·38	303	9·41287	2852·10
11–13	3	9·50	4·71	176·0	20	4·71	94·20	105	8·06	846·30	125	7·52400	940·50
13–15	4	8·53	4·23	..	2	4·23	8·46	6	4·23	25·38	8	4·23000	33·84
Total . .					1731		21872·66	1843		22924·98	3574		44797·64
Mean . .						12·635852			12·438947			12·534314	
$S(nt)$ and $\bar{t}$.					−656		−·378972	+163		+·088443	−493		−·137941

TABLE 2.01.3. RECOMBINATION DATA: SUMS OF SQUARES AND PRODUCTS

	$A[t^2]$	$B[t\phi]$	$C[\phi^2]$	b
Females	2335·4	−1359·9	4581·3	−·5823 } −·9830
Males	3674·6	−4548·0	7896·9	−1·2377 }
Sex difference	195·0	−82·2	34·6	
Total	6205·0	−5990·1	12512·8	−·9654

TABLE 2.01.4. RECOMBINATION DATA: ANALYSIS OF VARIANCE

	D.F.	S.S.	Differences: D.F.	Differences: S.S.	χ^2	Effect tested
D.F. and S.S. accounted for by:						
Single regression .	1	5782·6	1	5782·6	7·046	General regression
Parallel regressions .	2	5842·3	1	59·7	·073	Sex difference
Separate regressions .	3	6455·5	1	613·2	·747	Difference between regressions
Total	13	12512·8	10	6057·3	7·381	Residual variation

interpolation was used, but linear interpolation would give adequate accuracy for these data.) From these values and the observed numbers the working angles, ϕ_1 and ϕ_2, and totals, $n_1\phi_1$ and $n_2\phi_2$, are calculated: thus $(314\times 7{\cdot}02+20\times 119{\cdot}3)/314 = 14{\cdot}62$. The marginal totals and means are then entered as shown. The weights attached to the angles are proportional to the total numbers n in each class. Consequently the total sum of squares is given by $S\phi_1(n_1\phi_1)+S\phi_2(n_2\phi_2)-\bar{\phi}S(n_1\phi_1+n_2\phi_2)$.

The regressions of the separate sexes, and for both sexes together, on the age variate t are obtained by calculating, from the data for each sex separately, and from the pooled data for both sexes, the quantities

$$A = Sn(t-\bar{t})^2, \quad B = Sn\phi(t-\bar{t}) = St(n\phi) - \bar{t}Sn\phi$$

where $\bar{t}$ is the appropriately weighted mean of t. The values of A and B and of C, the corresponding sum of squares for ϕ, are shown in Table 2.01.3, together with the regression coefficients. These regression coefficients are given by B/A, and the sums of squares accounted for by B^2/A. If parallel regression lines are fitted to the two sexes the above sums of squares and products for the two sexes separately must be pooled, so that the regression coefficient is $(B_1+B_2)/(A_1+A_2)$ and the sum of squares accounted for is $(B_1+B_2)^2/(A_1+A_2)$. For the separate regressions, and for parallel regressions, the sum of squares due to the difference between the means of the sexes, namely $\bar{\phi}_1 Sn_1\phi_1+\bar{\phi}_2 Sn_2\phi_2-\bar{\phi}S(n_1\phi_1+n_2\phi_2) = 34{\cdot}6$, must be added to give the total sum of squares accounted for.

The sums of squares for testing the various items are found by taking the differences of the sums of squares accounted for. The corresponding χ^2 values are obtained by dividing by $820{\cdot}7$. These are shown in Table 2.01.4. The decrease with age is clearly significant, but the difference, $+{\cdot}2626$ in favour of males, after allowing for differences in mean age between the two sexes is substantially smaller than its standard error, nor is there any significant difference between the regressions for the two sexes. The residual variation also conforms with expectation.

Table XIII. Scores for Linkage Data from Intercrosses

Much linkage data with plants is obtained by selfing double heterozygotes, which may be in coupling or in repulsion. In some cases, especially with animals, single or double backcrosses may also be available. The compilation of such complex data, for the estimation of linkage, or the examination of heterogeneity, is much facilitated by efficient scores, based on the method of maximum likelihood, such as have been developed for human genetics.

Such scores possess the properties (i) that the scores are linear functions of the frequencies observed, (ii) that the expected score is zero if the trial value is correct, (iii) that the ratio of change of the expected score with change of trial value is equal to its sampling variance, and to the amount of information to be expected from such a body of data as is available. Estimation from a single body of data thus consists in finding the value for which the total score vanishes. The efficiently weighted combination of different lots of data consists simply in adding their scores. Homogeneity is tested by recognising that the square of the score divided by the amount of information is χ^2 for one degree of freedom. If m is the expectation of any distinguishable class, expressed as a function of the unknown parameter θ, the efficient score for cases observed in that class may be taken to be

$$\frac{1}{m}\,\frac{\partial m}{\partial \theta}.$$

For double heterozygotes intercrossed the two singly recessive classes have the same expectation and the same score. Table XIII gives the means of scoring such data in coupling and repulsion for each 1 per cent. of the recombination fraction, y, the quantities tabulated being:

	Coupling	Repulsion
Double dominant	$\dfrac{2(1-y)}{2+(1-y)^2}$	$-\dfrac{2y}{2+y^2}$
Single recessives	$\dfrac{1}{y}-\dfrac{1}{2-y}$	$\dfrac{1}{1-y}-\dfrac{1}{1+y}$
Double recessive	$\dfrac{2}{1-y}$	$-\dfrac{2}{y}$

The signs in the table are for $\theta = 1-y$, where y is the recombination fraction, in order that evidence for linkage ($y<{\cdot}5$) may be supplied by positive scores at 50 per cent.

Data should be scored at two adjacent values giving positive and negative scores. One hundred times the difference then gives the amount of information to a good approximation. For more accurate work, and especially for close linkage, four adjacent values with four-point interpolation may be used. The four-point value for the actual amount of information obtained at the value $x+\theta$ is

$$(u_{x+1}-u_x)-\frac{1}{6}(1-3\theta^2)\delta^2 u_{x+1}+\frac{1}{6}(1-3\phi^2)\delta^2 u_x,$$

where $\theta+\phi = 1$.

Since the Third Edition, similar tables have been calculated independently by D. J. Finney (57), who gives a good account of the origin and use of this mode of application of the method of maximum likelihood.

Example 7.2

For two factors in the Sweet Pea we have the data :

	Double dominant	Single recessives	Double recessive
Coupling	268	30	79
Repulsion	324	344	0

Scoring these at 6 per cent., 7 per cent., 8 per cent. and 9 per cent., we find

	6 per cent.		7 per cent.		8 per cent.		9 per cent.
Score	−119·701		−43·367		15·165		61·852
δ^1		+76·334		+58·532		+46·687	
δ^2			−17·802		−11·845		

To find the score at an intermediate value, such as 7·811 per cent., we find the inner interpolate, 4·102, and, since the linkage is close, the four-point value, 5·191. A method of calculating this is given in the section on *Interpolation* at the end of the introduction. The first approximation to the amount of information obtained is 5853·2, and the adjustment is −457·0, giving 5396·2 at this value.

Table XIII1. The Product-Ratio Method for Intercross Data

When one of the 3 : 1 ratios is disturbed, or both are disturbed independently, it is sometimes desirable to estimate the linkage, making allowance for differential viability between phenotypes ; although, since three of the four phenotypes are mixtures, such an allowance is somewhat insecure. In these cases the product ratios

bc/ad for coupling

and

ad/bc for repulsion

may make a satisfactory allowance.

Table XIII1 gives the theoretical values for these ratios for different recombination fractions, supposed the same for male and female gametogenesis. The significance of the estimate arrived at is best obtained from the χ^2 value of the fourfold table

$$\begin{matrix} a & c \\ b & d \end{matrix}$$

which tests the significance of linkage. The standard error may then be taken to be

$$\frac{50-y}{\chi}\,cM$$

if the significance is in doubt, where y is the recombination estimate expressed in centimorgans. The standard error for recombination fractions other than 50 per cent., may be obtained by dividing the second column of the table by $\sqrt{h}$, where h is the harmonic mean of the four frequencies observed. Since the method supplies no means of confirmation of the assumptions made, its use will generally be provisional pending the acquisition of data both in coupling and repulsion, or perhaps of backcross in place of intercross data.

Example 7.2.1. *Recombination fraction estimated from a cross-product ratio*

Kidwell, Gowen and Stadler (82) give the following data for intercrosses in repulsion between recessive pied *s* in the house-mouse, and a newly-discovered recessive, pugnose, *pn*

++	+*pn*	*s*+	*s pn*	Total
501	191	220	16	928

Suspecting some lack of viability among the pugnose, we may enter Table XIII with

$$ad/bc = 501 \times 16/191 \times 220 = \cdot 19077$$

giving (by four point interpolation) a recombination fraction 27·935 per cent. and the corresponding multiplier 23·147. The harmonic mean of the four frequencies is 53·852, $\sqrt{h} = 7\cdot 3384$; dividing by this, the standard error of estimation comes to 3·1542. The true value should lie between 21 per cent. and 34 per cent.

This publication gives no intercrosses in coupling, but a backcross of heterozygous females had

$++$	$+pn$	$s+$	$s\,pn$	Total
86	154	153	46	439

showing 30·068 per cent. recombinants. Although very likely the double recessive *s pn* has a particularly low viability, there is no significant difference between the two estimates of linkage, available at this early stage, before heterozygotes in coupling have been tested.

Table XIV. Segmental Functions, α, β, γ, δ

These functions are defined for all values of their argument as the sums of sub-series of the exponential series, containing every fourth term. The four sub-series having leading terms of degrees 0, 1, 2, and 3, define the functions α, β, γ, and δ.

They are closely connected with the hyperbolic and circular functions by the identities

$$\alpha+\gamma = \cosh, \qquad \beta+\delta = \sinh,$$
$$\alpha-\gamma = \cos, \qquad \beta-\delta = \sin.$$

Two elementary identities are found by eliminating the argument.

$$\alpha^2+\gamma^2-2\beta\delta = 1$$
$$\beta^2+\delta^2-2\alpha\gamma = 0.$$

Differentiation with respect to the argument merely permutes the symbols in the order $(\delta\gamma\beta\alpha)$.

The functions are peculiarly adapted to specifying the frequencies of non-recombinant and recombinant gametes of all types in terms of the metrical positions of the centromere, a series of any number of markers, and the terminus of the chromosome arm. The argument of each function is twice the metrical length of the segment. Thus with four markers and five segments the frequency of gametes non-recombinant in every segment is given by a simple combinatorial rule to be the sum of the following sixteen products,

$\alpha\alpha\alpha\alpha\alpha$	$\alpha\alpha\alpha\beta\delta$	$\beta\alpha\delta\alpha\alpha$	$\beta\alpha\alpha\alpha\delta$
$\alpha\beta\alpha\delta\alpha$	$\alpha\alpha\beta\alpha\delta$	$\beta\alpha\alpha\delta\alpha$	$\beta\alpha\delta\beta\delta$
$\alpha\alpha\beta\delta\alpha$	$\alpha\beta\alpha\alpha\delta$	$\beta\delta\alpha\alpha\alpha$	$\beta\delta\beta\alpha\delta$
$\alpha\beta\delta\alpha\alpha$	$\alpha\beta\delta\beta\delta$	$\beta\delta\beta\delta\alpha$	$\beta\delta\alpha\beta\delta$

in which, for the segment next the centromere the appropriate function is either α or β, while for the segment ending at the terminus it is either α or δ. Intervening segments may have α, β or δ, provided that β and δ appear in equal numbers and alternate.

The corresponding expressions for gametes having recombination in one or more of the five segments are found simply by making the substitution $(\alpha\gamma)(\beta\delta)$ in each corresponding column. In particular if as is usual, there is no marker at the centromere, or at the terminus, the eight observable classes in a four-point test will be specified by permutations in the three middle columns of the expression :—

$$(\alpha+\gamma)\left\{\begin{matrix}\alpha\alpha\alpha\\ \beta\alpha\delta\\ \alpha\beta\delta\\ \beta\delta\alpha\end{matrix}\right\}(\alpha+\gamma) \qquad (\beta+\delta)\left\{\begin{matrix}\alpha\delta\alpha\\ \alpha\alpha\delta\\ \delta\alpha\alpha\\ \delta\beta\delta\end{matrix}\right\}(\alpha+\gamma)$$

$$(\alpha+\gamma)\left\{\begin{matrix}\alpha\alpha\beta\\ \alpha\beta\alpha\\ \beta\alpha\alpha\\ \beta\delta\beta\end{matrix}\right\}(\beta+\delta) \qquad (\beta+\delta)\left\{\begin{matrix}\alpha\alpha\alpha\\ \alpha\delta\beta\\ \delta\beta\alpha\\ \delta\alpha\beta\end{matrix}\right\}(\beta+\delta).$$

With the aid of the tables these eight frequencies may be easily calculated for any assigned values of the five metrical lengths of the segments. Thus, starting from the left, we may calculate

$$P = (\beta+\delta)\alpha+(\alpha+\gamma)\beta,$$
$$Q = (\beta+\delta)\delta+(\alpha+\gamma)\alpha,$$

beside which we can calculate the parallel quantities,

$$P' = (\beta+\delta)\gamma+(\alpha+\gamma)\delta,$$
$$Q' = (\beta+\delta)\beta+(\alpha+\gamma)\gamma,$$

to be used in the cases in which recombination takes place in the second segment.

To bring in the third segment we make the similar calculation

$$R = P\alpha+Q\beta, \qquad R' = P\gamma+Q\delta,$$
$$S = P\delta+Q\alpha, \qquad S' = P\beta+Q\gamma,$$

with the same variation for recombination in the third segment, and, of course, using P', Q' in place of P and Q for the cases involving recombination in the second segment.

Equally, for the fourth segment, we take for each of the four pairs of values R, S

$$T = R\alpha+S\beta, \qquad T' = R\gamma+S\delta,$$
$$U = R\delta+S\alpha, \qquad U' = R\beta+S\gamma,$$

giving eight pairs of values for the eight distinguishable classes of gamete. Using the segmental functions of the fifth segment, we have from each of these pairs

$$T(\beta+\delta)+U(\alpha+\gamma)$$

the eight values of which should check to a total calculable in advance, namely the hyperbolic cosine of twice the total arm-length from centromere to terminus.

The expectations corresponding with assigned lengths of segments are then found by dividing each by $\cosh 2U$, where U now stands for the total metrical arm length. (References 58, 59, 60, 61).

Example 7.3

An arm is divided by four markers P, Q, R and S into five segments with metrical lengths 10, 25, 20, 15 and 5 cM, where the centimorgan is defined as 1 per cent. of the metrical unit. To apply the computational procedure outlined above, these lengths are first doubled to give the first two decimal places of the arguments of the segmental functions to be used. The first argument is therefore ·20 and the second ·50; we take then directly from the table :—

$$(\beta+\delta)(\cdot 20) = \cdot 2013360 \qquad \alpha(\cdot 50) = 1\cdot 0026043 \qquad \delta(\cdot 50) = \cdot 0208349$$
$$(\alpha+\gamma)(\cdot 20) = 1\cdot 0200668 \qquad \beta(\cdot 50) = \cdot 5002604 \qquad \gamma(\cdot 50) = \cdot 1250217$$

from which the pairs P, Q and P', Q' are calculated directly, as shown in Table 2.02 with the remaining calculations completed.

Since the differential coefficients of these functions are the same functions permuted, the rates of change of the eight frequencies for variations of the segment lengths are obtained by calculations involving the same tabulated values. Segment lengths may then be estimated from observed frequencies by the method of maximum likelihood. Estimates of unmarked segments will, naturally, have a very low precision.

TABLES XV AND XVI. LATIN SQUARES

Latin squares of size 7×7 and smaller have been enumerated (11), (22), (66). For 4×4 and 5×5 squares all the standard squares (*i.e.* squares with first row and first column in the standard order) are shown in Table XV, except that only one of each pair of conjugate 5×5 squares is given (two standard squares are conjugate if the rows of one are the columns of the other). From each 4×4 standard square 4 ! 3 !—1 non-standard squares, all different, may be generated by permuting all the rows except the first, and all columns. Thus each 4×4 standard square represents a set of 144 squares. There are consequently 4×144 or 576 4×4 Latin squares. Similarly, there are 56×2880 or 161,280 5×5 Latin squares. There is one 2×2 standard square and one 3×3 standard square, giving two 2×2 squares and twelve 3×3 squares in all.

For 6×6 squares one square of each transformation set (*i.e.* set generated from one of its members by permutation of rows, columns, and letters), or of each pair of conjugate transformation sets, is shown. Such permutation of rows, columns and letters will generate $(6\,!)^3$ or 373,248,000 squares, of which 6×6 ! or 4320 are standard squares. These squares are, however, not necessarily all different, the largest number actually obtained in any one 6×6 set being one-quarter of the possible number, *i.e.* 1080 standard squares, and the

TABLE 2.02

	$R, S.$	$T, U.$		Frequency per cent.	Segments recombining.
		1·4348573	1·1899512	50·584	None
	1·1237739	1·0410167			
	1·0356102				
		·0552312	·3912112	16·630	(4)
P ·7121594		·3837585			
Q 1·0269182					
		·1780864	·3871909	16·459	(3)
	·0679309	·3675134			
	·3670838				
		·0047088	·0375561	1·596	(34)
		·0368998			
		·2065418	·2515250	10·692	(2)
	·1377938	·2296870			
	·2289896				
		·0072313	·0526285	2·237	(24)
P' ·0464244		·0516457			
Q' ·2282509					
		·0172024	·0387828	1·649	(23)
	·0061490	·0368752			
	·0368351				
		·0004425	·0035643	·152	(234)
		·0035024			
			2·3524100	99·999	
		cosh 1·50	2·3524096		

smallest number one 216th, *i.e.* 20 standard squares. The actual numbers in each transformation set are shown in the table. There are in all 9408 standard 6×6 squares, giving 812,851,200 6×6 squares in all.

By permuting with each other the three categories, rows, columns, and letters, six sets are formed, not necessarily all different. These sets are said to be *adjugate*. In the case of the 6×6 squares each pair of conjugate sets and the set immediately following them in the table form an adjugate trio of sets. The remaining sets are self-adjugate.

Norton (22) has enumerated 562 sets of 7×7 squares, of 146 species, containing 16,927,968 standard squares. Later, A. Sade, using a different method of enumeration, obtained the value 16,942,080 for the number of standard 7×7 squares; he has subsequently discovered a 147th species (66) which accounts for the discrepancy of 14,112 standard squares. Four examples are shown in the table. Square (*a*) is a member of a transformation set in which the squares generated by the permutation of rows, columns and letters are all different. It thus contains $(7\,!)^3$ or 128,024,064,000 squares, of which $7 \times 7\,!$ or 35,280 are standard squares. The sets generated by the adjugate permutations are also all different. The example thus stands for an aggregate of 211,680 standard squares. No Graeco-Latin squares can be formed with any of these.

Square (*b*) differs from square (*a*) only by the interchange of two pairs of letters, B and C in the 5th and 6th columns. It belongs to a set of only 11,760 standard squares, which is one of an adjugate trio of sets, giving 35,280 standard squares in all. One-seventh (1680) of the standard squares of one of these sets are self-conjugate. Again no Graeco-Latin square can be formed.

The third illustration is a Graeco-Latin square, with the fourth category (Greek letters) represented by numeral suffixes. No fifth category can be added. The suffixes are standardised by assigning them in their natural order to the first row. The Latin square involved is one of a set of only 210 standard Latin squares, but the same transformation of the Graeco-Latin square yields a set of 1680 standard Graeco-Latin squares, containing eight different standard arrangements of the suffixes in each Latin square. These eight solutions are in fact all that exist, so that this set of Latin squares belongs to only one Graeco-Latin set. Permutation of the three categories, rows, columns, and letters, yields the same set, but if suffixes are interchanged with rows, columns, or letters, three more sets of 1680 standard Graeco-Latin squares are formed (6720 in all).

The three new sets all involve the cyclic Latin square shown as the fourth example. There are only 120 standard Latin squares in this set, so that each has 14 solutions in each of the Graeco-Latin sets, or 42 in all. Under orthogonal squares are given six 7×7 squares, of which the first is the above Latin square, while the rest are five more Graeco solutions which belong to a second set and are mutually orthogonal.

Until recently little was known about squares larger than 7×7, other than the orthogonal properties of the special sets shown in Table XVI. In Table XV a single example of each size from 8×8 to 12×12 is given. These examples will suffice (after permutation of rows, columns, and letters at random) for experimental arrangements requiring Latin squares of these sizes.

To select a square at random from all possible squares of a given size up to 6×6 proceed as follows:

(1) Select one of the given squares by using the key numbers printed below each square, selecting a number at random from all possible key numbers. If the key number falls in the second of two groups use the conjugate square.

(2) In the case of 3×3, 4×4, and 5×5 squares permute all rows except the first of the selected square, and all columns. Alternatively permute all rows except the first and assign the letters to the treatments at random. For 6×6 squares permute all rows and all columns at random, and then assign the letters to the treatments at random.

Table XVI gives complete sets of orthogonal Latin squares of sides 3, 4, 5, 7, 8, and 9. Two Latin squares are orthogonal to each other if, when they are superimposed, every letter of one square occurs once and once only with every letter of the other. Such a pair of squares (one square being written with Greek and the other with Latin letters) form a Graeco-Latin square.

If a complete set of $p-1$ orthogonal squares of side p exists, the p^2-1 degrees of freedom corresponding to the contrasts of the p^2 cells of the square split up into $p+1$ sets of $p-1$ degrees of freedom, corresponding to the contrasts between the rows, between the columns, and between the letters of each of the $p-1$ squares. A set of less than $p-1$ orthogonal squares gives a similar but incomplete partition.

Complete sets of orthogonal squares are known to exist for all prime numbers and powers of primes. (See *The Design of Experiments*.) The rule of formation for all prime numbers will be apparent from inspection of the sets for $p = 5$ and $p = 7$, each row being formed by moving the previous row 1, 2, 3 . . . places to the left. The existence of complete sets of orthogonal squares, for all prime-power values of p, was proved independently by R. C. Bose (63) and W. L. Stevens (64). See also R. C. Bose and K. R. Nair (65).

The rows of the 8×8 squares may be taken to represent the 8 combinations of three two-level factors A, B, and C, in the order (1), a, b, ab, c, ac, bc, abc, and the columns may be taken to represent the similar combinations of a further three factors D, E, and F. The 64 cells of each square will then represent the 64 combinations of the 6 factors. The numbers of each square will divide these combinations into 8 groups of 8, the contrasts among which will represent a set of 7 degrees of freedom orthogonal to A, B and C and their interactions, and to D, E, and F and their interactions, and also to the six other sets of seven degrees of freedom given by the other six squares. In the squares shown these degrees of freedom correspond to specific interactions between the six factors. The interactions corresponding to 3 degrees of freedom from each set are shown in the table, the remainder being given by the " generalised " interactions of these three. Note the law of formation of each set from the preceding one, corresponding to the cyclic interchange of all the rows of the square except the first.

The set of 9×9 squares possesses similar properties. The actual degrees of freedom confounded, when the cells of the squares represent the 81 combinations of four factors at three levels, are given in terms of the notation developed by Yates (3).

Besides their use in ordinary factorial design, complete sets of orthogonal squares serve as the basis of quasi-factorial and quasi-Latin square designs (balanced lattices and lattice squares) and also serve to generate other designs in balanced incomplete blocks. (See (15), (16), (17), and Tables XVII-XIX.) Pairs and larger incomplete sets of orthogonal squares (Græco- and hyper-Græco-Latin squares) also provide designs for varietal trials superseding the so-called semi-Latin squares (3).

An orthogonal pair of squares exists for all odd numbers, being formed of two " diagonal " squares (squares of the type of the 5×5 squares I and IV). Such a pair also exists for all numbers which are multiples of 4. The process of generation, which depends on the existence of orthogonal pairs of squares for factors of the number, may be illustrated for a pair of 12×12 squares. Let α and α' represent the 4×4 squares I and II, β and β', and γ and γ', the same squares with 1, 2, 3, 4 replaced by 5, 6, 7, 8 and 9, 10, 11, 12 respectively. Then

$$\begin{matrix}\alpha & \beta & \gamma\\ \beta & \gamma & \alpha\\ \gamma & \alpha & \beta\end{matrix} \quad \text{and} \quad \begin{matrix}\alpha' & \beta' & \gamma'\\ \gamma' & \alpha' & \beta'\\ \beta' & \gamma' & \alpha'\end{matrix}$$

form an orthogonal pair of 12×12 squares. The process will generate larger orthogonal sets if such sets exist for both factors of the number. Thus there will be a set of three squares of side 20.

Bose *et al.* recently investigated the existence of larger orthogonal sets of 12×12 squares, using a computer. They found two sets of 5 squares (76) and a class of sets of three squares (77). Either set of 5 squares can be used, for example, to construct a seven-fold lattice for 144 varieties in 84 blocks of 12 plots each. Both sets of 5 squares can be generated from the standard 12×12 square shown in Table 2.03 below by rearranging the rows so as to give one of the following sets of first columns

1	3	2	11	10	12	4	6	5	8	7	9
1	11	5	9	3	7	10	2	8	6	12	4
1	6	10	3	11	8	9	5	12	2	4	7
1	7	6	8	12	5	3	9	11	4	2	10

1	3	2	8	7	9	10	12	11	5	4	6
1	8	5	12	3	10	4	11	2	9	6	7
1	9	4	3	11	8	6	5	12	2	7	10
1	10	12	2	6	11	9	3	8	7	5	4

It may be noted that the standard square can be transformed by intramutation into the square

$$\begin{matrix} \alpha & \beta & \gamma \\ \beta & \gamma & \alpha' \\ \gamma & \alpha' & \beta' \end{matrix}$$

where α etc. are as defined above.

It was shown independently by Tarry in 1900 (72) and by Fisher and Yates in 1934 (11) that no orthogonal pair of 6×6 squares exists. Euler had conjectured, and several later mathematicians had purported to prove, that this was true also for larger squares of side $4s+2$. An event of importance was the development in 1959 by Bose, Shrikhande and Parker (73, 74) of a method for forming orthogonal pairs of squares having sides of some numbers, such as 10 and 22, of the form $4s+2$. They also proved (75) that such pairs exist for all $s>1$. The 10×10 square given in Table 2.03 is a transformation of the first 10×10 square of this kind to be published (73). This form has an intercalated 3×3 Græco-Latin square on the leading diagonal, flanked by two 7×3 patterns which on inspection are seen to be Youden squares both in their letters and their suffices. The remaining 7×7 portion has diagonals in the letters *ABC* and the suffices 123. The seventh diagonal is compounded of all the other letters and figures. The square possesses seven-fold symmetry by the intramutation

$$\begin{pmatrix} D & E & F & G & H & I & J \\ 4 & 5 & 6 & 7 & 8 & 9 & 10 \end{pmatrix}$$

TABLE 2.03. ORTHOGONAL 10×10 SQUARE AND ONE OF A SET OF FIVE 12×12 SQUARES

A_1	B_2	C_3	D_4	E_5	F_6	G_7	H_8	I_9	J_{10}
B_3	C_1	A_2	G_9	H_{10}	I_4	J_5	D_6	E_7	F_8
C_2	A_3	B_1	F_5	G_6	H_7	I_8	J_9	D_{10}	E_4
D_8	H_4	I_{10}	A_7	B_9	E_1	C_6	F_3	J_2	G_5
E_9	I_5	J_4	H_6	A_8	B_{10}	F_1	C_7	G_3	D_2
F_{10}	J_6	D_5	E_2	I_7	A_9	B_4	G_1	C_8	H_3
G_4	D_7	E_6	I_3	F_2	J_8	A_{10}	B_5	H_1	C_9
H_5	E_8	F_7	C_{10}	J_3	G_2	D_9	A_4	B_6	I_1
I_6	F_9	G_8	J_1	C_4	D_3	H_2	E_{10}	A_5	B_7
J_7	G_{10}	H_9	B_8	D_1	C_5	E_3	I_2	F_4	A_6

1	2	3	4	5	6	7	8	9	10	11	12
2	3	4	5	6	1	8	9	10	11	12	7
3	4	5	6	1	2	9	10	11	12	7	8
4	5	6	1	2	3	10	11	12	7	8	9
5	6	1	2	3	4	11	12	7	8	9	10
6	1	2	3	4	5	12	7	8	9	10	11
7	8	9	10	11	12	1	2	3	4	5	6
8	9	10	11	12	7	2	3	4	5	6	1
9	10	11	12	7	8	3	4	5	6	1	2
10	11	12	7	8	9	4	5	6	1	2	3
11	12	7	8	9	10	5	6	1	2	3	4
12	7	8	9	10	11	6	1	2	3	4	5

TABLES XVII TO XIX. BALANCED INCOMPLETE BLOCKS

If v varieties (or treatments) are to be compared in randomized blocks of k experimental units (k being less than v), block differences can be simply eliminated if the arrangement is such that every two varieties occur together in the same number (λ) of blocks. If r is the number of replicates, and b the number of blocks, the number of experimental units is $rv(=bk)$, and $\lambda = r(k-1)/(v-1)$. Examples of experimental material and processes in which this type of arrangement, called *balanced incomplete blocks*, is likely to be of use are: litters of animals, monozygotic twins and other paired material, laboratory and technological processes in which the nature of the process or the apparatus available imposes a definite limit on the number of treatments in a group, agricultural experiments on commercial farms when each farm is only prepared to undertake a small number of treatments, and varietal trials involving a large number of varieties. (See also Yates (16).)

Arrangements satisfying the required conditions are clearly provided by taking all possible combinations of the v varieties, k at a time, but if v is at all large, the number of replicates required will be very large. Arrangements involving fewer replicates are therefore of importance. Tables XVIII and XIX give indexes, by number of replications (r), and by number of units in a block (k) respectively, of the arrangements requiring ten or less replications which are known to exist, and of all arithmetically possible arrangements, the existence of which had not been disproved at the date of the first edition. Since then the non-existence of solutions of Nos. 10 and 14 has been demonstrated by Q. M. Hussain (44, 45), and No. 8 has been eliminated by H. K. Nandi (46). These cases are marked by an asterisk. Cases not yet solved are marked by a dagger. General discussions of the construction of incomplete block designs have been given by R. C. Bose (31), and C. R. Rao (47). The solutions for Nos. 26 and 27 and those for Nos. 17 and 20 are due to K. N. Bhattacharya (50, 32). Schützenberger (48) has since proved that if $b = v$, and they are even, then a solution is impossible unless $r-\lambda$ is a perfect square. This confirms Hussain's rejection of No. 10, and excludes No. 30. Theorems of S. Chowla and H. J. Ryser (53) and S. S. Shrikhande (49) confirm the impossibility of No. 14, and exclude other cases (beyond the range of the table) with $b = v$, when they are odd. W. S. Connor (51) has found that designs Nos. 12 and 28 are impossible.

The combinatorial solutions which are referred to by numbers are shown in Table XVII. Those involving all combinations are indicated by u (unreduced), and the two series derivable from sets of orthogonal squares (described below) by *o.s.* When $k>\frac{1}{2}v$ arrangements can be obtained from arrangements for the same number of varieties in blocks of $v-k$ units, by replacing each block by its complement, *i.e.* by a block containing all the varieties missing from the original block.

Sometimes the blocks may be arranged in groups so as to minimise the inter-block variance. Each such group must contain an equal number of replicates of each variety. In general it is not possible so to arrange things that each group of blocks contains only a single replication. Such groups as are possible in the designs of Table XVII are indicated in the table itself. Other arrangements possessing this property are possible in certain cases when a multiple of the number of replications required for the designs of the table are used. Thus when $k = \frac{1}{2}v$ a balanced design and its complement will give such a grouping. Within each group the arrangement of the blocks should be at random.

Some of the combinatorial solutions are derivable very simply by cyclic substitutions. In such cases one block only for each family is given in Table XVII. In solution number 2, for example, the block *aefgi* is given. The contents of the other 10 blocks are obtained progressively by replacing each letter by the next in order, with the convention that the last letter k is followed by a. Thus the blocks are *aefgi*, *bfghj*, *cghik*, *dhija*, etc.

For the dicyclic solutions a similar procedure holds. In solution number 5, for example, the 16 varieties are represented by 4 letters in combination with 4 suffixes. The 16 blocks are generated by performing the cyclic substitution on the letters (thus obtaining 4 blocks) and then performing the cyclic substitution on the suffixes of each of these 4 blocks.

Some of the cyclic and dicyclic solutions consist of more than one family, each family being generated by a cyclic or dicyclic substitution of the above type.

Two series of arrangements and their complements are derivable from complete sets of orthogonal Latin squares (Table XVI). In the first, for p^2 varieties in $p(p+1)$ blocks of p plots ($r=p+1$, $\lambda = 1$), each of the p^2 cells of the set of squares is taken to represent a variety, the blocks being determined by the rows, the columns and the numbers of the $p-1$ squares. In the second, for p^2+p+1 varieties in p^2+p+1 blocks of $p+1$ plots ($r=p+1$, $\lambda = 1$), p^2 of the varieties are divided as in the first arrangement, each of the $p+1$ additional varieties being allotted to one of the $p+1$ sets of blocks so formed, while the last block consists entirely of these additional varieties. Thus the set of 4×4 squares gives the following arrangements for 16 and 21 varieties (the latter including the bracketed varieties):

a b c d (*q*)	*a e i m* (*r*)	*a f k p* (*s*)	*a g l n* (*t*)	*a h j o* (*u*)	(*q r s t u*)
e f g h (*q*)	*b f j n* (*r*)	*b e l o* (*s*)	*b h k m* (*t*)	*b g i p* (*u*)	
i j k l (*q*)	*c g k o* (*r*)	*c h i n* (*s*)	*c e j p* (*t*)	*c f l m* (*u*)	
m n o p (*q*)	*d h l p* (*r*)	*d g j m* (*s*)	*d f i o* (*t*)	*d e k n* (*u*)	

The second of these series, at least up to 10 replications, may also be generated by the cyclic substitutions shown in Table 2.1. For convenience the varieties are here represented by numbers instead of letters. The first series may then be obtained from the second by deleting any set of varieties occurring in the same block.

Tables XVII1 and XVIII1 extend Tables XVII and XVIII to arrangements requiring fifteen or less replications. These tables are based on tables by C. R. Rao (79). Since their publication Rao has supplied solutions for No. 64 (due to himself), for No. 80 (P. K. Menon), and for Nos. 46 and 49 (S. S. Shrikhande). The last two were derived from the partially balanced incomplete block design No. S 1.14 given on p. 222 of (80). The procedure for No. 49 is to combine all blocks containing the number 1, omitting this number, and so on for the 45 numbers. No. 46 is derivable from No. 49 by omitting one block and all the numbers in that

TABLE 2.1. CYCLIC SOLUTIONS FOR BALANCED INCOMPLETE BLOCKS

Problem.	Modulus (v).	Selected numbers.
o. s. 2	7	1, 2, 4
o. s. 3	13	1, 2, 4, 10
2	11	1, 5, 6, 7, 9
o. s. 4	21	1, 4, 5, 10, 12
o. s. 5	31	1, 2, 4, 11, 15, 27
9	15	1, 2, 3, 5, 6, 9, 11
o. s. 7	57	1, 4, 6, 14, 15, 21, 33, 37
19	19	1, 3, 5, 6, 7, 8, 11, 14, 15
23	37	1, 2, 4, 8, 18, 25, 26, 30, 36
o. s. 8	73	1, 2, 4, 8, 16, 32, 37, 55, 64
o. s. 9	91	1, 2, 7, 11, 24, 27, 35, 42, 54, 56

To these may be added the unreduced (u) designs with $k = v-1$. See also the dicyclic solutions in one family (No. 5), in two families (No. 13), and the cyclic solutions in two families (Nos. 4, 11, 29) and in three families (No. 18) of Table XVII. The unreduced (u) design with $k = 2$ and v odd has an obvious cyclic solution in $\frac{1}{2}(v-1)$ families, and by taking complements a corresponding result is obtained when $k = v-2$.

block from the other blocks. These two designs are not given in Table XVII1, since no compact representation has yet been found. The solution given by Rao for No. 56 is not completely balanced and has therefore been omitted.

The reference system in Table XVIII1 is that given by Rao. P and E indicate designs that are derivable from projective and Euclidean finite geometrical configurations. $P(2, x) : 1$ and $E(2, x) : 1$ correspond to the solutions (*o.s.*) based on orthogonal squares of side x of Table XVIII.

In Table XVII1 numbers are used instead of letters to denote the varieties, as in Table 2.1 above. Mod (x) indicates monocyclic permutation over the range 1 to x; mod (x, y) dicyclic permutation over the ranges 1 to x and 1 to y of the components. In monocyclic solutions in one or more families an invariant variety is assigned the number $x+1$, so that, for example, in No. 32 the varieties are numbered consecutively from 1 to 12. In dicyclic solutions an invariant variety, additional to the dicylic set, is denoted by I. Numbers or components which are not to be permuted are printed in bold type.

Youden's Squares

It has been shown (Youden, 21; Smith and Hartley, 52) that when $b = v$ it is possible so to arrange the order of the varieties in the blocks that each variety occurs once in each position in the block, and it is thus possible to eliminate experimental differences between these positions, each of which comprises a complete replication.

For example, experimental material consisting of 13 plants each with 4 corresponding leaves may be assigned to 13 treatments so that each treatment occurs once on each type of leaf. At the same time the 13 plants constitute randomized blocks of 4 treatments each, in which each treatment occurs on the same plant once with every other treatment. Similarly a varietal trial of 13 varieties in blocks of 4 may be arranged on the ground in a 4×13 rectangle of plots so that both row and column differences are eliminated from the varietal comparisons.

In drawing up an actual experimental arrangement of this type the blocks should be assigned to the plants (or arranged on the ground) at random, and the experimental treatments (or varieties) should be similarly assigned to the letters.

A cyclic or dicyclic substitution which gives an incomplete block solution will also give the corresponding Youden square, provided that each block is written in the order in which it is generated. The complementary Youden square solution may be obtained by performing the same cyclic substitution on the block which is complementary to the given block. Of the remaining solutions of Table XVII for which $b = v$, No. 20 is arranged in the form of a Youden square, and a Youden square form of the solution for No. 27 is shown in Table 2.2.

Cyclic and dicyclic solutions in two or more families provide an extension of Youden squares. In these the differences corresponding to the positions in the blocks of each family may be eliminated, *e.g.* 9 degrees of freedom in solution No. 29 for 10 replicates of 41 varieties. Solution 15 is also given in the form of an extended Youden design. Each variety occurs 3 times in each position in the blocks, and the 2 degrees of freedom between the three positions can be eliminated. This form of No. 15 (and Youden solutions of Nos. 20 and 27) was given by S. S. Shrikhande (54) in a study of designs for two-way elimination so heterogeneity.

TABLE 2.2. YOUDEN SQUARE SOLUTION ($b = v = 31$, $r = k = 10$)

a b c d e f g h i j	*i m t g u j B q v z*	*q r k n j D b c x u*	*y e j B k z r a E s*
b l q u p a t y e o	*j y h f t n m o w r*	*r f l i B C v b D y*	*z w v e q k f d l n*
c k d y x o i t s v	*k o b s h w q C B g*	*s t n D b v z A a h*	*A x e m r s l i h q*
d q m w A c D B y a	*l C r t a g w z c x*	*t p i k D e C j A w*	*B i a x s p u w n f*
e B x b m E d n t C	*m z f p g y A x b k*	*u h B c E A k l f t*	*C a u v d r h k p m*
f s C o c m e u z D	*n A s C y u j g d l*	*v j w E l b s p m c*	*D E g q f t p s r d*
g D E a n l o m k i	*o n p r v B c e g A*	*w u z A i d E r o b*	*E g D h w x y v u e*
h c y z C i n E q p	*p d o l z h x D j B*	*x v A j o q a f C E*	

Besides balanced incomplete blocks and Youden squares a number of different types of non-balanced arrangement are also available, namely quasi-factorial arrangements, described in (25) and (26), lattice squares (27), and partially balanced arrangements (28). The various types of arrangement are reviewed in (3) and (29). The experimenter should acquaint himself with their advantages and disadvantages before deciding which is most appropriate for his own needs.

Analysis of Balanced Incomplete Block Experiments

In the analysis of incomplete block experiments the block differences may be eliminated entirely from the varietal comparisons (Yates, 16), or, as has been shown later (Yates, 23), the information contributed by these comparisons may be combined with that provided by the intra-block comparisons, with weights depending on the relative accuracy of the two sets of comparisons (as estimated from the experimental results).

If the choice of blocks has been successful in substantially reducing the variability of the material, then the amount of information contributed by the inter-block comparisons will be small, but in the limiting case when the inter-block and intra-block comparisons are of equal accuracy, the fraction of the total information contained in the inter-block comparisons is given by the efficiency factor $E = \frac{1-1/k}{1-1/v} = \frac{v\lambda}{rk}$, which approaches the value $\frac{1}{2}$ when $k = 2$ and v is large.

In general therefore, it is worth recovering the inter-block information, except in small experiments with less than 10 degrees of freedom for blocks, in which the relative weights cannot be determined with sufficient accuracy. If the method described below is followed the amount of additional computation is not great.

If V_s be the sum of all the yields of variety s, T_s the sum of all the block totals of blocks containing variety s, T'_s the sum of all the remaining block totals, and G the total yield of all plots, the estimates of the varietal differences derived from the intra-block comparisons are obtained from the quantities

$$Q_s = V_s - T_s/k,$$

or, as is more convenient when $k > \frac{1}{2}v$,

$$Q'_s = V_s + T'_s/k.$$

The actual differences in units of the total yield of the r replicates are given by the differences of

$$Q_s/E \quad \text{or} \quad Q'_s/E,$$

the sum of the first set being zero, and of the second set $rG/\lambda v$.

The error variance of these latter sets of quantities is r/Ew, where $1/w$ is the intra-block error variance.

The estimates of the varietal differences derived from the inter-block comparisons are similarly given by the differences of

$$rT_s/(r-\lambda) \quad \text{or} \quad rT'_s/(r-\lambda),$$

in units of the total yield of r replicates. The error variance of these sets of quantities is $kr^2/(r-\lambda)w'$, where $1/w'$ is the error variance of the inter-block comparisons, in units of a single plot.

The most efficient estimates of the varietal differences are given by the quantities

$$Y_s = V_s + \mu W_s$$

where

$$W_s = (v-k)V_s - (v-1)T_s + (k-1)G,$$

and

$$\mu = \frac{w - w'}{wv(k-1) + w'(v-k)}.$$

The error variance of the Y's is

$$\frac{kr(v-1)}{wv(k-1)+w'(v-k)}.$$

The analysis of variance follows the lines shown in the accompanying table, in which dev^2 indicates the sum of squares of the deviations, y the individual yields and B the block totals. If recovery of inter-block information is not required then only the part (*a*) of the analysis of variance is required. There is also no need to subdivide the sum of squares for blocks, ignoring varieties. The comparison of the mean square (*a*) for varieties with intra-block error gives an exact test of significance for the intra-block estimates.

Method (*a*).	D.F.	S.S. (*a*).		S.S. (*b*).	Method (*b*).
Blocks (ignoring varieties) :					Blocks (eliminating varieties) :
Varietal component	$v-1$	$\frac{dev^2\, T}{k(r-\lambda)}$		$\frac{dev^2\, W}{rv(v-k)\,(k-1)}$	Varietal component
Remainder	$b-v$	†	→	†	Remainder
Total	$b-1$	$\frac{dev^2\, B^*}{k}$		†	Total
Varieties (eliminating blocks)	$v-1$	$\frac{dev^2\, kQ}{k^2\, rE}$		$\frac{dev^2\, V}{r}$	Varieties (ignoring blocks)
Intra-block error	$rv-v-b+1$	†	←→	†	Intra-block error
Total	$rv-1$	$dev^2\, y^*$	←→	$dev^2\, y^*$	Total

* Requires checking. † Calculated by addition or subtraction.

If the mean square for intra-block error is M and the mean square (*b*) for the $b-1$ degrees of freedom for blocks, eliminating varieties, is M'', w and w' are estimated from the equations

$$w=\frac{1}{M},\quad w'=\frac{v(r-1)}{k(b-1)M''-(v-k)M}.$$

If M'' is less than M, w' may ordinarily be taken as equal to w.

If the structure of the design is such that the blocks fall into r groups, each containing a single replication of all the varieties (as in solutions Nos. 6, 7, 21 and 25), and each of these groups of blocks is itself arranged in a large block, then the sum of squares corresponding to the $r-1$ degrees of freedom for complete replications must be eliminated from the sums of squares for blocks. The formula for w' will then become

$$w'=\frac{r-1}{rM''-M}.$$

In the more general case in which there are c such groups of blocks, each containing r/c replications, we have

$$w'=\frac{v(r-1)-k(c-1)}{k(b-c)M''-(v-k)M}.$$

In the case of a Youden square the sum of squares corresponding to the $k-1$ degrees of freedom for the k series (representing leaf position, etc.) must be eliminated from the intra-block error.

Example 8

Table 3 shows the scores of 18 litters of 4 rats in a discrimination test, the values shown being the square roots of the numbers of trials prior to 12 consecutive successful trials (the data are a random selection from scores published by F. A. E. Crew (10)). A dummy trial of 9 treatments (*e.g.* dietary treatments) in blocks of 4, corresponding to litters, is superimposed on these scores. (Design 11 of Table XVII, $r=8$, $v=9$, $k=4$, $b=18$, $\lambda=3$.)

Table 4 shows the values of V, T, $4Q$ and W for the nine treatments *a*-*i*. The analysis of variance is shown in Table 5.

TABLE 3. SCORES OF 18 LITTERS OF RATS

f 2·6	*f* 5·9	*a* 7·0	*i* 2·4	*i* 5·0	*d* 10·1	*b* 3·9	*h* 4·0	*b* 2·8
d 9·7	*g* 2·6	*f* 4·6	*d* 4·0	*h* 7·4	*a* 9·7	*d* 4·1	*f* 6·1	*f* 2·6
c 5·4	*i* 5·9	*i* 4·9	*g* 3·0	*e* 10·3	*f* 5·7	*e* 6·4	*g* 4·4	*e* 2·8
e 6·9	*b* 6·3	*c* 3·3	*f* 2·4	*c* 9·4	*h* 7·5	*i* 6·3	*c* 3·3	*h* 3·3
24·6	20·7	19·8	11·8	32·1	33·0	20·7	17·8	11·5

b 5·7	*b* 4·7	*a* 3·0	*c* 7·5	*c* 3·7	*i* 3·0	*d* 4·5	*g* 2·6	*b* 7·3
h 9·3	*g* 6·6	*h* 1·4	*g* 2·2	*a* 5·2	*g* 2·6	*b* 6·0	*e* 4·9	*e* 5·4
c 5·4	*a* 5·5	*i* 4·2	*e* 2·6	*d* 2·4	*a* 4·7	*g* 4·6	*d* 6·0	*f* 5·7
i 6·1	*h* 5·3	*d* 2·8	*a* 4·4	*b* 2·4	*e* 2·4	*c* 3·3	*h* 4·6	*a* 4·4
26·5	22·1	11·4	16·7	13·7	12·7	18·4	18·1	22·8

Grand Total: 354·4. Mean: 4·9$\dot{2}$.

TABLE 4. CALCULATION OF ADJUSTED SCORES IN DISCRIMINATION TEST

	V.	T.	4Q = 4V−T.	Q/E+$\frac{1}{9}$ G.	W = 5V−8T+3G.	Y = V+μW.
a	43·9	152·2	+23·4	46·3	+65·1	45·7
b	39·1	156·4	0	39·4	+7·5	39·3
c	41·3	169·6	−4·4	38·1	−87·1	38·9
d	43·6	151·7	+22·7	46·1	+67·6	45·4
e	41·7	159·2	+7·6	41·6	−1·9	41·6
f	35·6	162·0	−19·6	33·6	−54·8	34·1
g	28·6	138·3	−23·9	32·3	+99·8	31·3
h	42·8	172·5	−1·3	39·0	−102·8	40·0
i	37·8	155·7	−4·5	38·0	+6·6	38·0
Divisor	354·4 8	1417·6 4·5 = 20	0 4²·8·27/32 = 108	354·4	0 8·9·5·3 = 1080	354·3

TABLE 5. ANALYSIS OF VARIANCE, DISCRIMINATION TEST

	D.F.	S.S. (*a*).	S.S. (*b*).	M.S.
Blocks:				
Varietal component .	8	41·4684	37·0634	4·6329 (*b*)
Remainder . .	9	138·2011	138·2011	15·3557
Total . . .	17	179·6695	175·2645	10·3097 (*b*)
Varieties . . .	8	19·6044	24·0094	2·4506 (*a*)
Error	46	119·4506	119·4506	2·5968
Total	71	318·7245	318·7245	—

From the results of the analysis of variance we obtain

$$w = \frac{1}{2\cdot5968} = 0\cdot3851, \quad w' = \frac{63}{68\times10\cdot3097-5\times2\cdot5968} = 0\cdot0916,$$

$$\mu = \frac{0\cdot3851-0\cdot0916}{27\times0\cdot3851+5\times0\cdot0916} = \frac{0\cdot2935}{10\cdot8557} = 0\cdot02704.$$

The final adjusted scores in terms of the total scores of eight rats are given in the last column of Table 4. The standard error of these scores is

$$\sqrt{\frac{256}{10\cdot8557}} = \sqrt{23\cdot58} = 4\cdot86.$$

The scores based on intra-block information only (which are equal to $Q/E+\frac{1}{9}G$) are also shown in Table 4. Their standard errors are

$$\sqrt{(8\times2\cdot5968\times32/27)} = \sqrt{24\cdot62} = 4\cdot96.$$

Thus the gain in information from the recovery of the inter-block information is $24\cdot62/23\cdot58-1$ or 4·4 per cent. (excluding losses due to inaccuracy of weighting). The gain is here trivial, but if inter-litter and intra-litter comparisons had been of equal accuracy, the gain would have been 18·5 per cent.

The litter totals can be adjusted to allow for the effects of the treatments by deducting one-eighth the scores based on intra-litter differences only of each of the four treatments contained in the litter concerned from the total of that litter (adding four times the general mean if desired). Thus the adjusted total of the first litter is $24\cdot6-4\cdot20-5\cdot76-4\cdot76-5\cdot20+4\times4\cdot92 = 24\cdot36$.

Tables XX and XXI. Ordinal (or Ranked) Data

It is often necessary to draw statistical conclusions from data giving the order of a number of magnitudes without knowledge of their quantitative values. Thus in tests of psychological preference subjects can often express preferences, without being able to assign numerical values to the force with which the preference is felt. Not infrequently, also, an experimenter who possesses quantitative values may suspect that the metric used is unsuitable to the comparisons he wishes to make, and prefer to draw conclusions only from the order of the magnitudes observed.

The analysis of such data is greatly facilitated by Table XX, which gives the average deviate of the rth largest of samples of n observations drawn from a normal distribution having unit variance. Symbolically this comes to

$$\mu_r = \int_{-\infty}^{\infty} \frac{n!}{(r-1)!(n-r)!} p^{n-r} q^{r-1} xz\, dx,$$

where z is the ordinate of the normal curve, and p and q are the probabilities respectively of falling short of and exceeding x.

From the order in which any subject places a series of objects scores may now be assigned. With a number of different subjects, the experimenter is then in a position to test, for example, whether the scores assigned to any one object significantly exceed those assigned to another; or, more generally, to apply an analysis of variance to determine whether the variance in score among objects exceeds the remainder due to differences in the order chosen by different subjects. More elaborate analyses afford tests of differentiation in preference between classes of subjects of different sex or age. The labour of summing the squares of individual scores is saved by using the sums of squares given in Table XXI.

The scoring of ties causes little difficulty. If, for example, the fifth and sixth objects out of 13 are judged to be equal, instead of scoring one ·39 and the other ·19, each is given the same score ·29. In such cases the sum of squares needs adjustment. When only two objects are judged to be equal it is easy to see that the reduction is $\frac{1}{2}(\cdot39-\cdot19)^2 = \cdot02$, so that the sum of squares is 10·7904 in place of 10·8104 shown by the table.

The table can also be used to obtain the factor by which the range of a sample of size n should be multiplied to give an estimate of the standard deviation of an observation. The factor is in fact $1/2\mu_1$. Thus the range of a sample of 8 should be multiplied by $1/(2\times1\cdot42) = 0\cdot352$. Similarly if the mean of the difference between the largest and smallest observations, and between the second largest and second smallest is taken, the appropriate factor is $1/\{2\times\frac{1}{2}(\mu_1+\mu_2)\}$.

In a normal distribution the range provides quite an efficient estimate of the standard deviation if the sample size is small, but the efficiency becomes low as the size of the sample is increased, as is shown by the following values on page 32.

Size.	Efficiency.	Size.	Efficiency.
2	100	15	76·8
4	97·5	20	69·9
6	93·3	25	64·7
8	89·0	50	48·9
10	85·0	100	34·8

Moreover if the distribution is not normal the use of the factors appropriate to the normal distribution will give biased estimates of the standard deviation, whereas the ordinary estimate of the variance based on the sum of the squares of the deviations will be unbiased whatever the form of the distribution. The magnitude of the bias increases as the size of the sample increases.

On both grounds, therefore, the range should not be used with large samples. Nor has its use with a single small sample much to recommend it, since in this case the calculation of the ordinary estimate is not laborious. Similar objections apply to the use of a modified *t* test based on the range. If a large number of small samples of the same size (say not greater than 10) from a distribution are available, however, the mean range will provide a reasonably efficient estimate of the standard deviation and one that is not likely to be seriously biased unless the distribution departs markedly from normality. The range can also be appropriately used in quality control work to provide a control on the variability, though in this case there is no point in converting to the standard deviation, provided the sample size is constant.

TABLE XXII. THE LEADING DIFFERENCES OF POWERS OF NATURAL NUMBERS

The table gives values of $\Delta^r 0^s/r!$, where $\Delta^r 0^s$ denotes the leading rth difference of the sth powers of the natural numbers commencing at zero. Its application to combinatorial problems arises from the identity

$$\frac{\Delta^r 0^s}{r!} \equiv \sum_{s/r} \frac{s!}{(p_1!)^{\pi_1}\pi_1!(p_2!)^{\pi_2}\pi_2! \ldots}$$

where $(p_1^{\pi_1}\, p_2^{\pi_2} \ldots)$ is a partition of the number s into r parts, *i.e.*

$$\Sigma \pi p = s, \qquad \Sigma \pi = r.$$

The partitional function on the right represents the number of ways of dividing s persons into the r parties specified by the partition. The summation proceeds over all partitions of s into r parts.

Hence $\Delta^s 0^s/r\,!$ is the number of ways of dividing s persons into r parties of all possible sizes. For example, with $s = 4$, $r = 2$, the table gives 7 ways, which are in fact,

$$(ab)(cd),\ (ac)(bd),\ (ad)(bc),\ (abc)(d),\ (abd)(c),\ (acd)(b),\ (bcd)(a).$$

If s objects are equally likely to fall into any of n classes, the probability that they will fall into exactly r of the possible n classes can be shown to be

$$\frac{1}{n^s}\,\frac{n!}{(n-r)!}\,\frac{\Delta^r 0^s}{r!}.$$

Hence the distribution of r can be derived from the table. Thus the table is useful for testing whether objects fall together in classes too often or too rarely. For practical applications see Stevens (12), (13).

Example 9

In the expression for π it may be noticed that in the first 20 decimal places all digits save zero occur at least once. Zero is the only digit absent. It is not obvious how frequently 0, 1, 2,. . . absent digits would be found in a series chosen at random.

Putting $s = 20$, $n = 10$, $r = 9$ in the expression shown we find

$$\frac{10!}{10^{20}} \times 1201{\cdot}13 \times 10^{10}$$
$$= 3{\cdot}6288 \times {\cdot}120113 = {\cdot}4359.$$

Hence in 43·59 per cent. of trials we should expect exactly one absent digit; similarly we should expect no digit absent in 21·47 per cent. of such trials; consequently two or more digits will be absent in the remaining 34·94 per cent. The absence of one digit out of ten is the commonest event in such trials.

The table may be extended by using the recurrence relation

$$\frac{\Delta^{r+1}0^{s+1}}{(r+1)!} = (r+1)\frac{\Delta^{r+1}0^{s}}{(r+1)!} + \frac{\Delta^{r}0^{s}}{r!}.$$

This equation may be written

$$\frac{\Delta^{r+1}0^{s+1}}{(r+1)!} = \frac{\Delta^{r}(\Delta+1)0^{s}}{r!} = \frac{\Delta^{r}1^{s}}{r!}$$

showing that the entries derived from powers of natural numbers commencing at *unity*, may be read from this same table, with r and s both increased by one.

Cumulants of the binomial distribution

The first four cumulants of the binomial distribution are well known

$$\kappa_1 = Np \qquad \kappa_3 = Npq(q-p)$$
$$\kappa_2 = Npq \qquad \kappa_4 = Npq(1-6pq)$$

Table XXII makes the higher cumulants readily available in the form

$$\frac{\kappa_s}{N} = \sum_{1}^{s} (-)^{r-1}\frac{p^r}{r}\,\Delta^r 0^s$$

and, what is more directly useful,

$$\frac{\kappa_s}{Npq} = \sum_{1}^{s-1} (-)^{r-1}p^{r-1}\Delta^r 0^{s-1}.$$

Example 9.1

To obtain κ_6/Npq take the series for $s = 5$, supply the initial unit entry, and multiply by $r!$ to give

$$1-30p+150p^2-240p^3+120p^4$$

or, more compactly

$$1-30pq+120p^2q^2.$$

For odd numbers the value is divisible by $(q-p)$ or $\sqrt{(1-4pq)}$; so, for κ_7/Npq, we find from the series for $s = 6$

$$\begin{aligned} &1-62p+540p^2-1560p^3+1800p^4-720p^5 \\ &= (1-2p)(1-60p+420p^2-720p^3+360p^4) \\ &= (q-p)(1-60pq+360p^2q^2). \end{aligned}$$

The coefficients of powers of pq are really central differences at zero, with different formulæ for odd and even cumulants. Their derivation by the method exemplified above is more expeditious and more easily remembered. Authors of standard textbooks seem not to know this method.

Table XXIII. Orthogonal Polynomials

If a polynomial regression line

$$Y = a+bx+cx^2+dx^3+\ldots$$

is to be fitted to a series of n' observations at equal intervals of x, we may conveniently fit instead the equivalent line

$$Y = A+B\xi_1+C\xi_2+D\xi_3+\ldots$$

where ξ_1, ξ_2, ξ_3, . . . are orthogonal polynomials of degree 1, 2, 3 . . . ($A = \bar{y}$ and $\xi_1 = x-\bar{x}$). A method of determining the coefficients by successive addition, and also of calculating the polynomial values by the same process, is given in *Statistical Methods*. If the numerical values of the ξ's (or multiples $\xi' = \lambda\xi$ of them) are known for each value of x an alternative procedure (set out below) is available. Table XXIII shows these numerical values for all ξ' from ξ'_1 to ξ'_5, and all n' from 3 to 75. In each case the smallest value of λ is chosen which gives integral values to ξ'. This value is shown at the foot of the corresponding column, and is the coefficient of the highest power of x in ξ'. The sum of squares of each set of ξ''s is also shown. Above $n' = 8$ only the values for positive $x-\bar{x}$ are given. More extensive tables are available elsewhere (38, 39). The first reference (38) gives values of ξ'_1 to ξ'_5 for all n' to 104, the second (39) gives values of ξ'_1 to ξ'_9 for all n' to 52.

The coefficient of each ξ' in the regression equation is found by calculating the sum of the products of the observed values y and the corresponding values of ξ', and dividing by the sum of squares of ξ'. Thus the coefficient of ξ'_2 is

$$C' = \frac{S(y\xi'_2)}{S(\xi'^2_2)}.$$

The corresponding sum of squares in the analysis of variance is

$$C'.S(y\xi'_2) = \frac{\{S(y\xi'_2)\}^2}{S(\xi'^2_2)}$$

and the estimated standard error of C' is $s/\sqrt{S(\xi'^2_2)}$.

If any or all of the values of Y corresponding to the n' observations are required they may be calculated directly from the values of the coefficients and the ξ''s, using the equation

$$Y = A + B'\,\xi'_1 + C'\,\xi'_2 + D'\,\xi'_3 + \ldots$$

This procedure is particularly useful when only a few values are required.

If the regression equation is required in terms of powers of x the formulæ for the ξ's in terms of powers of x given in *Statistical Methods* (editions 3-6), or the recurrence formula

$$\xi_{r+1} = \xi_1\xi_r - \frac{r^2(n'^2-r^2)}{4(4r^2-1)}\,\xi_{r-1}$$

($\xi_0 = 1$) may be used, together with $\xi' = \lambda\xi$. Alternatively the coefficients of the powers of x in each ξ' may be deduced from the tabulated values of the ξ''s, as illustrated in the example below (remembering that the even ξ''s contain only even powers of x, and the odd ξ''s odd powers of x).

The relative advantages of the above method and the alternative process of successive addition depend on the computing facilities available, but with a good multiplying machine, and no printing adding machine, the use of the ξ''s will be found to be decidedly more expeditious, especially when n' is small. If polynomials of the fourth or fifth degree are to be fitted the sums and differences of the pairs of observations (commencing from the centre) should be tabulated, thereby halving the number of multiplications. At the same time the values may frequently be reduced in magnitude by the deduction of a constant amount, and if necessary by rounding off. If polynomials have to be fitted to a number of sets of observations with the same n' it is worth tabulating $\xi'_1+\xi'_3+\xi'_5$ and $\xi'_2+\xi'_4$, and using the checks $Sy(\xi'_1+\xi'_3+\xi'_5) = Sy\xi'_1+Sy\xi'_3+Sy\xi'_5$ and $Sy(\xi'_2+\xi'_4) = Sy\xi'_2+Sy\xi'_4$.

Example 10

Table 6 shows the yields (bushels per acre) of plots 9 and 7 of Broadbalk wheat field for the 30 years 1855-1884. The only difference in manurial treatment was that plot 9 received nitrate of soda and 7 an equivalent quantity of nitrogen as sulphate of ammonia. Investigate the slow changes in the difference between the two plots by fitting a polynomial of the fifth degree. (This example is discussed in *Statistical Methods*, Sections 26 and 28.1. The values there given have been rounded off to one decimal place.)

TABLE 6

Harvest Year.	9.	7.	9-7.	Harvest Year.	9.	7.	9-7.	Harvest Year.	9.	7.	9-7.
1855	29·6	33·0	−3·4	1865	44·1	40·2	+3·9	1875	30·5	26·6	+3·9
1856	32·4	36·9	−4·5	1866	32·5	29·9	+2·6	1876	33·3	25·5	+7·8
1857	43·8	44·8	−1·0	1867	29·1	22·2	+6·9	1877	40·1	19·1	+21·0
1858	37·6	38·9	−1·3	1868	47·8	39·2	+8·6	1878	37·2	32·2	+5·0
1859	30·0	34·7	−4·7	1869	39·0	28·2	+10·8	1879	21·9	17·2	+4·7
1860	32·6	27·7	+4·9	1870	45·5	41·4	+4·1	1880	34·1	34·3	−·2
1861	33·8	34·9	−1·1	1871	34·4	22·3	+12·1	1881	35·4	26·1	+9·3
1862	43·4	35·9	+7·5	1872	40·7	29·1	+11·6	1882	31·8	34·8	−3·0
1863	55·6	53·7	+1·9	1873	35·8	22·8	+13·0	1883	43·4	36·3	+7·1
1864	51·1	45·8	+5·3	1874	38·2	39·6	−1·4	1884	40·4	37·8	+2·6

The computations are given in Table 7. The first step is to form the sums and differences of pairs of values, working outwards from the centre pair.

Thus $$+14\cdot9 = +4\cdot1 + 10\cdot8 \quad\text{and}\quad -6\cdot7 = +4\cdot1 - 10\cdot8.$$

The totals of the sums and the differences check against the sum and difference of the totals of the first fifteen and last fifteen of the original values. The total sum of squares of the deviations from the mean, 1016·99, can be calculated from the sums and differences, and, as a check, from the original values, at the same time as these totals are obtained.

TABLE 7

Years.	Sum.	Difference.
'70, '69	+14·9	−6·7
'71, '68	+20·7	+3·5
'72, '67	+18·5	+4·7
'73, '66	+15·6	+10·4
'74, '65	+2·5	−5·3
'75, '64	+9·2	−1·4
'76, '63	+9·7	+5·9
'77, '62	+28·5	+13·5
'78, '61	+3·9	+6·1
'79, '60	+9·6	−0·2
'80, '59	−4·9	+4·5
'81, '58	+8·0	+10·6
'82, '57	−4·0	−2·0
'83, '56	+2·6	+11·6
'84, '55	−0·8	+6·0
	+134·0 4·46	+61·2

Sums of Products.

Term.	$S\xi'y$.	Regression Coefficient.	Sum of Squares.
1	+1191·6	+0·1325473	157·9434
2	−8990·0	−0·02976190	267·5595
3	−8772·2	$-0\cdot0^{3}4106789$	3·6026
4	+148509·0	$+0\cdot0^{4}4044817$	6·0069
5	+72400·4	$+0\cdot0^{4}3374157$	2·4429

Analysis of Variance.

	Degrees of Freedom.	Sum of Squares.	Mean Square.
Polynomial terms .	5	437·55	87·51
Remainder . .	24	579·44	24·14
Total . .	29	1016·99	

The sums of products with the ξ' values are next calculated, and at the same time the regression coefficients and the sum of squares accounted for by each term. The sums of products may be checked by direct recalculation or by the method described above. Finally the analysis of variance can be completed, giving the residual mean square of 24·14. As will be seen, the first two terms account for a substantial part of the variation, but the mean squares of the remaining three terms are all below the residual mean square. Thus a parabola adequately describes the slow changes.

The equation of the parabola is

$$Y = 4\cdot4667 + 0\cdot13255\,\xi'_1 - 0\cdot029762\,\xi'_2 \quad \text{. . . . . . .} \quad (1)$$

which can be converted into an equation in x, if this is required, by substitution for ξ'_1 and ξ'_2. If x is measured from the centre point of the series $\xi'_1 = 2x$ and $\xi'_2 = \frac{3}{2}x^2 + C$, the coefficients 2 and $\frac{3}{2}$ of the highest powers of x being given in Table XXIII. Further, since $\xi'_2 = -112$ when $x = +\frac{1}{2}$, we have immediately $C = -112\cdot375$.

Alternatively the recurrence formula gives

$$\xi'_2 = \frac{3}{2}\xi_2 = \frac{3}{2}\left[\xi_1 \,.\, \xi_1 - \frac{1^2 \,.\, 899}{4\cdot3} \,.\, 1\right] = \frac{3}{2}x^2 - 112\cdot375.$$

This gives the equation

$$Y = 7\cdot8112 + 0\cdot26510x - 0\cdot044643x^2 \quad \text{. . . . . . .} \quad (2)$$

The value of Y for any given year can be calculated from (1). Thus for the year 1855

$$Y = 4{\cdot}4667 + 0{\cdot}13255\ (-29) - 0{\cdot}029762\ (+203) = -5{\cdot}4189.$$

Similarly for the year 1884 the value is $+2{\cdot}2690$. Equation (2) will, of course, give the same values.

The parabola has a maximum at $x = +2{\cdot}969$, *i.e.* at the date 1872·469, Y then having the value $+8{\cdot}2048$. The improvement of plot 9 relative to plot 7 therefore appears to have reached a climax at about this date, there being some evidence of a reduction in the difference subsequently.

As a further exercise polynomials may be fitted to the yields of the two plots separately. The difference of these two polynomials clearly gives the polynomial already determined.

Table XXIV. Calculation of Integrals from equally spaced Ordinates

The table supplies data useful for the mechanical integration of functions calculated at not very closely spaced ordinates. Its use may be illustrated by the series of reciprocals of numbers from 15 to 19. In this case the integral is known to be $\log_e 19/15$, or ·23638,87780,66; the average ordinate over the four units in the panel is, therefore, ·05909,71945,16. An accurate calculation can be obtained from the series of reciprocals and their even differences. Taking enough to use the eighth difference, these are as follows to ten places of decimals :—

Number.	Reciprocal.	Differences.			
		2nd.	4th.	6th.	8th.
13	·07692,30769				
14	·07142,85714	73,26008			
15	·06666,66667	59,52380	3,23209		
16	·06250,00000	49,01961	2,33426	28353	
17	·05882,35294	40,84968	1,71996	18438	3745
18	·05555,55556	34,39971	1,29004	12268	
19	·05263,15789	29,23978	98280		
20	·05000,00000	25,06265			
21	·04761,90476				

Using the coefficients for 4 sub-panels in the table for central ordinate and even differences, the successive approximations obtained are as follows :—

	Correction.	Value.	Error.
No difference		·05882,35294,00	−27,36651,16
2nd ,,	+27,23312,00	·05909,58606,00	−13339,16
4th ,,	+13377,47	·05909,71983,47	+38,31
6th ,,	−39,02	·05909,71944,45	−71
8th ,,	+86	·05909,71945,31	+15

To show the adequacy of the adjustments two additional figures have been retained. The evaluation is satisfactory using only 6th differences, although the 8th difference is a four-figure number.

The formulæ using integral multipliers in the first part of the table (4 sub-panels) give exactly equivalent values to the use of 4th and 6th differences.

Tables XXV to XXXII and XXXIV. Logarithms, Trigonometrical Functions, Etc.

These tables comprise the standard functions that are likely to be required by the statistician. Five-figure accuracy has been adopted in most of the tables, since four-figure tables are already sufficiently plentiful, and five figures bridge the gap between the slide-rule (roughly three-figure accuracy) and seven-figure standard tables.

In addition to the ordinary table of logarithms to base 10 of numbers 1·00 to 9·99 (Table XXV) eight pages of natural logarithms ($\log_e$ ·0001—$\log_e$ 99·9) are included (Table XXVI). This table will prove useful to those who find it convenient to work with natural logarithms. Exact squares of all three-figure numbers (Table XXVII) will enable an occasional analysis of variance to be performed without a machine. Square roots (Table XXVIII) are so tabulated that the square roots of each three-figure number and of ten times that number appear together. This facilitates reference to the table, and serves as a reminder of the necessity of considering the position of the decimal point before taking out a square root. Reciprocals (Table XXIX) are given to six figures, since interpolation is not likely to be required in this table when a machine is available, and the additional figure is useful when division is being performed on a machine by means of reciprocals.

Table XXX gives the factorials of the numbers 1-300 to six significant figures, and their logarithms to seven decimal places. Of the trigonometrical functions only natural sines (Table XXXI) and tangents (Table XXXII) have been included. By tabulating $(90°-\theta) \tan \theta$ instead of $\tan \theta$ for values of θ greater than 60° five-figure accuracy of interpolated values is assured, even for high angles.

In Table XXXIV will be found a collection of constants, conversion factors, etc. Only the English, American and metric systems of weights and measures have been included, the more uncommon measures of these systems being given in their most easily remembered forms.

The tabulation of the above tables to 5 and 6 figure accuracy does not preclude their use to four-figure accuracy only when the circumstances warrant it. For this purpose they will be found little slower than the ordinary four-figure tables, since the amount of page-turning is the same. When using logarithms to four-figure accuracy it will frequently be found most expeditious, as well as slightly more accurate, to retain 5 figures in the logarithmic calculations, without making any attempt to interpolate to more than four-figure accuracy. General instructions for interpolation are given in the last section of the Introduction.

Table XXXIII. Random Numbers

This table contains 7500 two-figure random numbers arranged on six pages. The method of construction of the table, and the tests made on it, are described in the first edition. Methods of using the table may best be made clear by a couple of examples.

Example 11. Required to select a series of numbers at random from the numbers 1-16.

The simplest way of doing this is to select a row, column or diagonal of two-figure numbers (such selection being made without previous inspection of the numbers themselves), and pick out all numbers between 1 and 16 as they occur. With a number like 16, however, such a procedure is tedious and wasteful of numbers. As an alternative we may divide each two-figure number by 16 and take the remainder (0 counting as 16). It is clear that to give all numbers an equal chance of selection the numbers 97, 98, 99 and 00 must be rejected. A modification of this process, which somewhat lightens the arithmetic, is to use the divisor 20 instead of 16, rejecting any number which gives 00, 17, 18 or 19. Thus 73 would be read as 13.

Example 12. Required to arrange 8 treatments, numbered 1-8, in random order.

This operation can be performed by selecting one of the treatments at random from the eight, then selecting a second from the seven that remain, and so on. When the number of treatments is at all large, however, this procedure is tiresome, since each treatment must be deleted from a list as it is selected and a fresh count made for each further selection. To avoid this C. R. Rao (78) has proposed an alternative method. The one-dimensional variant of this, suitable for the present example, consists of taking 10 cells numbered 0-9, and allocating the numbers 1-8 to these according to a sequence of 8 single digit random numbers. Thus, using the first column of Table XXXIII (I), which begins 0, 9, 1, 1, 5, 1, 8, 6 we allocate 1 to cell 0, 2 to cell 9, 3 to cell 1, 4 to cell 1, etc., the complete allocation being

Cell:	0	1	2	3	4	5	6	7	8	9
	1	3, 4, 6	—	—	—	5	8	—	7	2

The three numbers in cell 1 must now be permuted. This can be done by the same process, using the next three random numbers 3, 5, 1 to give the order 6, 3, 4, so that the final permutation is 1, 6, 3, 4, 5, 8, 7, 2. Alternatively this secondary permutation can be done as the numbers are written down, *e.g.* by using the random digit in the adjoining column. For 4 the number is 2, which divided by 2 gives remainder 0; this can be taken to indicate that 4 is placed first, *i.e.* 4, 3. For 6 the number is 6, which divided by 3 again gives remainder 0, *i.e.* the order is 6, 4, 3. (In this latter case the number 9, if it occurs, must be rejected.) A further alternative to secondary permutations is to reject the random number if the cell is already occupied, but this will, on the average, lead to a substantial proportion of rejections if the number of treatments is greater than half the number of cells, and the use of secondary permutations, at least for pairs, is generally preferable.

Example 12.1. Arrange the 40 symbols *A-Z*, *a-n*, in random order.

Rao's method can again be followed, using two-digit random numbers to define a two-way 10×10 table of cells. Thus, using the first two columns of Table XXXIII (II) and the third column for secondary permutations, the first number is 53, so that *A* is written in row 5, column 3, etc. The full permutation is

O F j K b N M c m V X J C g a h T R k U f e A d I L D B G P Y n l Z H Q i W S E

Tables XXXIII1 and XXXIII2. Random Permutations

Table XXXIII1 gives 750 random permutations, numbered 0-749, of the numbers 0-9. By its aid random permutations of up to 10 objects can be obtained immediately, instead of by the procedure given in Example 12. Table XXXIII2 gives 200 similar permutations of the numbers 0-19. It can also be used to give additional permutations of the numbers 0-9 which are independent of one another and of those given in Table XXXIII1, by taking separately the numbers 0-9 and 10-19 of each permutation. These tables are useful in such processes as the allocation of treatments to the plots of a randomized block experiment, the randomization of the rows, columns and letters of a Latin square, etc.

Each permutation of 10 is arranged in two groups of 5 printed one beneath the other, and each permutation of 20 is similarly arranged in a 4×5 pattern. The reading of the permutation in a different order, *e.g.* down instead of across, will give another random permutation, but not one that is fully independent of the first. For this reason, when large numbers of random and strictly independent permutations are required they should be generated from a table of random numbers, or by some other fully random process. For the purpose of experimental design, however, Tables XXXIII1 and XXXIII2 will provide all that is normally required.

Numbering the permutations serves two purposes. It enables permutations to be selected from the table at random by selecting a number at random in the appropriate range from Table XXXIII. It also enables the permutations which are to be used in particular experiments to be specified, thus ensuring that the arrangement used is really random, without drawing up the detailed plans.

The tables were generated by Mr B. M. Church from Table XXXIII by the aid of Hollerith equipment. Tables of 1000 permutations of 9 and 1000 of 16 have previously been published by Cochran and Cox (3b), following a suggestion by G. W. Snedecor.

Interpolation

Many of the tables are provided with special aids to rapid interpolation, such as proportional parts, mean differences, and tabular differences printed in the tables. In other cases linear interpolation is generally sufficient. Thus if the table gives values for u_x and u_{x+1}, and the value is required for $u_{x+\theta}$, we may take

$$u' = \theta u_{x+1} + \phi u_x,$$

where ϕ stands for $1-\theta$. This form is convenient for machine work; mentally it is often quicker to use the difference between the tabular values and calculate

$$u' = u_x + \theta(u_{x+1} - u_x).$$

Modern work, especially that of Jordan (20) and Aitken (18), (19), has much facilitated the use of interpolation, without the aid of differences, in cases for which linear interpolation is thought not to be sufficiently accurate. A very useful improvement on the linear interpolate, based on four instead of two adjacent tabular values, is obtained by supplementing the linear interpolate from the two nearest entries by that from the next two which enclose them. This latter is

$$u'' = \tfrac{1}{3}\{(1+\theta)u_{x+2} + (1+\phi)u_{x-1}\}.$$

Curvature of the entries is shown by u' differing from u'', the error of the former being generally of the same sign as that of the latter, though smaller. To complete the adjustment calculate the product $\theta\phi$, and use

$$u = \tfrac{1}{2}\{(2+\theta\phi)u' - \theta\phi u''\},$$

which is the correct 4-point interpolate, obtained without the use of differences.

This procedure can be used also for inverse interpolation. The inverse linear interpolate $x+\theta$ corresponding to a value u is given by

$$\theta = (u-u_x)/(u_{x+1}-u_x).$$

An improved value of θ (though not the exact 4-point interpolate) is then given by

$$\theta_1 = \theta+\tfrac{1}{2}\theta\phi(u''-u)/(u_{x+1}-u_x).$$

Alternatively, exact 4- and 6-point inverse interpolates can be obtained by the method of divided differences.

In the tables in which mean proportional parts are given for each line of the table, only the proportional parts for 1 to 5 units are included, instead of, as is customary in tables of this type, for 1 to 9 units, it being intended that subtraction from the next higher tabular value should be used when interpolating for fractions above 0·5. The exact procedure will be clear from the examples below. The advantages of this innovation are two-fold: firstly the differences which have to be dealt with are smaller, thus facilitating the arithmetic (which should be done mentally), and secondly the errors arising from the use of *mean* proportional parts for the whole of each line are reduced.

The actual tables in which this type of arrangement has been adopted, for the whole or the greater part of the table, comprise: II (ordinates of the normal distribution), IX (probits), XXV and XXVI (logarithms), XXVIII (square roots), XXXI (sines), and XXXII (tangents). In tables XXV and XXVI (logarithms) the common practice is followed of tabulating the values for the numbers 10-19 on 20 lines instead of 10, so that five values instead of ten occur on each line, thus reducing the range covered by each set of mean proportional parts (which are here changing rather rapidly) to five units of the interval of tabulation.

In Table XXIX (reciprocals) mean differences of consecutive tabular values, instead of mean proportional parts, are given for each line of the table (thus allowing space for the tabulation of six figures instead of five). These mean differences correspond to ten times the mean proportional parts for 1 unit given in the other tables. This course has also been followed for Table VII (transformation of r to z).

In those parts of Table IX (probits) and Table XXIX (reciprocals) where the first differences are changing so rapidly that the mean proportional parts or mean differences are too inaccurate to be of value, the actual first differences between each tabular value and the next are given.

In Table XXVII (squares) no differences are given. Squares of four-figure numbers may be easily derived by the use of the expansion of $(a+b)^2$, as indicated in the note below the table.

In all the above tables, except the last two lines of the table of probits and the first 20 lines (1·0-2·9) of the table of reciprocals, linear interpolation will give results accurate to one unit in the last tabulated place. The use of mean proportional parts or mean differences will, of course, give errors greater than this in those parts of the tables where the differences are changing rapidly. The actual magnitude of the error may be assessed by seeing how rapidly the proportional parts are changing from line to line, as illustrated in Example 13 below. The mean proportional parts should therefore be used with discretion when full tabular accuracy is required.

Example 13

Find the natural logarithm of 11·06. We have $\log_e$ 11·1 = 2·40695. The proportional part for 4 is 357. Thus $\log_e$ 11·06 = 2·40338. The true value = 2·40333,5, the discrepancy being due to the use of the mean proportional part instead of the correct proportional part, which is 0·4 (2·40695—2·39790), *i.e.* 362. As will be seen by inspection the proportional parts are here changing rather rapidly. A closer approximation would have been obtained by taking this change into account, interpolating roughly in the required column of proportional parts, *i.e.* between 357 and 374.

Example 14

Find the square root of 29·426. The square root of 29·4 is 5·4222, and since 2·6 = 3—0·4 the required proportional part is 28—3·7 or 24, giving the value 5·4246. (True value = 5·42457.)

Example 15

Find the angle whose tangent is 0·79134. The nearest tabular value, 0·79259, is that for 38° 24′. The difference is —125. To the nearest minute, therefore, the angle is 38° 21′, a closer approximation being 38° 21·4′. (True value = 38° 21·36′.)

Example 16

Find the angle whose tangent is 2·82. Ordinary inverse interpolation is here impossible, since $(90°-\theta)$ tan θ is tabulated, but any approximation may be rapidly improved by dividing the tabular value by the given tangent, and subtracting from 90°. First find a rough approximation by interpolation in the auxiliary marginal table of tangents. This gives a value between 70·4° and 70·5°, *i.e.* between 70° 24′ and 70° 30′.

We then obtain

Angle.	Tabular Value (x).	$x/2{\cdot}82$.	*Plus* Original Angle.
70·4	55·043	19·5188	89·9188
70·5	55·066	19·5270	90·0270

The second column gives the tabular values from the table, the third is the result of dividing these by the given tangent 2·82. Subtracting either of these from 90° gives a better approximation. For full accuracy we use the fourth column, which is the sum of these quotients and the original angles. Inverse linear interpolation between the last two values for the value 90 will now give the required angle, namely 70·4750, *i.e.* 70° 28.50′. True value = 70° 28·499′.)

REFERENCES

Books and Monographs

1. R. A. FISHER (1950) . . . Statistical methods for research workers (11th edition). Oliver and Boyd, Edinburgh.
2. R. A. FISHER (1949) . . . The design of experiments (5th edition). Oliver and Boyd, Edinburgh.
3. F. YATES (1937) . . . The design and analysis of factorial experiments. Imperial Bureau of Soil Science, Harpenden.

3a. D. J. FINNEY (1951) . . . Probit analysis (2nd edition). Cambridge University Press.

3b. W. G. COCHRAN and G. M. COX (1950) Experimental designs. Wiley, New York.

Papers

4. C. I. BLISS (1935) . . . The calculation of the dosage-mortality curve. Annals of Applied Biology, xxii, 134-167.
5. C. I. BLISS (1935) . . . The comparison of dosage-mortality data. Annals of Applied Biology, xxii, 307-333.
6. C. I. BLISS (1937) . . . The calculation of the time-mortality curve. Annals of Applied Biology, xxiv, 815-852.
7. C. I. BLISS (1938) . . . The determination of the dosage-mortality curve from small numbers. Quarterly Journal of Pharmacy and Pharmacology, xi, 192-216.
8. W. G. COCHRAN (1938) . . Some difficulties in the statistical analysis of replicated experiments. Empire Journal of Experimental Agriculture, vi, 157-175.
9. C. G. COLCORD and L. S. DEMING (1935) The one-tenth per cent. level of "z." Sankhyā, ii, 423-424.
10. F. A. E. CREW (1932) . . . Inheritance of educability. Proceedings of the Sixth International Congress of Genetics, i, 121-134.
11. R. A. FISHER and F. YATES (1934) The six by six Latin squares. Proceedings of the Cambridge Philosophical Society, xxx, 492-507.
12. W. L. STEVENS (1937) . . Significance of grouping. Annals of Eugenics, viii, 57-69.
13. W. L. STEVENS (1937) . . A test for uniovular twins in mice. Annals of Eugenics, viii, 70-73.
14. F. YATES (1934) . . . Contingency tables involving small numbers and the χ^2 test. Journal of the Royal Statistical Society, Supplement, i, 217-235.
15. F. YATES (1936) . . . A new method of arranging variety trials involving a large number of varieties. Journal of Agricultural Science, xxvi, 424-455.
16. F. YATES (1936) . . . Incomplete randomized blocks. Annals of Eugenics, vii, 121-140.
17. F. YATES (1937) . . . A further note on the arrangement of variety trials: quasi-Latin squares. Annals of Eugenics, vii, 319-332.
18. A. C. AITKEN (1932) . . . On interpolation by iteration of proportional parts, without the use of differences. Proceedings of the Edinburgh Mathematical Society, Series 2, iii, 56-76.
19. A. C. AITKEN, G. J. LIDSTONE, and J. WISHART (1932) Interpolation without printed differences: Jordan's and Aitken's formulæ. Mathematical Gazette, xvi, 14-25.
20. C. JORDAN (1932) . . . Sur une formule d'interpolation. Atti del Congresso Internazionale dei Matematici, Bologna, vi, 157-177.
21. W. J. YOUDEN (1937) . . . Use of incomplete block replications in estimating tobacco-mosaic virus. Contributions from Boyce Thompson Institute, ix, 41-48.
22. H. W. NORTON (1939) . . The 7×7 squares. Annals of Eugenics, ix, 269-307.
23. F. YATES (1940) . . . The recovery of inter-block information in balanced incomplete block designs. Annals of Eugenics, x, 317-325.
24. W. L. STEVENS (1942) . . Accuracy of mutation rates. Journal of Genetics, xliii, 301-307.
25. F. YATES (1939) . . . The recovery of inter-block information in variety trials arranged in three dimensional lattices. Annals of Eugenics, ix, 136-156.
26. G. M. COX, R. C. ECKHARDT and W. G. COCHRAN (1940) The analysis of lattice and triple lattice experiments in corn varietal tests. Iowa Agricultural Experiment Station Research Bulletin, 281.

27. F. YATES (1940) . . . Lattice squares. Journal of Agricultural Science, xxx, 672-687.

28. R. C. BOSE and K. R. NAIR (1939) Partially balanced incomplete block designs. Sankhyā, iv, 337-372.

29. F. YATES (1940) . . . Modern experimental design and its function in plant selection. Empire Journal of Experimental Agriculture, viii, 223-230.

30. R. A. FISHER (1921) . . . On the mathematical foundations of theoretical statistics. Philosophical Transactions of the Royal Society of London, ccxxii, 309-368.

31. R. C. BOSE (1939) . . . On the construction of balanced incomplete block designs. Annals of Eugenics, ix, 353-399.

32. K. N. BHATTACHARYA (1944) . On a new symmetrical balanced incomplete block design. Bulletin of the Calcutta Mathematical Society, xxxvi, 91-96.

33. W. -V. BEHRENS (1929) . . Ein Beitrag zur Fehlerberechnung bei wenigen Beobachtungen. Landwirtschaftliche Jahrbücher, lxviii, 807-837.

34. R. A. FISHER (1939) . . . The comparison of samples with possibly unequal variances. Annals of Eugenics, ix, 174-180.

35. R. A. FISHER (1941) . . . The asymptotic approach to Behrens's integral, with further tables for the *d* test of significance. Annals of Eugenics, ii, 141-172.

36. P. V. SUKHATME (1938) . . On Fisher and Behrens's test of significance for the difference in means of two normal samples. Sankhyā, iv, 39-48.

37. D. J. FINNEY (1944) . . . The application of the probit method to toxicity test data adjusted for mortality in the controls. Annals of Applied Biology, xxxi, 68-74.

38. R. L. ANDERSON and E. E. HOUSEMAN (1942) Tables of orthogonal polynomial values extended to N = 104. Research Bulletin 292, Ames, Iowa.

39. D. VAN DER REYDEN (1943) . Curve fitting by the orthogonal polynomials of least squares. Onderstepoort Journal of Veterinary Science and Animal Industry, xxviii, 355-404.

40. C. M. THOMPSON (1941) . . Tables of percentage points of the incomplete Beta function. Biometrika, xxxii, 151-181.

41. M. MERRINGTON and C. M. THOMPSON (1943) Tables of percentage points of the inverted Beta (F) distribution. Biometrika, xxxiii, 73-88.

42. V. G. PANSE and G. R. AYACHIT (1944) Ten per cent. probability of z and the variance ratio. Indian Journal of Agricultural Science, xiv, 244-247.

43. J. T. MARTIN (1940) . . . The problem of the evaluation of rotenone-containing plants. V. The relative toxicities of different species of derris. Annals of Applied Biology, xxvii, 274-294.

44 Q. M. HUSSAIN (1946) . . Impossibility of the symmetrical incomplete block design with $\lambda = 2$, $k = 7$. Sankhyā, vii, 317-322.

45. Q. M. HUSSAIN (1945) . . Symmetrical incomplete block designs with $\lambda = 2$, $k = 8$ or 9. Bulletin of the Calcutta Mathematical Society, xxxvii, 115-123.

46. H. K. NANDI (1945) . . . On the relation between certain types of tactical configurations. Bulletin of the Calcutta Mathematical Society, xxxvii, 92-94.

47. C. R. RAO (1946) . . . Difference sets and combinatorial arrangements derivable from finite geometrics. Proceedings of the National Institute of Sciences of India, xii, 123-135.

48. M. P. SCHÜTZENBERGER (1949) . A non-existence theorem for an infinite family of symmetrical block designs. Annals of Eugenics, xiv, 286-287.

49. S. S. SHRIKHANDE (1950) . . The impossibility of certain symmetrical balanced incomplete block designs. Annals of Mathematical Statistics, xxi, 106-111.

50. K. N. BHATTACHARYA (1946) . A new solution in symmetrical balanced incomplete block designs. Sankhyā, vii, 423-424.

51. W. S. CONNOR (1951) . . . The structure of balanced incomplete block designs, and the impossibility of certain unsymmetrical cases. (Abstract) Annals of Mathematical Statistics, xxii, 312-313.

52. C. A. B. SMITH and H. O. HARTLEY (1948) The construction of Youden squares. Journal of the Royal Statistical Society, B, x, 262-263.

53. S. CHOWLA and H. J. RYSER (1950) Combinatorial problems. Canadian Journal of Mathematics, ii, 93-99.

54. S. S. SHRIKHANDE (1951) . . Designs for two-way elimination of heterogeneity. Annals of Mathematical Statistics, xxii, 235-247.

55. R. A. FISHER (1929) . . . Tests of significance in harmonic analysis. Proceedings of the Royal Society of London, Series A, cxxv, 54-59.

56. R. A. FISHER (1950) . . . Gene frequencies in a cline determined by selection and diffusion. Biometrics, vi, 353-361.

57. D. J. FINNEY (1949) . . . The estimation of the frequency of recombinations. I. Matings of known phase. Journal of Genetics, xlix, 159-176.

58. R. A. FISHER, M. F. LYON and A. R. G. OWEN (1947) The sex chromosome in the house mouse. Heredity, i, 335-365.

59. R. A. FISHER (1948) . . . A quantitative theory of genetic recombination and chiasma formation. Biometrics, iv, 1-13.

60. A. R. G. OWEN (1949) . . The theory of genetical recombination. I. Long-chromosome arms. Proceedings of the Royal Society of London, Series B, cxxxvi, 67-94.

61. A. R. G. OWEN (1950) . . The theory of genetical recombination. Advances in Genetics, iii, 117-157.

62. K. MATHER (1949) . . . The analysis of extinction time data in bioassay. Biometrics, v, 127-143.

63. R. C. BOSE (1938) . . . On the application of the properties of Galois fields to the problem of construction of Hyper-Græco-Latin squares. Sankhyā, iii, 323-338.

64. W. L. STEVENS (1939) . . The completely orthogonalized Latin square. Annals of Eugenics, ix, 82-93.

65. R. C. BOSE and K. R. NAIR (1941) On complete sets of Latin squares. Sankhyā, v, 361-382.

66. A. SADE (1951) . . . . An omission in Norton's list of 7×7 squares. Annals of Mathematical Statistics, xxii, 306-307.

67. R. A. FISHER and M. J. R. HEALY (1956) New tables of Behrens' test of significance. Journal of the Royal Statistical Society, Series B, xviii, 212-216.

68. R. A. FISHER (1949) . . . A preliminary linkage test with *Agouti* and *Undulated* mice. Heredity, iii, 229-241.

69. E. A. CORNISH and R. A. FISHER (1937) Moments and cumulants in the specification of distributions. Revue de l'Institut International de Statistique, iv, 1-14.

70. E. A. CORNISH (1960) . . . Fiducial limits for parameters in compound hypotheses. Australian Journal of Statistics, ii, 32-40.

71. R. A. FISHER (1935) . . . The fiducial argument in statistical inference. Annals of Eugenics, vi, 391-398.

72. G. TARRY (1900) . . . Le problème des 36 officiers. Association française pour l'Avancement des Sciences, xxix, 170-203.

73. E. T. PARKER (1959) . . Orthogonal Latin squares. Proceedings of the National Academy of Sciences of the United States of America, xlv, 859-862.

74. R. C. BOSE and S. S. SHRIKHANDE (1959) On the falsity of Euler's conjecture about the non-existence of two orthogonal Latin squares of order $4t+2$. Proceedings of the National Academy of Sciences of the United States of America, xlv, 734-737.

75. R. C. BOSE, S. S. SHRIKHANDE and E. T. PARKER (1960) Further results on the construction of mutually orthogonal Latin squares and the falsity of Euler's conjecture. Canadian Journal of Mathematics, xii, 189-203.

76. R. C. BOSE, I. M. CHAKRAVARTI and D. E. KNUTH (1960) On methods of constructing sets of mutually orthogonal Latin squares using a computer. I. Technometrics, ii, 507-516.

77. R. C. BOSE, I. M. CHAKRAVARTI and D. E. KNUTH (1961) On methods of constructing sets of mutually orthogonal Latin squares using a computer. II. Technometrics, iii, 111-117.

78. C. R. RAO (1961) . . . Generation of random permutations of given number of elements using random sampling numbers. Sankhyā, xxiii, 305-307.

79. C. R. RAO (1961) . . . A study of BIB designs with replications 11 to 15. Sankhyā, xxiii, 117-127.

80. R. C. BOSE, W. H. CLATWORTHY and S. S. SHRIKHANDE (1954) Tables of partially balanced designs with two associate classes. North Carolina Agricultural Experiment Station, Technical Bulletin No. 107.

81. M. J. R. HEALY (1963) . . Fiducial limits for a variance component. J. R. Statist. Soc. B.

82. J. F. KIDWELL, J. W. GOWEN and J. STADLER (1961) Pugnose—a recessive mutation in linkage group 3 of mice. Journal of Heredity, lii, 145-148.

Table I. The Normal Distribution

P	·00	·01	·02	·03	·04	·05	·06	·07	·08	·09
·0	∞	2·575829	2·326348	2·170090	2·053749	1·959964	1·880794	1·811911	1·750686	1·695398
·1	1·644854	1·598193	1·554774	1·514102	1·475791	1·439531	1·405072	1·372204	1·340755	1·310579
·2	1·281552	1·253565	1·226528	1·200359	1·174987	1·150349	1·126391	1·103063	1·080319	1·058122
·3	1·036433	1·015222	·994458	·974114	·954165	·934589	·915365	·896473	·877896	·859617
·4	·841621	·823894	·806421	·789192	·772193	·755415	·738847	·722479	·706303	·690309
·5	·674490	·658838	·643345	·628006	·612813	·597760	·582842	·568051	·553385	·538836
·6	·524401	·510073	·495850	·481727	·467699	·453762	·439913	·426148	·412463	·398855
·7	·385320	·371856	·358459	·345126	·331853	·318639	·305481	·292375	·279319	·266311
·8	·253347	·240426	·227545	·214702	·201893	·189118	·176374	·163658	·150969	·138304
·9	·125661	·113039	·100434	·087845	·075270	·062707	·050154	·037608	·025069	·012533

P	·002	·001	·000,1	·000,01	·000,001	·000,000,1	·000,000,01	·000,000,001
x	3·090232	3·29053	3·89059	4·41717	4·89164	5·32672	5·73073	6·10941

The value of P for each entry is found by adding the column heading to the value in the left-hand margin. The corresponding value of x is the deviation such that the probability of an observation falling outside the range from $-x$ to $+x$ is P. For example, P = ·03 for $x = 2·170090$; so that 3 per cent. of normally distributed values will have positive or negative deviations exceeding the standard deviation in the ratio 2·170090 at least.

The table of probits (Table IX) provides a more extensive (P, x) table. The probit table refers to a single tail of the distribution, and the P's derived from that table must therefore be multiplied by 2 to bring them into line with the P's of Table I.

Table II. Ordinates of the Normal Distribution

x	·00	·01	·02	·03	·04	·05	·06	·07	·08	·09	1	2	3	4	5
0·0	·3989	·3989	·3989	·3988	·3986	·3984	·3982	·3980	·3977	·3973	0	0	−1	−1	−1
0·1	·3970	·3965	·3961	·3956	·3951	·3945	·3939	·3932	·3925	·3918	−1	−1	−2	−2	−3
0·2	·3910	·3902	·3894	·3885	·3876	·3867	·3857	·3847	·3836	·3825	−1	−2	−3	−4	−5
0·3	·3814	·3802	·3790	·3778	·3765	·3752	·3739	·3725	·3712	·3697	−1	−3	−4	−5	−6
0·4	·3683	·3668	·3653	·3637	·3621	·3605	·3589	·3572	·3555	·3538	−2	−3	−5	−6	−8
0·5	·3521	·3503	·3485	·3467	·3448	·3429	·3410	·3391	·3372	·3352	−2	−4	−6	−8	−9
0·6	·3332	·3312	·3292	·3271	·3251	·3230	·3209	·3187	·3166	·3144	−2	−4	−6	−8	−10
0·7	·3123	·3101	·3079	·3056	·3034	·3011	·2989	·2966	·2943	·2920	−2	−5	−7	−9	−11
0·8	·2897	·2874	·2850	·2827	·2803	·2780	·2756	·2732	·2709	·2685	−2	−5	−7	−9	−12
0·9	·2661	·2637	·2613	·2589	·2565	·2541	·2516	·2492	·2468	·2444	−2	−5	−7	−10	−12
1·0	·2420	·2396	·2371	·2347	·2323	·2299	·2275	·2251	·2227	·2203	−2	−5	−7	−10	−12
1·1	·2179	·2155	·2131	·2107	·2083	·2059	·2036	·2012	·1989	·1965	−2	−5	−7	−10	−12
1·2	·1942	·1919	·1895	·1872	·1849	·1826	·1804	·1781	·1758	·1736	−2	−5	−7	−9	−11
1·3	·1714	·1691	·1669	·1647	·1626	·1604	·1582	·1561	·1539	·1518	−2	−4	−7	−9	−11
1·4	·1497	·1476	·1456	·1435	·1415	·1394	·1374	·1354	·1334	·1315	−2	−4	−6	−8	−10
1·5	·1295	·1276	·1257	·1238	·1219	·1200	·1182	·1163	·1145	·1127	−2	−4	−6	−7	−9
1·6	·1109	·1092	·1074	·1057	·1040	·1023	·1006	·0989	·0973	·0957	−2	−3	−5	−7	−8
1·7	·0940	·0925	·0909	·0893	·0878	·0863	·0848	·0833	·0818	·0804	−2	−3	−5	−6	−8
1·8	·0790	·0775	·0761	·0748	·0734	·0721	·0707	·0694	·0681	·0669	−1	−3	−4	−5	−7
1·9	·0656	·0644	·0632	·0620	·0608	·0596	·0584	·0573	·0562	·0551	−1	−2	−4	−5	−6
2·0	·0540	·0529	·0519	·0508	·0498	·0488	·0478	·0468	·0459	·0449	−1	−2	−3	−4	−5
2·1	·0440	·0431	·0422	·0413	·0404	·0396	·0387	·0379	·0371	·0363	−1	−2	−3	−3	−4
2·2	·0355	·0347	·0339	·0332	·0325	·0317	·0310	·0303	·0297	·0290	−1	−1	−2	−3	−4
2·3	·0283	·0277	·0270	·0264	·0258	·0252	·0246	·0241	·0235	·0229	−1	−1	−2	−2	−3
2·4	·0224	·0219	·0213	·0208	·0203	·0198	·0194	·0189	·0184	·0180	0	−1	−1	−2	−2
2·5	·0175	·0171	·0167	·0163	·0158	·0154	·0151	·0147	·0143	·0139	0	−1	−1	−2	−2
2·6	·0136	·0132	·0129	·0126	·0122	·0119	·0116	·0113	·0110	·0107	0	−1	−1	−1	−2
2·7	·0104	·0101	·0099	·0096	·0093	·0091	·0088	·0086	·0084	·0081	0	−1	−1	−1	−1
2·8	·0079	·0077	·0075	·0073	·0071	·0069	·0067	·0065	·0063	·0061	0	0	−1	−1	−1
2·9	·0060	·0058	·0056	·0055	·0053	·0051	·0050	·0048	·0047	·0046	0	0	0	−1	−1

x	·0	·1	·2	·3	·4	·5	·6	·7	·8	·9
3·0	·0044	·0033	·0024	·0017	·0012	·0009	·0006	·0004	·0003	·0002

TABLE II1. THE NORMAL PROBABILITY INTEGRAL

x		0	1	2	3	4	5	6	7	8	9
0·0	0·	50000	49601	49202	48803	48405	48006	47608	47210	46812	46414
0·1		46017	45620	45224	44828	44433	44038	43644	43251	42858	42465
0·2		42074	41683	41294	40905	40517	40129	39743	39358	38974	38591
0·3		38209	37828	37448	37070	36693	36317	35942	35569	35197	34827
0·4		34458	34090	33724	33360	32997	32636	32276	31918	31561	31207
0·5		30854	30503	30153	29806	29460	29116	28774	28434	28096	27760
0·6		27425	27093	26763	26435	26109	25785	25463	25143	24825	24510
0·7		24196	23885	23576	23270	22965	22663	22363	22065	21770	21476
0·8		21186	20897	20611	20327	20045	19766	19489	19215	18943	18673
0·9		18406	18141	17879	17619	17361	17106	16853	16602	16354	16109
1·0		15866	15625	15386	15151	14917	14686	14457	14231	14007	13786
1·1		13567	13350	13136	12924	12714	12507	12302	12100	11900	11702
1·2		11507	11314	11123	10935	10749	10565	10383	10204	10027	**98525**
1·3	0·0	96800	95098	93418	91759	90123	88508	86915	85343	83793	82264
1·4		80757	79270	77804	76359	74934	73529	72145	70781	69437	68112
1·5		66807	65522	64255	63008	61780	60571	59380	58208	57053	55917
1·6		54799	53699	52616	51551	50503	49471	48457	47460	46479	45514
1·7		44565	43633	42716	41815	40930	40059	39204	38364	37538	36727
1·8		35930	35148	34380	33625	32884	32157	31443	30742	30054	29379
1·9		28717	28067	27429	26803	26190	25588	24998	24419	23852	23295
2·0		22750	22216	21692	21178	20675	20182	19699	19226	18763	18309
2·1		17864	17429	17003	16586	16177	15778	15386	15003	14629	14262
2·2		13903	13553	13209	12874	12545	12224	11911	11604	11304	11011
2·3		10724	10444	10170	**99031**	**96419**	**93867**	**91375**	**88940**	**86563**	**84242**
2·4	$0·0^2$	81975	79763	77603	75494	73436	71428	69469	67557	65691	63872
2·5		62097	60366	58677	57031	55426	53861	52336	50849	49400	47988
2·6		46612	45271	43965	42692	41453	40246	39070	37926	36811	35726
2·7		34670	33642	32641	31667	30720	29798	28901	28028	27179	26354
2·8		25551	24771	24012	23274	22557	21860	21182	20524	19884	19262
2·9		18658	18071	17502	16948	16411	15889	15382	14890	14412	13949
3·0		13499	13062	12639	12228	11829	11442	11067	10703	10350	10008
3·1	$0·0^3$	96760	93544	90426	87403	84474	81635	78885	76219	73638	71136
3·2		68714	66367	64095	61895	59765	57703	55706	53774	51904	50094
3·3		48342	46648	45009	43423	41889	40406	38971	37584	36243	34946
3·4		33693	32481	31311	30179	29086	28029	27009	26023	25071	24151
3·5		23263	22405	21577	20778	20006	19262	18543	17849	17180	16534
3·6		15911	15310	14730	14171	13632	13112	12611	12128	11662	11213
3·7		10780	10363	**99611**	**95740**	**92010**	**88417**	**84957**	**81624**	**78414**	**75324**
3·8	$0·0^4$	72348	69483	66726	64072	61517	59059	56694	54418	52228	50122
3·9		48096	46148	44274	42473	40741	39076	37475	35936	34458	33037
4·0		31671	30359	29099	27888	26726	25609	24536	23507	22518	21569
4·1		20658	19783	18944	18138	17365	16624	15912	15230	14575	13948
4·2		13346	12769	12215	11685	11176	10689	10221	**97736**	**93447**	**89337**
4·3	$0·0^5$	85399	81627	78015	74555	71241	68069	65031	62123	59340	56675
4·4		54125	51685	49350	47117	44979	42935	40980	39110	37322	35612
4·5		33977	32414	30920	29492	28127	26823	25577	24386	23249	22162
4·6		21125	20133	19187	18283	17420	16597	15810	15060	14344	13660
4·7		13008	12386	11792	11226	10686	10171	**96796**	**92113**	**87648**	**83391**
4·8	$0·0^6$	79333	75465	71779	68267	64920	61731	58693	55799	53043	50418
4·9		47918	45538	43272	41115	39061	37107	35247	33476	31792	30190

TABLE III. DISTRIBUTION OF t

Probability.

n	·9	·8	·7	·6	·5	·4	·3	·2	·1	·05	·02	·01	·001
1	·158	·325	·510	·727	1·000	1·376	1·963	3·078	6·314	12·706	31·821	63·657	636·619
2	·142	·289	·445	·617	·816	1·061	1·386	1·886	2·920	4·303	6·965	9·925	31·598
3	·137	·277	·424	·584	·765	·978	1·250	1·638	2·353	3·182	4·541	5·841	12·924
4	·134	·271	·414	·569	·741	·941	1·190	1·533	2·132	2·776	3·747	4·604	8·610
5	·132	·267	·408	·559	·727	·920	1·156	1·476	2·015	2·571	3·365	4·032	6·869
6	·131	·265	·404	·553	·718	·906	1·134	1·440	1·943	2·447	3·143	3·707	5·959
7	·130	·263	·402	·549	·711	·896	1·119	1·415	1·895	2·365	2·998	3·499	5·408
8	·130	·262	·399	·546	·706	·889	1·108	1·397	1·860	2·306	2·896	3·355	5·041
9	·129	·261	·398	·543	·703	·883	1·100	1·383	1·833	2·262	2·821	3·250	4·781
10	·129	·260	·397	·542	·700	·879	1·093	1·372	1·812	2·228	2·764	3·169	4·587
11	·129	·260	·396	·540	·697	·876	1·088	1·363	1·796	2·201	2·718	3·106	4·437
12	·128	·259	·395	·539	·695	·873	1·083	1·356	1·782	2·179	2·681	3·055	4·318
13	·128	·259	·394	·538	·694	·870	1·079	1·350	1·771	2·160	2·650	3·012	4·221
14	·128	·258	·393	·537	·692	·868	1·076	1·345	1·761	2·145	2·624	2·977	4·140
15	·128	·258	·393	·536	·691	·866	1·074	1·341	1·753	2·131	2·602	2·947	4·073
16	·128	·258	·392	·535	·690	·865	1·071	1·337	1·746	2·120	2·583	2·921	4·015
17	·128	·257	·392	·534	·689	·863	1·069	1·333	1·740	2·110	2·567	2·898	3·965
18	·127	·257	·392	·534	·688	·862	1·067	1·330	1·734	2·101	2·552	2·878	3·922
19	·127	·257	·391	·533	·688	·861	1·066	1·328	1·729	2·093	2·539	2·861	3·883
20	·127	·257	·391	·533	·687	·860	1·064	1·325	1·725	2·086	2·528	2·845	3·850
21	·127	·257	·391	·532	·686	·859	1·063	1·323	1·721	2·080	2·518	2·831	3·819
22	·127	·256	·390	·532	·686	·858	1·061	1·321	1·717	2·074	2·508	2·819	3·792
23	·127	·256	·390	·532	·685	·858	1·060	1·319	1·714	2·069	2·500	2·807	3·767
24	·127	·256	·390	·531	·685	·857	1·059	1·318	1·711	2·064	2·492	2·797	3·745
25	·127	·256	·390	·531	·684	·856	1·058	1·316	1·708	2·060	2·485	2·787	3·725
26	·127	·256	·390	·531	·684	·856	1·058	1·315	1·706	2·056	2·479	2·779	3·707
27	·127	·256	·389	·531	·684	·855	1·057	1·314	1·703	2·052	2·473	2·771	3·690
28	·127	·256	·389	·530	·683	·855	1·056	1·313	1·701	2·048	2·467	2·763	3·674
29	·127	·256	·389	·530	·683	·854	1·055	1·311	1·699	2·045	2·462	2·756	3·659
30	·127	·256	·389	·530	·683	·854	1·055	1·310	1·697	2·042	2·457	2·750	3·646
40	·126	·255	·388	·529	·681	·851	1·050	1·303	1·684	2·021	2·423	2·704	3·551
60	·126	·254	·387	·527	·679	·848	1·046	1·296	1·671	2·000	2·390	2·660	3·460
120	·126	·254	·386	·526	·677	·845	1·041	1·289	1·658	1·980	2·358	2·617	3·373
∞	·126	·253	·385	·524	·674	·842	1·036	1·282	1·645	1·960	2·326	2·576	3·291

TABLE IV. DISTRIBTION OF χ^2

Probability.

n	·99	·98	·95	·90	·80	·70	·50	·30	·20	·10	·05	·02	·01	·001
1	·0^{3}157	·0^{3}628	·00393	·0158	·0642	·148	·455	1·074	1·642	2·706	3·841	5·412	6·635	10·827
2	·0201	·0404	·103	·211	·446	·713	1·386	2·408	3·219	4·605	5·991	7·824	9·210	13·815
3	·115	·185	·352	·584	1·005	1·424	2·366	3·665	4·642	6·251	7·815	9·837	11·345	16·266
4	·297	·429	·711	1·064	1·649	2·195	3·357	4·878	5·989	7·779	9·488	11·668	13·277	18·467
5	·554	·752	1·145	1·610	2·343	3·000	4·351	6·064	7·289	9·236	11·070	13·388	15·086	20·515
6	·872	1·134	1·635	2·204	3·070	3·828	5·348	7·231	8·558	10·645	12·592	15·033	16·812	22·457
7	1·239	1·564	2·167	2·833	3·822	4·671	6·346	8·383	9·803	12·017	14·067	16·622	18·475	24·322
8	1·646	2·032	2·733	3·490	4·594	5·527	7·344	9·524	11·030	13·362	15·507	18·168	20·090	26·125
9	2·088	2·532	3·325	4·168	5·380	6·393	8·343	10·656	12·242	14·684	16·919	19·679	21·666	27·877
10	2·558	3·059	3·940	4·865	6·179	7·267	9·342	11·781	13·442	15·987	18·307	21·161	23·209	29·588
11	3·053	3·609	4·575	5·578	6·989	8·148	10·341	12·899	14·631	17·275	19·675	22·618	24·725	31·264
12	3·571	4·178	5·226	6·304	7·807	9·034	11·340	14·011	15·812	18·549	21·026	24·054	26·217	32·909
13	4·107	4·765	5·892	7·042	8·634	9·926	12·340	15·119	16·985	19·812	22·362	25·472	27·688	34·528
14	4·660	5·368	6·571	7·790	9·467	10·821	13·339	16·222	18·151	21·064	23·685	26·873	29·141	36·123
15	5·229	5·985	7·261	8·547	10·307	11·721	14·339	17·322	19·311	22·307	24·996	28·259	30·578	37·697
16	5·812	6·614	7·962	9·312	11·152	12·624	15·338	18·418	20·465	23·542	26·296	29·633	32·000	39·252
17	6·408	7·255	8·672	10·085	12·002	13·531	16·338	19·511	21·615	24·769	27·587	30·995	33·409	40·790
18	7·015	7·906	9·390	10·865	12·857	14·440	17·338	20·601	22·760	25·989	28·869	32·346	34·805	42·312
19	7·633	8·567	10·117	11·651	13·716	15·352	18·338	21·689	23·900	27·204	30·144	33·687	36·191	43·820
20	8·260	9·237	10·851	12·443	14·578	16·266	19·337	22·775	25·038	28·412	31·410	35·020	37·566	45·315
21	8·897	9·915	11·591	13·240	15·445	17·182	20·337	23·858	26·171	29·615	32·671	36·343	38·932	46·797
22	9·542	10·600	12·338	14·041	16·314	18·101	21·337	24·939	27·301	30·813	33·924	37·659	40·289	48·268
23	10·196	11·293	13·091	14·848	17·187	19·021	22·337	26·018	28·429	32·007	35·172	38·968	41·638	49·728
24	10·856	11·992	13·848	15·659	18·062	19·943	23·337	27·096	29·553	33·196	36·415	40·270	42·980	51·179
25	11·524	12·697	14·611	16·473	18·940	20·867	24·337	28·172	30·675	34·382	37·652	41·566	44·314	52·620
26	12·198	13·409	15·379	17·292	19·820	21·792	25·336	29·246	31·795	35·563	38·885	42·856	45·642	54·052
27	12·879	14·125	16·151	18·114	20·703	22·719	26·336	30·319	32·912	36·741	40·113	44·140	46·963	55·476
28	13·565	14·847	16·928	18·939	21·588	23·647	27·336	31·391	34·027	37·916	41·337	45·419	48·278	56·893
29	14·256	15·574	17·708	19·768	22·475	24·577	28·336	32·461	35·139	39·087	42·557	46·693	49·588	58·302
30	14·953	16·306	18·493	20·599	23·364	25·508	29·336	33·530	36·250	40·256	43·773	47·962	50·892	59·703
32	16·362	17·783	20·072	22·271	25·148	27·373	31·336	35·665	38·466	42·585	46·194	50·487	53·486	62·487
34	17·789	19·275	21·664	23·952	26·938	29·242	33·336	37·795	40·676	44·903	48·602	52·995	56·061	65·247
36	19·233	20·783	23·269	25·643	28·735	31·115	35·336	39·922	42·879	47·212	50·999	55·489	58·619	67·985
38	20·691	22·304	24·884	27·343	30·537	32·992	37·335	42·045	45·076	49·513	53·384	57·969	61·162	70·703
40	22·164	23·838	26·509	29·051	32·345	34·872	39·335	44·165	47·269	51·805	55·759	60·436	63·691	73·402
42	23·650	25·383	28·144	30·765	34·157	36·755	41·335	46·282	49·456	54·090	58·124	62·892	66·206	76·084
44	25·148	26·939	29·787	32·487	35·974	38·641	43·335	48·396	51·639	56·369	60·481	65·337	68·710	78·750
46	26·657	28·504	31·439	34·215	37·795	40·529	45·335	50·507	53·818	58·641	62·830	67·771	71·201	81·400
48	28·177	30·080	33·098	35·949	39·621	42·420	47·335	52·616	55·993	60·907	65·171	70·197	73·683	84·037
50	29·707	31·664	34·764	37·689	41·449	44·313	49·335	54·723	58·164	63·167	67·505	72·613	76·154	86·661
52	31·246	33·256	36·437	39·433	43·281	46·209	51·335	56·827	60·332	65·422	69·832	75·021	78·616	89·272
54	32·793	34·856	38·116	41·183	45·117	48·106	53·335	58·930	62·496	67·673	72·153	77·422	81·069	91·872
56	34·350	36·464	39·801	42·937	46·955	50·005	55·335	61·031	64·658	69·919	74·468	79·815	83·513	94·461
58	35·913	38·078	41·492	44·696	48·797	51·906	57·335	63·129	66·816	72·160	76·778	82·201	85·950	97·039
60	37·485	39·699	43·188	46·459	50·641	53·809	59·335	65·227	68·972	74·397	79·082	84·580	88·379	99·607
62	39·063	41·327	44·889	48·226	52·487	55·714	61·335	67·322	71·125	76·630	81·381	86·953	90·802	102·166
64	40·649	42·960	46·595	49·996	54·336	57·620	63·335	69·416	73·276	78·860	83·675	89·320	93·217	104·716
66	42·240	44·599	48·305	51·770	56·188	59·527	65·335	71·508	75·424	81·085	85·965	91·681	95·626	107·258
68	43·838	46·244	50·020	53·548	58·042	61·436	67·335	73·600	77·571	83·308	88·250	94·037	98·028	109·791
70	45·442	47·893	51·739	55·329	59·898	63·346	69·334	75·689	79·715	85·527	90·531	96·388	100·425	112·317

For odd values of n between 30 and 70 the mean of the tabular values for $n-1$ and $n+1$ may be taken. For larger values of n, the expression $\sqrt{2\chi^2}-\sqrt{2n-1}$ may be used as a normal deviate with unit variance, remembering that the probability for χ^2 corresponds with that of a single tail of the normal curve. (For fuller formulæ see Introduction.)

Table V. Distribution of z

20 Per Cent. Points (H. W. Norton)

n_2 \ n_1	1	2	3	4	5	6	8	12	24	∞
1	1·1242	1·2425	1·2849	1·3067	1·3198	1·3286	1·3397	1·3508	1·3619	1·3730
2	·6343	·6931	·7125	·7219	·7275	·7313	·7360	·7406	·7453	·7500
3	·4933	·5299	·5385	·5418	·5435	·5444	·5453	·5461	·5465	·5467
4	·4274	·4526	·4551	·4547	·4537	·4528	·4512	·4491	·4464	·4431
5	·3892	·4075	·4062	·4032	·4004	·3982	·3947	·3905	·3853	·3791
6	·3645	·3780	·3740	·3692	·3651	·3618	·3569	·3510	·3438	·3350
7	·3471	·3573	·3512	·3450	·3399	·3358	·3298	·3224	·3135	·3025
8	·3342	·3419	·3342	·3269	·3211	·3163	·3093	·3008	·2904	·2774
9	·3243	·3300	·3211	·3130	·3064	·3011	·2933	·2838	·2721	·2573
10	·3164	·3206	·3107	·3018	·2946	·2889	·2805	·2701	·2572	·2407
11	·3100	·3129	·3021	·2926	·2850	·2790	·2699	·2588	·2449	·2268
12	·3047	·3065	·2951	·2850	·2770	·2706	·2611	·2493	·2345	·2149
13	·3002	·3011	·2891	·2786	·2703	·2636	·2536	·2412	·2255	·2046
14	·2964	·2965	·2840	·2732	·2644	·2575	·2471	·2342	·2178	·1956
15	·2931	·2926	·2795	·2684	·2594	·2522	·2415	·2281	·2110	·1876
16	·2903	·2891	·2757	·2642	·2550	·2476	·2366	·2228	·2050	·1805
17	·2877	·2860	·2723	·2605	·2511	·2436	·2322	·2180	·1997	·1741
18	·2855	·2833	·2692	·2572	·2476	·2399	·2283	·2138	·1949	·1682
19	·2835	·2809	·2665	·2543	·2445	·2367	·2249	·2100	·1905	·1629
20	·2817	·2787	·2641	·2517	·2418	·2338	·2217	·2066	·1866	·1581
21	·2800	·2767	·2619	·2493	·2392	·2311	·2189	·2034	·1831	·1536
22	·2786	·2750	·2599	·2471	·2369	·2287	·2163	·2006	·1798	·1495
23	·2772	·2733	·2581	·2452	·2348	·2265	·2139	·1980	·1767	·1457
24	·2760	·2719	·2564	·2433	·2329	·2245	·2117	·1956	·1740	·1421
25	·2749	·2705	·2549	·2417	·2311	·2226	·2097	·1933	·1715	·1388
26	·2738	·2692	·2534	·2401	·2295	·2209	·2079	·1913	·1691	·1357
27	·2729	·2681	·2521	·2387	·2280	·2193	·2062	·1894	·1669	·1328
28	·2720	·2670	·2509	·2374	·2266	·2178	·2046	·1876	·1648	·1300
29	·2711	·2660	·2498	·2362	·2253	·2165	·2031	·1860	·1629	·1274
30	·2703	·2650	·2487	·2350	·2241	·2152	·2017	·1844	·1611	·1250
40	·2647	·2582	·2411	·2267	·2152	·2058	·1915	·1731	·1477	·1062
60	·2592	·2514	·2334	·2184	·2063	·1965	·1813	·1616	·1338	·0848
120	·2536	·2446	·2258	·2101	·1974	·1870	·1710	·1499	·1193	·0582
∞	·2481	·2379	·2183	·2018	·1885	·1776	·1606	·1379	·1041	0

For high values of n_1 and n_2, z (20 per cent.) $= \dfrac{0{\cdot}8416}{\sqrt{h-0{\cdot}6}} - 0{\cdot}4514\left(\dfrac{1}{n_1}-\dfrac{1}{n_2}\right)$ approximately, where $\dfrac{2}{h} = \dfrac{1}{n_1} + \dfrac{1}{n_2}$. In other parts of the table interpolation is approximately linear if reciprocals of n_1 and n_2 are taken.

Table V. Variance Ratio

20 Per Cent. Points of e^{2z} (H. W. Norton)

n_2 \ n_1	1	2	3	4	5	6	8	12	24	∞
1	9·47	12·00	13·06	13·64	14·01	14·26	14·58	14·90	15·24	15·58
2	3·56	4·00	4·16	4·24	4·28	4·32	4·36	4·40	4·44	4·48
3	2·68	2·89	2·94	2·96	2·97	2·97	2·98	2·98	2·98	2·98
4	2·35	2·47	2·48	2·48	2·48	2·47	2·47	2·46	2·44	2·43
5	2·18	2·26	2·25	2·24	2·23	2·22	2·20	2·18	2·16	2·13
6	2·07	2·13	2·11	2·09	2·08	2·06	2·04	2·02	1·99	1·95
7	2·00	2·04	2·02	1·99	1·97	1·96	1·93	1·91	1·87	1·83
8	1·95	1·98	1·95	1·92	1·90	1·88	1·86	1·83	1·79	1·74
9	1·91	1·94	1·90	1·87	1·85	1·83	1·80	1·76	1·72	1·67
10	1·88	1·90	1·86	1·83	1·80	1·78	1·75	1·72	1·67	1·62
11	1·86	1·87	1·83	1·80	1·77	1·75	1·72	1·68	1·63	1·57
12	1·84	1·85	1·80	1·77	1·74	1·72	1·69	1·65	1·60	1·54
13	1·82	1·83	1·78	1·75	1·72	1·69	1·66	1·62	1·57	1·51
14	1·81	1·81	1·76	1·73	1·70	1·67	1·64	1·60	1·55	1·48
15	1·80	1·79	1·75	1·71	1·68	1·66	1·62	1·58	1·53	1·46
16	1·79	1·78	1·74	1·70	1·67	1·64	1·61	1·56	1·51	1·43
17	1·78	1·77	1·72	1·68	1·65	1·63	1·59	1·55	1·49	1·42
18	1·77	1·76	1·71	1·67	1·64	1·62	1·58	1·53	1·48	1·40
19	1·76	1·75	1·70	1·66	1·63	1·61	1·57	1·52	1·46	1·39
20	1·76	1·75	1·70	1·65	1·62	1·60	1·56	1·51	1·45	1·37
21	1·75	1·74	1·69	1·65	1·61	1·59	1·55	1·50	1·44	1·36
22	1·75	1·73	1·68	1·64	1·61	1·58	1·54	1·49	1·43	1·35
23	1·74	1·73	1·68	1·63	1·60	1·57	1·53	1·49	1·42	1·34
24	1·74	1·72	1·67	1·63	1·59	1·57	1·53	1·48	1·42	1·33
25	1·73	1·72	1·66	1·62	1·59	1·56	1·52	1·47	1·41	1·32
26	1·73	1·71	1·66	1·62	1·58	1·56	1·52	1·47	1·40	1·31
27	1·73	1·71	1·66	1·61	1·58	1·55	1·51	1·46	1·40	1·30
28	1·72	1·71	1·65	1·61	1·57	1·55	1·51	1·46	1·39	1·30
29	1·72	1·70	1·65	1·60	1·57	1·54	1·50	1·45	1·39	1·29
30	1·72	1·70	1·64	1·60	1·57	1·54	1·50	1·45	1·38	1·28
40	1·70	1·68	1·62	1·57	1·54	1·51	1·47	1·41	1·34	1·24
60	1·68	1·65	1·59	1·55	1·51	1·48	1·44	1·38	1·31	1·18
120	1·66	1·63	1·57	1·52	1·48	1·45	1·41	1·35	1·27	1·12
∞	1·64	1·61	1·55	1·50	1·46	1·43	1·38	1·32	1·23	1·00

Lower 20 per cent. points are found by interchange of n_1 and n_2, *i.e.* n_1 must always correspond with the greater mean square.

Table V. Distribution of *z*—*continued*

10 Per Cent. Points

n_2 \ n_1	1	2	3	4	5	6	8	12	24	∞
1	1·8427	1·9510	1·9907	2·0112	2·0236	2·0320	2·0425	2·0530	2·0636	2·0742
2	1·0716	1·0986	1·1075	1·1120	1·1146	1·1164	1·1186	1·1208	1·1230	1·1252
3	·8558	·8489	·8423	·8379	·8347	·8324	·8293	·8258	·8221	·8179
4	·7570	·7322	·7164	·7064	·6994	·6944	·6875	·6799	·6716	·6623
5	·7006	·6648	·6432	·6293	·6196	·6125	·6029	·5921	·5801	·5665
6	·6643	·6211	·5953	·5786	·5669	·5583	·5465	·5332	·5181	·5007
7	·6390	·5905	·5615	·5427	·5295	·5197	·5061	·4907	·4730	·4523
8	·6203	·5678	·5364	·5160	·5015	·4907	·4757	·4585	·4386	·4148
9	·6060	·5504	·5171	·4953	·4798	·4682	·4520	·4333	·4114	·3849
10	·5947	·5366	·5017	·4788	·4624	·4502	·4330	·4130	·3893	·3602
11	·5855	·5253	·4892	·4653	·4483	·4355	·4173	·3962	·3710	·3395
12	·5779	·5160	·4788	·4541	·4365	·4231	·4043	·3821	·3555	·3219
13	·5715	·5082	·4701	·4447	·4265	·4127	·3932	·3702	·3422	·3066
14	·5661	·5015	·4626	·4366	·4180	·4038	·3836	·3598	·3308	·2931
15	·5614	·4957	·4561	·4296	·4106	·3961	·3754	·3508	·3207	·2813
16	·5572	·4907	·4504	·4235	·4041	·3893	·3681	·3429	·3118	·2706
17	·5537	·4863	·4455	·4181	·3984	·3833	·3617	·3359	·3038	·2611
18	·5505	·4823	·4411	·4134	·3933	·3780	·3560	·3296	·2967	·2524
19	·5476	·4788	·4371	·4091	·3887	·3732	·3508	·3240	·2904	·2445
20	·5451	·4757	·4336	·4052	·3846	·3689	·3462	·3189	·2846	·2373
21	·5427	·4728	·4304	·4017	·3809	·3650	·3420	·3143	·2793	·2307
22	·5407	·4703	·4275	·3986	·3776	·3615	·3382	·3101	·2744	·2245
23	·5388	·4679	·4248	·3957	·3745	·3582	·3347	·3062	·2700	·2188
24	·5370	·4657	·4224	·3931	·3717	·3553	·3315	·3027	·2659	·2135
25	·5354	·4638	·4201	·3906	·3691	·3526	·3286	·2994	·2621	·2086
26	·5339	·4620	·4181	·3884	·3667	·3500	·3258	·2964	·2585	·2039
27	·5326	·4603	·4162	·3863	·3645	·3477	·3233	·2936	·2553	·1996
28	·5313	·4587	·4144	·3844	·3624	·3455	·3210	·2910	·2522	·1955
29	·5301	·4572	·4128	·3826	·3605	·3435	·3188	·2885	·2493	·1916
30	·5290	·4559	·4112	·3809	·3587	·3416	·3167	·2863	·2466	·1880
40	·5211	·4461	·4001	·3688	·3458	·3280	·3019	·2696	·2268	·1599
60	·5133	·4363	·3891	·3567	·3328	·3142	·2868	·2526	·2063	·1279
120	·5054	·4266	·3781	·3446	·3198	·3005	·2717	·2354	·1848	·0881
∞	·4976	·4170	·3671	·3326	·3069	·2866	·2565	·2178	·1622	0

For high values of n_1 and n_2, z (10 per cent.) $= \dfrac{1{\cdot}2816}{\sqrt{h-0{\cdot}8}} - 0{\cdot}6071\left(\dfrac{1}{n_1}-\dfrac{1}{n_2}\right)$ approximately, where $\dfrac{2}{h}=\dfrac{1}{n_1}+\dfrac{1}{n_2}$. In other parts of the table interpolation is approximately linear if reciprocals of n_1 and n_2 are taken.

TABLE V. VARIANCE RATIO—*continued*

10 Per Cent. Points of e^{2z}

n_2 \ n_1	1	2	3	4	5	6	8	12	24	∞
1	39·86	49·50	53·59	55·83	57·24	58·20	59·44	60·70	62·00	63·33
2	8·53	9·00	9·16	9·24	9·29	9·33	9·37	9·41	9·45	9·49
3	5·54	5·46	5·39	5·34	5·31	5·28	5·25	5·22	5·18	5·13
4	4·54	4·32	4·19	4·11	4·05	4·01	3·95	3·90	3·83	3·76
5	4·06	3·78	3·62	3·52	3·45	3·40	3·34	3·27	3·19	3·10
6	3·78	3·46	3·29	3·18	3·11	3·05	2·98	2·90	2·82	2·72
7	3·59	3·26	3·07	2·96	2·88	2·83	2·75	2·67	2·58	2·47
8	3·46	3·11	2·92	2·81	2·73	2·67	2·59	2·50	2·40	2·29
9	3·36	3·01	2·81	2·69	2·61	2·55	2·47	2·38	2·28	2·16
10	3·28	2·92	2·73	2·61	2·52	2·46	2·38	2·28	2·18	2·06
11	3·23	2·86	2·66	2·54	2·45	2·39	2·30	2·21	2·10	1·97
12	3·18	2·81	2·61	2·48	2·39	2·33	2·24	2·15	2·04	1·90
13	3·14	2·76	2·56	2·43	2·35	2·28	2·20	2·10	1·98	1·85
14	3·10	2·73	2·52	2·39	2·31	2·24	2·15	2·05	1·94	1·80
15	3·07	2·70	2·49	2·36	2·27	2·21	2·12	2·02	1·90	1·76
16	3·05	2·67	2·46	2·33	2·24	2·18	2·09	1·99	1·87	1·72
17	3·03	2·64	2·44	2·31	2·22	2·15	2·06	1·96	1·84	1·69
18	3·01	2·62	2·42	2·29	2·20	2·13	2·04	1·93	1·81	1·66
19	2·99	2·61	2·40	2·27	2·18	2·11	2·02	1·91	1·79	1·63
20	2·97	2·59	2·38	2·25	2·16	2·09	2·00	1·89	1·77	1·61
21	2·96	2·57	2·36	2·23	2·14	2·08	1·98	1·88	1·75	1·59
22	2·95	2·56	2·35	2·22	2·13	2·06	1·97	1·86	1·73	1·57
23	2·94	2·55	2·34	2·21	2·11	2·05	1·95	1·84	1·72	1·55
24	2·93	2·54	2·33	2·19	2·10	2·04	1·94	1·83	1·70	1·53
25	2·92	2·53	2·32	2·18	2·09	2·02	1·93	1·82	1·69	1·52
26	2·91	2·52	2·31	2·17	2·08	2·01	1·92	1·81	1·68	1·50
27	2·90	2·51	2·30	2·17	2·07	2·00	1·91	1·80	1·67	1·49
28	2·89	2·50	2·29	2·16	2·06	2·00	1·90	1·79	1·66	1·48
29	2·89	2·50	2·28	2·15	2·06	1·99	1·89	1·78	1·65	1·47
30	2·88	2·49	2·28	2·14	2·05	1·98	1·88	1·77	1·64	1·46
40	2·84	2·44	2·23	2·09	2·00	1·93	1·83	1·71	1·57	1·38
60	2·79	2·39	2·18	2·04	1·95	1·87	1·77	1·66	1·51	1·29
120	2·75	2·35	2·13	1·99	1·90	1·82	1·72	1·60	1·45	1·19
∞	2·71	2·30	2·08	1·94	1·85	1·77	1·67	1·55	1·38	1·00

Lower 10 per cent. points are found by interchange of n_1 and n_2, *i.e.* n_1 must always correspond with the greater mean square.

TABLE V. DISTRIBUTION OF *z*—*continued*

5 Per Cent. Points

n_2 \ n_1	1	2	3	4	5	6	8	12	24	∞
1	2·5421	2·6479	2·6870	2·7071	2·7194	2·7276	2·7380	2·7484	2·7588	2·7693
2	1·4592	1·4722	1·4765	1·4787	1·4800	1·4808	1·4819	1·4830	1·4840	1·4851
3	1·1577	1·1284	1·1137	1·1051	1·0994	1·0953	1·0899	1·0842	1·0781	1·0716
4	1·0212	·9690	·9429	·9272	·9168	·9093	·8993	·8885	·8767	·8639
5	·9441	·8777	·8441	·8236	·8097	·7997	·7862	·7714	·7550	·7368
6	·8948	·8188	·7798	·7558	·7394	·7274	·7112	·6931	·6729	·6499
7	·8606	·7777	·7347	·7080	·6896	·6761	·6576	·6369	·6134	·5862
8	·8355	·7475	·7014	·6725	·6525	·6378	·6175	·5945	·5682	·5371
9	·8163	·7242	·6757	·6450	·6238	·6080	·5862	·5613	·5324	·4979
10	·8012	·7058	·6553	·6232	·6009	·5843	·5611	·5346	·5035	·4657
11	·7889	·6909	·6387	·6055	·5822	·5648	·5406	·5126	·4795	·4387
12	·7788	·6786	·6250	·5907	·5666	·5487	·5234	·4941	·4592	·4156
13	·7703	·6682	·6134	·5783	·5535	·5350	·5089	·4785	·4419	·3957
14	·7630	·6594	·6036	·5677	·5423	·5233	·4964	·4649	·4269	·3782
15	·7568	·6518	·5950	·5585	·5326	·5131	·4855	·4532	·4138	·3628
16	·7514	·6451	·5876	·5505	·5241	·5042	·4760	·4428	·4022	·3490
17	·7466	·6393	·5811	·5434	·5166	·4964	·4676	·4337	·3919	·3366
18	·7424	·6341	·5753	·5371	·5099	·4894	·4602	·4255	·3827	·3253
19	·7386	·6295	·5701	·5315	·5040	·4832	·4535	·4182	·3743	·3151
20	·7352	·6254	·5654	·5265	·4986	·4776	·4474	·4116	·3668	·3057
21	·7322	·6216	·5612	·5219	·4938	·4725	·4420	·4055	·3599	·2971
22	·7294	·6182	·5574	·5178	·4894	·4679	·4370	·4001	·3536	·2892
23	·7269	·6151	·5540	·5140	·4854	·4636	·4325	·3950	·3478	·2818
24	·7246	·6123	·5508	·5106	·4817	·4598	·4283	·3904	·3425	·2749
25	·7225	·6097	·5478	·5074	·4783	·4562	·4244	·3862	·3376	·2685
26	·7205	·6073	·5451	·5045	·4752	·4529	·4209	·3823	·3330	·2625
27	·7187	·6051	·5427	·5017	·4723	·4499	·4176	·3786	·3287	·2569
28	·7171	·6030	·5403	·4992	·4696	·4471	·4146	·3752	·3248	·2516
29	·7155	·6011	·5382	·4969	·4671	·4444	·4117	·3720	·3211	·2466
30	·7141	·5994	·5362	·4947	·4648	·4420	·4090	·3691	·3176	·2419
40	·7037	·5866	·5217	·4789	·4479	·4242	·3897	·3475	·2920	·2057
60	·6933	·5738	·5073	·4632	·4311	·4064	·3702	·3255	·2654	·1644
120	·6830	·5611	·4930	·4475	·4143	·3885	·3506	·3032	·2376	·1131
∞	·6729	·5486	·4787	·4319	·3974	·3706	·3309	·2804	·2085	0

For high values of n_1 and n_2, z (5 per cent.) $= \dfrac{1·6449}{\sqrt{h-1}} - 0·7843 \left(\dfrac{1}{n_1} - \dfrac{1}{n_2}\right)$ approximately, where $\dfrac{2}{h} = \dfrac{1}{n_1} + \dfrac{1}{n_2}$. In other parts of the table interpolation is approximately linear if reciprocals of n_1 and n_2 are taken.

Table V. Variance Ratio—*continued*

5 Per Cent. Points of e^{2z}

n_2 \ n_1	1	2	3	4	5	6	8	12	24	∞
1	161·4	199·5	215·7	224·6	230·2	234·0	238·9	243·9	249·0	254·3
2	18·51	19·00	19·16	19·25	19·30	19·33	19·37	19·41	19·45	19·50
3	10·13	9·55	9·28	9·12	9·01	8·94	8·84	8·74	8·64	8·53
4	7·71	6·94	6·59	6·39	6·26	6·16	6·04	5·91	5·77	5·63
5	6·61	5·79	5·41	5·19	5·05	4·95	4·82	4·68	4·53	4·36
6	5·99	5·14	4·76	4·53	4·39	4·28	4·15	4·00	3·84	3·67
7	5·59	4·74	4·35	4·12	3·97	3·87	3·73	3·57	3·41	3·23
8	5·32	4·46	4·07	3·84	3·69	3·58	3·44	3·28	3·12	2·93
9	5·12	4·26	3·86	3·63	3·48	3·37	3·23	3·07	2·90	2·71
10	4·96	4·10	3·71	3·48	3·33	3·22	3·07	2·91	2·74	2·54
11	4·84	3·98	3·59	3·36	3·20	3·09	2·95	2·79	2·61	2·40
12	4·75	3·88	3·49	3·26	3·11	3·00	2·85	2·69	2·50	2·30
13	4·67	3·80	3·41	3·18	3·02	2·92	2·77	2·60	2·42	2·21
14	4·60	3·74	3·34	3·11	2·96	2·85	2·70	2·53	2·35	2·13
15	4·54	3·68	3·29	3·06	2·90	2·79	2·64	2·48	2·29	2·07
16	4·49	3·63	3·24	3·01	2·85	2·74	2·59	2·42	2·24	2·01
17	4·45	3·59	3·20	2·96	2·81	2·70	2·55	2·38	2·19	1·96
18	4·41	3·55	3·16	2·93	2·77	2·66	2·51	2·34	2·15	1·92
19	4·38	3·52	3·13	2·90	2·74	2·63	2·48	2·31	2·11	1·88
20	4·35	3·49	3·10	2·87	2·71	2·60	2·45	2·28	2·08	1·84
21	4·32	3·47	3·07	2·84	2·68	2·57	2·42	2·25	2·05	1·81
22	4·30	3·44	3·05	2·82	2·66	2·55	2·40	2·23	2·03	1·78
23	4·28	3·42	3·03	2·80	2·64	2·53	2·38	2·20	2·00	1·76
24	4·26	3·40	3·01	2·78	2·62	2·51	2·36	2·18	1·98	1·73
25	4·24	3·38	2·99	2·76	2·60	2·49	2·34	2·16	1·96	1·71
26	4·22	3·37	2·98	2·74	2·59	2·47	2·32	2·15	1·95	1·69
27	4·21	3·35	2·96	2·73	2·57	2·46	2·30	2·13	1·93	1·67
28	4·20	3·34	2·95	2·71	2·56	2·44	2·29	2·12	1·91	1·65
29	4·18	3·33	2·93	2·70	2·54	2·43	2·28	2·10	1·90	1·64
30	4·17	3·32	2·92	2·69	2·53	2·42	2·27	2·09	1·89	1·62
40	4·08	3·23	2·84	2·61	2·45	2·34	2·18	2·00	1·79	1·51
60	4·00	3·15	2·76	2·52	2·37	2·25	2·10	1·92	1·70	1·39
120	3·92	3·07	2·68	2·45	2·29	2·17	2·02	1·83	1·61	1·25
∞	3·84	2·99	2·60	2·37	2·21	2·10	1·94	1·75	1·52	1·00

Lower 5 per cent. points are found by interchange of n_1 and n_2, *i.e.* n_1 must always correspond with the greater mean square.

TABLE V. DISTRIBUTION OF *z*—*continued*

1 Per Cent. Points

n_2 \ n_1	1	2	3	4	5	6	8	12	24	∞
1	4·1535	4·2585	4·2974	4·3175	4·3297	4·3379	4·3482	4·3585	4·3689	4·3794
2	2·2950	2·2976	2·2984	2·2988	2·2991	2·2992	2·2994	2·2997	2·2999	2·3001
3	1·7649	1·7140	1·6915	1·6786	1·6703	1·6645	1·6569	1·6489	1·6404	1·6314
4	1·5270	1·4452	1·4075	1·3856	1·3711	1·3609	1·3473	1·3327	1·3170	1·3000
5	1·3943	1·2929	1·2449	1·2164	1·1974	1·1838	1·1656	1·1457	1·1239	1·0997
6	1·3103	1·1955	1·1401	1·1068	1·0843	1·0680	1·0460	1·0218	·9948	·9643
7	1·2526	1·1281	1·0672	1·0300	1·0048	·9864	·9614	·9335	·9020	·8658
8	1·2106	1·0787	1·0135	·9734	·9459	·9259	·8983	·8673	·8319	·7904
9	1·1786	1·0411	·9724	·9299	·9006	·8791	·8494	·8157	·7769	·7305
10	1·1535	1·0114	·9399	·8954	·8646	·8419	·8104	·7744	·7324	·6816
11	1·1333	·9874	·9136	·8674	·8354	·8116	·7785	·7405	·6958	·6408
12	1·1166	·9677	·8919	·8443	·8111	·7864	·7520	·7122	·6649	·6061
13	1·1027	·9511	·8737	·8248	·7907	·7652	·7295	·6882	·6386	·5761
14	1·0909	·9370	·8581	·8082	·7732	·7471	·7103	·6675	·6159	·5500
15	1·0807	·9249	·8448	·7939	·7582	·7314	·6937	·6496	·5961	·5269
16	1·0719	·9144	·8331	·7814	·7450	·7177	·6791	·6339	·5786	·5064
17	1·0641	·9051	·8229	·7705	·7335	·7057	·6663	·6199	·5630	·4879
18	1·0572	·8970	·8138	·7607	·7232	·6950	·6549	·6075	·5491	·4712
19	1·0511	·8897	·8057	·7521	·7140	·6854	·6447	·5964	·5366	·4560
20	1·0457	·8831	·7985	·7443	·7058	·6768	·6355	·5864	·5253	·4421
21	1·0408	·8772	·7920	·7372	·6984	·6690	·6272	·5773	·5150	·4294
22	1·0363	·8719	·7860	·7309	·6916	·6620	·6196	·5691	·5056	·4176
23	1·0322	·8670	·7806	·7251	·6855	·6555	·6127	·5615	·4969	·4068
24	1·0285	·8626	·7757	·7197	·6799	·6496	·6064	·5545	·4890	·3967
25	1·0251	·8585	·7712	·7148	·6747	·6442	·6006	·5481	·4816	·3872
26	1·0220	·8548	·7670	·7103	·6699	·6392	·5952	·5422	·4748	·3784
27	1·0191	·8513	·7631	·7062	·6655	·6346	·5902	·5367	·4685	·3701
28	1·0164	·8481	·7595	·7023	·6614	·6303	·5856	·5316	·4626	·3624
29	1·0139	·8451	·7562	·6987	·6576	·6263	·5813	·5269	·4570	·3550
30	1·0116	·8423	·7531	·6954	·6540	·6226	·5773	·5224	·4519	·3481
40	·9949	·8223	·7307	·6712	·6283	·5956	·5481	·4901	·4138	·2952
60	·9784	·8025	·7086	·6472	·6028	·5687	·5189	·4574	·3746	·2352
120	·9622	·7829	·6867	·6234	·5774	·5419	·4897	·4243	·3339	·1612
∞	·9462	·7636	·6651	·5999	·5522	·5152	·4604	·3908	·2913	0

For high values of n_1 and n_2, z (1 per cent.) $= \dfrac{2\cdot3263}{\sqrt{h-1\cdot4}} - 1\cdot235\left(\dfrac{1}{n_1}-\dfrac{1}{n_2}\right)$ approximately, where $\dfrac{2}{h}=\dfrac{1}{n_1}+\dfrac{1}{n_2}$. In other parts of the table interpolation is approximately linear if reciprocals of n_1 and n_2 are taken.

TABLE V. VARIANCE RATIO—*continued*

1 Per Cent. Points of e^{2z}

n_2 \ n_1	1	2	3	4	5	6	8	12	24	∞
1	4052	4999	5403	5625	5764	5859	5982	6106	6234	6366
2	98·50	99·00	99·17	99·25	99·30	99·33	99·37	99·42	99·46	99·50
3	34·12	30·82	29·46	28·71	28·24	27·91	27·49	27·05	26·60	26·12
4	21·20	18·00	16·69	15·98	15·52	15·21	14·80	14·37	13·93	13·46
5	16·26	13·27	12·06	11·39	10·97	10·67	10·29	9·89	9·47	9·02
6	13·74	10·92	9·78	9·15	8·75	8·47	8·10	7·72	7·31	6·88
7	12·25	9·55	8·45	7·85	7·46	7·19	6·84	6·47	6·07	5·65
8	11·26	8·65	7·59	7·01	6·63	6·37	6·03	5·67	5·28	4·86
9	10·56	8·02	6·99	6·42	6·06	5·80	5·47	5·11	4·73	4·31
10	10·04	7·56	6·55	5·99	5·64	5·39	5·06	4·71	4·33	3·91
11	9·65	7·20	6·22	5·67	5·32	5·07	4·74	4·40	4·02	3·60
12	9·33	6·93	5·95	5·41	5·06	4·82	4·50	4·16	3·78	3·36
13	9·07	6·70	5·74	5·20	4·86	4·62	4·30	3·96	3·59	3·16
14	8·86	6·51	5·56	5·03	4·69	4·46	4·14	3·80	3·43	3·00
15	8·68	6·36	5·42	4·89	4·56	4·32	4·00	3·67	3·29	2·87
16	8·53	6·23	5·29	4·77	4·44	4·20	3·89	3·55	3·18	2·75
17	8·40	6·11	5·18	4·67	4·34	4·10	3·79	3·45	3·08	2·65
18	8·28	6·01	5·09	4·58	4·25	4·01	3·71	3·37	3·00	2·57
19	8·18	5·93	5·01	4·50	4·17	3·94	3·63	3·30	2·92	2·49
20	8·10	5·85	4·94	4·43	4·10	3·87	3·56	3·23	2·86	2·42
21	8·02	5·78	4·87	4·37	4·04	3·81	3·51	3·17	2·80	2·36
22	7·94	5·72	4·82	4·31	3·99	3·76	3·45	3·12	2·75	2·31
23	7·88	5·66	4·76	4·26	3·94	3·71	3·41	3·07	2·70	2·26
24	7·82	5·61	4·72	4·22	3·90	3·67	3·36	3·03	2·66	2·21
25	7·77	5·57	4·68	4·18	3·86	3·63	3·32	2·99	2·62	2·17
26	7·72	5·53	4·64	4·14	3·82	3·59	3·29	2·96	2·58	2·13
27	7·68	5·49	4·60	4·11	3·78	3·56	3·26	2·93	2·55	2·10
28	7·64	5·45	4·57	4·07	3·75	3·53	3·23	2·90	2·52	2·06
29	7·60	5·42	4·54	4·04	3·73	3·50	3·20	2·87	2·49	2·03
30	7·56	5·39	4·51	4·02	3·70	3·47	3·17	2·84	2·47	2·01
40	7·31	5·18	4·31	3·83	3·51	3·29	2·99	2·66	2·29	1·80
60	7·08	4·98	4·13	3·65	3·34	3·12	2·82	2·50	2·12	1·60
120	6·85	4·79	3·95	3·48	3·17	2·96	2·66	2·34	1·95	1·38
∞	6·64	4·60	3·78	3·32	3·02	2·80	2·51	2·18	1·79	1·00

Lower 1 per cent. points are found by interchange of n_1 and n_2, *i.e.* n_1 must always correspond with the greater mean square.

Table V. Distribution of *z*—*continued*

0·1 Per Cent. Points (Colcord and Deming)

n_2 \ n_1	1	2	3	4	5	6	8	12	24	∞
1	6·4562	6·5612	6·6000	6·6201	6·6323	6·6405	6·6508	6·6611	6·6715	6·6819
2	3·4531	3·4534	3·4535	3·4535	3·4535	3·4535	3·4536	3·4536	3·4536	3·4536
3	2·5591	2·5003	2·4748	2·4603	2·4511	2·4446	2·4361	2·4272	2·4179	2·4080
4	2·1529	2·0574	2·0143	1·9892	1·9728	1·9612	1·9459	1·9294	1·9118	1·8927
5	1·9270	1·8071	1·7513	1·7184	1·6964	1·6808	1·6598	1·6370	1·6121	1·5845
6	1·7849	1·6479	1·5828	1·5438	1·5175	1·4986	1·4730	1·4449	1·4136	1·3783
7	1·6879	1·5384	1·4662	1·4224	1·3927	1·3711	1·3417	1·3090	1·2721	1·2296
8	1·6177	1·4587	1·3809	1·3333	1·3008	1·2770	1·2443	1·2077	1·1659	1·1169
9	1·5646	1·3982	1·3160	1·2653	1·2304	1·2047	1·1694	1·1293	1·0830	1·0279
10	1·5232	1·3509	1·2650	1·2116	1·1748	1·1475	1·1098	1·0668	1·0165	·9557
11	1·4900	1·3128	1·2238	1·1683	1·1297	1·1012	1·0614	1·0157	·9619	·8957
12	1·4627	1·2814	1·1900	1·1326	1·0926	1·0628	1·0213	·9733	·9162	·8450
13	1·4400	1·2553	1·1616	1·1026	1·0614	1·0306	·9875	·9374	·8774	·8014
14	1·4208	1·2332	1·1376	1·0772	1·0348	1·0031	·9586	·9066	·8439	·7635
15	1·4043	1·2141	1·1169	1·0553	1·0119	·9795	·9336	·8800	·8147	·7301
16	1·3900	1·1976	1·0989	1·0362	·9920	·9588	·9119	·8567	·7891	·7005
17	1·3775	1·1832	1·0832	1·0195	·9745	·9407	·8927	·8361	·7664	·6740
18	1·3665	1·1704	1·0693	1·0047	·9590	·9246	·8757	·8178	·7462	·6502
19	1·3567	1·1591	1·0569	·9915	·9452	·9103	·8605	·8014	·7279	·6285
20	1·3480	1·1489	1·0458	·9798	·9329	·8974	·8469	·7867	·7115	·6086
21	1·3401	1·1398	1·0358	·9691	·9217	·8858	·8346	·7734	·6965	·5904
22	1·3329	1·1315	1·0268	·9595	·9116	·8753	·8234	·7612	·6828	·5738
23	1·3264	1·1240	1·0186	·9507	·9024	·8657	·8132	·7501	·6704	·5583
24	1·3205	1·1171	1·0111	·9427	·8939	·8569	·8038	·7400	·6589	·5440
25	1·3151	1·1108	1·0041	·9354	·8862	·8489	·7953	·7306	·6483	·5307
26	1·3101	1·1050	·9978	·9286	·8791	·8415	·7873	·7220	·6385	·5183
27	1·3055	1·0997	·9920	·9223	·8725	·8346	·7800	·7140	·6294	·5066
28	1·3013	1·0947	·9866	·9165	·8664	·8282	·7732	·7066	·6209	·4957
29	1·2973	1·0901	·9815	·9112	·8607	·8223	·7669	·6997	·6129	·4853
30	1·2936	1·0859	·9768	·9061	·8554	·8168	·7610	·6932	·6056	·4756
40	1·2672	1·0552	·9431	·8701	·8174	·7771	·7184	·6463	·5513	·4016
60	1·2413	1·0250	·9100	·8345	·7798	·7377	·6760	·5992	·4955	·3184
120	1·2159	·9954	·8773	·7994	·7426	·6986	·6338	·5519	·4381	·2170
∞	1·1910	·9663	·8453	·7648	·7059	·6599	·5917	·5044	·3786	0

For high values of n_1 and n_2, $z(0\cdot1 \text{ per cent.}) = \dfrac{3\cdot0902}{\sqrt{h-2\cdot1}} - 1\cdot925\left(\dfrac{1}{n_1} - \dfrac{1}{n_2}\right)$ approximately, where $\dfrac{2}{h} = \dfrac{1}{n_1} + \dfrac{1}{n_2}$. In other parts of the table interpolation is approximately linear if reciprocals of n_1 and n_2 are taken.

TABLE V. VARIANCE RATIO—*continued*

0·1 Per Cent. Points of e^{2z}

$n_2 \backslash n_1$	1	2	3	4	5	6	8	12	24	∞
1	405284	500000	540379	562500	576405	585937	598144	610667	623497	636619
2	998·5	999·0	999·2	999·2	999·3	999·3	999·4	999·4	999·5	999·5
3	167·0	148·5	141·1	137·1	134·6	132·8	130·6	128·3	125·9	123·5
4	74·14	61·25	56·18	53·44	51·71	50·53	49·00	47·41	45·77	44·05
5	47·18	37·12	33·20	31·09	29·75	28·84	27·64	26·42	25·14	23·78
6	35·51	27·00	23·70	21·92	20·81	20·03	19·03	17·99	16·89	15·75
7	29·25	21·69	18·77	17·19	16·21	15·52	14·63	13·71	12·73	11·69
8	25·42	18·49	15·83	14·39	13·49	12·86	12·04	11·19	10·30	9·34
9	22·86	16·39	13·90	12·56	11·71	11·13	10·37	9·57	8·72	7·81
10	21·04	14·91	12·55	11·28	10·48	9·92	9·20	8·45	7·64	6·76
11	19·69	13·81	11·56	10·35	9·58	9·05	8·35	7·63	6·85	6·00
12	18·64	12·97	10·80	9·63	8·89	8·38	7·71	7·00	6·25	5·42
13	17·81	12·31	10·21	9·07	8·35	7·86	7·21	6·52	5·78	4·97
14	17·14	11·78	9·73	8·62	7·92	7·43	6·80	6·13	5·41	4·60
15	16·59	11·34	9·34	8·25	7·57	7·09	6·47	5·81	5·10	4·31
16	16·12	10·97	9·00	7·94	7·27	6·81	6·19	5·55	4·85	4·06
17	15·72	10·66	8·73	7·68	7·02	6·56	5·96	5·32	4·63	3·85
18	15·38	10·39	8·49	7·46	6·81	6·35	5·76	5·13	4·45	3·67
19	15·08	10·16	8·28	7·26	6·62	6·18	5·59	4·97	4·29	3·52
20	14·82	9·95	8·10	7·10	6·46	6·02	5·44	4·82	4·15	3·38
21	14·59	9·77	7·94	6·95	6·32	5·88	5·31	4·70	4·03	3·26
22	14·38	9·61	7·80	6·81	6·19	5·76	5·19	4·58	3·92	3·15
23	14·19	9·47	7·67	6·69	6·08	5·65	5·09	4·48	3·82	3·05
24	14·03	9·34	7·55	6·59	5·98	5·55	4·99	4·39	3·74	2·97
25	13·88	9·22	7·45	6·49	5·88	5·46	4·91	4·31	3·66	2·89
26	13·74	9·12	7·36	6·41	5·80	5·38	4·83	4·24	3·59	2·82
27	13·61	9·02	7·27	6·33	5·73	5·31	4·76	4·17	3·52	2·75
28	13·50	8·93	7·19	6·25	5·66	5·24	4·69	4·11	3·46	2·70
29	13·39	8·85	7·12	6·19	5·59	5·18	4·64	4·05	3·41	2·64
30	13·29	8·77	7·05	6·12	5·53	5·12	4·58	4·00	3·36	2·59
40	12·61	8·25	6·60	5·70	5·13	4·73	4·21	3·64	3·01	2·23
60	11·97	7·76	6·17	5·31	4·76	4·37	3·87	3·31	2·69	1·90
120	11·38	7·32	5·79	4·95	4·42	4·04	3·55	3·02	2·40	1·54
∞	10·83	6·91	5·42	4·62	4·10	3·74	3·27	2·74	2·13	1·00

Lower 0·1 per cent. points are found by interchange of n_1 and n_2, *i.e.* n_1 must always correspond with the greater mean square.

Table VI. Fiducial Limits for a Variance Component
(M. J. R. Healy)

	n_2	$F = 0·5$	1·0	2·0	4·0	8·0	16·0	∞
$P = 0·95$	6	0	0	0	0	0·254	0·372	0·477
	8	0	0	0	0·060	0·288	0·386	0·477
$n_1 = 6$	12	0	0	0	0·134	0·315	0·397	0·477
	24	0	0	0	0·187	0·336	0·407	0·477
	∞	0	0	0	0·227	0·352	0·415	0·477
	6	0	0	0	0	0·288	0·410	0·516
	8	0	0	0	0·088	0·325	0·425	0·516
$n_1 = 8$	12	0	0	0	0·168	0·353	0·436	0·516
	24	0	0	0	0·225	0·375	0·446	0·516
	∞	0	0	0·016	0·266	0·391	0·454	0·516
	6	0	0	0	0	0·333	0·462	0·571
	8	0	0	0	0·125	0·375	0·479	0·571
$n_1 = 12$	12	0	0	0	0·212	0·405	0·491	0·571
	24	0	0	0	0·276	0·428	0·501	0·571
	∞	0	0	0·071	0·321	0·446	0·509	0·571
	6	0	0	0	0·034	0·404	0·546	0·659
	8	0	0	0	0·179	0·453	0·565	0·659
$n_1 = 24$	12	0	0	0	0·284	0·489	0·579	0·659
	24	0	0	0·006	0·358	0·515	0·589	0·659
	∞	0	0	0·159	0·409	0·534	0·597	0·659
	6	0	0	0	0·083	0·542	0·734	1·000
	8	0	0	0	0·268	0·634	0·817	1·000
$n_1 = ∞$	12	0	0	0	0·426	0·713	0·856	1·000
	24	0	0	0·134	0·567	0·784	0·892	1·000
	∞	0	0	0·500	0·750	0·875	0·938	1·000
$P = 0·05$	6	1·608	2·502	3·030	3·329	3·492	3·577	3·669
	8	1·620	2·540	3·066	3·354	3·508	3·586	3·669
$n_1 = 6$	12	1·633	2·580	3·101	3·378	3·521	3·594	3·669
	24	1·649	2·623	3·136	3·399	3·533	3·601	3·669
	∞	1·669	2·669	3·169	3·419	3·544	3·607	3·669
	6	1·027	1·831	2·317	2·598	2·754	2·837	2·928
	8	1·010	1·851	2·343	2·619	2·768	2·846	2·928
$n_1 = 8$	12	0·989	1·873	2·370	2·639	2·781	2·853	2·928
	24	0·963	1·898	2·399	2·659	2·792	2·860	2·928
	∞	0·928	1·928	2·428	2·678	2·803	2·866	2·928
	6	0·588	1·291	1·724	1·982	2·128	2·208	2·296
	8	0·545	1·291	1·739	1·998	2·139	2·215	2·296
$n_1 = 12$	12	0·489	1·292	1·757	2·014	2·151	2·222	2·296
	24	0·413	1·294	1·775	2·030	2·162	2·228	2·296
	∞	0·296	1·296	1·796	2·046	2·171	2·234	2·296
	6	0·269	0·861	1·222	1·444	1·575	1·648	1·733
	8	0·205	0·841	1·225	1·453	1·583	1·654	1·733
$n_1 = 24$	12	0·118	0·816	1·227	1·462	1·590	1·660	1·733
	24	0	0·783	1·229	1·470	1·600	1·666	1·733
	∞	0	0·733	1·233	1·483	1·608	1·671	1·733
	6	0·046	0·523	0·762	0·881	0·960	0·970	1·000
	8	0	0·484	0·742	0·871	0·936	0·968	1·000
$n_1 = ∞$	12	0	0·429	0·714	0·857	0·929	0·964	1·000
	24	0	0·341	0·640	0·835	0·918	0·959	1·000
	∞	0	0	0·500	0·750	0·875	0·938	1·000

TABLE VI. FIDUCIAL LIMITS FOR A VARIANCE COMPONENT—*continued*

	n_2	$F = 0{\cdot}5$	1·0	2·0	4·0	8·0	16·0	∞
$P = 0{\cdot}99$	6	0	0	0	0	0	0·220	0·357
	8	0	0	0	0	0·100	0·252	0·357
$n_1 = 6$	12	0	0	0	0	0·170	0·273	0·357
	24	0	0	0	0·034	0·208	0·286	0·357
	∞	0	0	0	0·107	0·232	0·295	0·357
	6	0	0	0	0	0	0·256	0·398
	8	0	0	0	0	0·135	0·293	0·398
$n_1 = 8$	12	0	0	0	0	0·209	0·314	0·398
	24	0	0	0	0·074	0·249	0·327	0·398
	∞	0	0	0	0·148	0·273	0·336	0·398
	6	0	0	0	0	0·025	0·312	0·458
	8	0	0	0	0	0·182	0·350	0·458
$n_1 = 12$	12	0	0	0	0	0·265	0·372	0·458
	24	0	0	0	0·129	0·308	0·386	0·458
	∞	0	0	0	0·208	0·333	0·396	0·458
	6	0	0	0	0	0·070	0·394	0·558
	8	0	0	0	0	0·255	0·444	0·558
$n_1 = 24$	12	0	0	0	0·041	0·355	0·471	0·558
	24	0	0	0	0·219	0·406	0·486	0·558
	∞	0	0	0·058	0·308	0·433	0·496	0·558
	6	0	0	0	0	0·140	0·570	1·000
	8	0	0	0	0	0·393	0·696	1·000
$n_1 = \infty$	12	0	0	0	0·160	0·580	0·790	1·000
	24	0	0	0	0·447	0·724	0·862	1·000
	∞	0	0	0·500	0·750	0·875	0·938	1·000
$P = 0{\cdot}01$	6	4·612	5·628	6·209	6·531	6·697	6·775	6·880
	8	4·670	5·690	6·257	6·560	6·715	6·784	6·880
$n_1 = 6$	12	4·734	5·753	6·306	6·586	6·731	6·793	6·880
	24	4·804	5·817	6·343	6·608	6·745	6·799	6·880
	∞	4·880	5·880	6·380	6·630	6·755	6·818	6·880
	6	2·798	3·693	4·221	4·520	4·682	4·764	4·859
	8	2·809	3·730	4·256	4·544	4·697	4·774	4·859
$n_1 = 8$	12	2·822	3·770	4·291	4·568	4·711	4·782	4·859
	24	2·839	3·813	4·325	4·589	4·723	4·789	4·859
	∞	2·859	3·859	4·359	4·609	4·734	4·797	4·859
	6	1·566	2·312	2·770	3·039	3·191	3·271	3·361
	8	1·532	2·322	2·790	3·057	3·203	3·279	3·361
$n_1 = 12$	12	1·491	2·332	2·812	3·076	3·215	3·287	3·361
	24	1·437	2·346	2·836	3·094	3·226	3·293	3·361
	∞	1·361	2·361	2·861	3·111	3·236	3·299	3·361
	6	0·761	1·335	1·698	1·920	2·052	2·126	2·211
	8	0·695	1·316	1·700	1·929	2·060	2·132	2·211
$n_1 = 24$	12	0·605	1·291	1·703	1·939	2·069	2·138	2·211
	24	0·469	1·258	1·706	1·949	2·077	2·143	2·211
	∞	0·211	1·211	1·711	1·961	2·086	2·149	2·211
	6	0·286	0·643	0·822	0·911	0·955	0·978	1·000
	8	0·204	0·602	0·801	0·900	0·950	0·975	1·000
$n_1 = \infty$	12	0·085	0·542	0·771	0·886	0·943	0·972	1·000
	24	0	0·442	0·721	0·860	0·930	0·965	1·000
	∞	0	0	0·500	0·750	0·875	0·938	1·000

TABLE VI. SIGNIFICANCE OF DIFFERENCE BETWEEN TWO MEANS
(P. V. Sukhatme)

	n_1	0°	15°	30°	45°	60°	75°	90°
5 per cent. points	6	2·447	2·440	2·435	2·435	2·435	2·440	2·447
	8	2·447	2·430	2·398	2·364	2·331	2·310	2·306
$n_2 = 6$	12	2·447	2·423	2·367	2·301	2·239	2·193	2·179
	24	2·447	2·418	2·342	2·247	2·156	2·088	2·064
	∞	2·447	2·413	2·322	2·201	2·082	1·993	1·960
	6	2·306	2·310	2·331	2·364	2·398	2·430	2·447
	8	2·306	2·300	2·294	2·292	2·294	2·300	2·306
$n_2 = 8$	12	2·306	2·292	2·262	2·229	2·201	2·183	2·179
	24	2·306	2·286	2·236	2·175	2·118	2·077	2·064
	∞	2·306	2·281	2·215	2·128	2·044	1·982	1·960
	6	2·179	2·193	2·239	2·301	2·367	2·423	2·447
	8	2·179	2·183	2·201	2·229	2·262	2·292	2·306
$n_2 = 12$	12	2·179	2·175	2·169	2·167	2·169	2·175	2·179
	24	2·179	2·168	2·142	2·112	2·085	2·069	2·064
	∞	2·179	2·163	2·120	2·064	2·011	1·973	1·960
	6	2·064	2·088	2·156	2·247	2·342	2·418	2·447
	8	2·064	2·077	2·118	2·175	2·236	2·286	2·306
$n_2 = 24$	12	2·064	2·069	2·085	2·112	2·142	2·168	2·179
	24	2·064	2·062	2·058	2·056	2·058	2·062	2·064
	∞	2·064	2·056	2·035	2·009	1·983	1·966	1·960
	6	1·960	1·993	2·082	2·201	2·322	2·413	2·447
	8	1·960	1·982	2·044	2·128	2·215	2·281	2·306
$n_2 = \infty$	12	1·960	1·973	2·011	2·064	2·120	2·163	2·179
	24	1·960	1·966	1·983	2·009	2·035	2·056	2·064
	∞	1·960	1·960	1·960	1·960	1·960	1·960	1·960
1 per cent. points	6	3·707	3·654	3·557	3·514	3·557	3·654	3·707
	8	3·707	3·643	3·495	3·363	3·307	3·328	3·355
$n_2 = 6$	12	3·707	3·636	3·453	3·246	3·104	3·053	3·055
	24	3·707	3·631	3·424	3·158	2·938	2·822	2·797
	∞	3·707	3·626	3·402	3·093	2·804	2·627	2·576
	6	3·355	3·328	3·307	3·363	3·495	3·643	3·707
	8	3·355	3·316	3·239	3·206	3·239	3·316	3·355
$n_2 = 8$	12	3·355	3·307	3·192	3·083	3·032	3·039	3·055
	24	3·355	3·301	3·158	2·988	2·862	2·805	2·797
	∞	3·355	3·295	3·132	2·916	2·723	2·608	2·576
	6	3·055	3·053	3·104	3·246	3·453	3·636	3·707
	8	3·055	3·039	3·032	3·083	3·192	3·307	3·355
$n_2 = 12$	12	3·055	3·029	2·978	2·954	2·978	3·029	3·055
	24	3·055	3·020	2·938	2·853	2·803	2·793	2·797
	∞	3·055	3·014	2·909	2·775	2·661	2·595	2·576
	6	2·797	2·822	2·938	3·158	3·424	3·631	3·707
	8	2·797	2·805	2·862	2·988	3·158	3·301	3·355
$n_2 = 24$	12	2·797	2·793	2·803	2·853	2·938	3·020	3·055
	24	2·797	2·785	2·759	2·747	2·759	2·785	2·797
	∞	2·797	2·777	2·726	2·664	2·613	2·585	2·576
	6	2·576	2·627	2·804	3·093	3·402	3·626	3·707
	8	2·576	2·608	2·723	2·916	3·132	3·295	3·355
$n_2 = \infty$	12	2·576	2·595	2·661	2·775	2·909	3·014	3·055
	24	2·576	2·585	2·613	2·664	2·726	2·777	2·797
	∞	2·576	2·576	2·576	2·576	2·576	2·576	2·576

TABLE VI₁. SIGNIFICANCE OF DIFFERENCE BETWEEN TWO MEANS

Behrens' Test—Odd degrees of freedom

n_1, n_2	P	0°	15°	30°	45°	60°	75°	90°
1, 1	10 per cent.	6·31375	7·73273	8·62474	8·92899	8·62474	7·73273	6·31375
	5 "	12·70620	15·56186	17·35700	17·96929	17·35700	15·56186	12·70620
	2 "	31·82052	38·97201	43·46763	45·00101	43·46763	38·97201	31·82052
	1 "	63·65674	77·96326	86·95672	90·02423	86·95672	77·96326	63·65674
1, 3	10 per cent.	2·35336	3·23550	4·12401	4·95809	5·65934	6·13872	6·31375
	5 "	3·18245	4·95977	7·12269	9·30312	11·11208	12·29429	12·70620
	2 "	4·54070	9·52070	16·30672	22·64110	27·60237	30·74479	31·82052
	1 "	5·84091	17·28287	32·03669	45·08352	55·15098	61·49196	63·65674
1, 5	10 per cent.	2·01505	2·81232	3·72983	4·70125	5·55092	6·11690	6·31375
	5 "	2·57058	4·21758	6·63646	9·09012	11·04305	12·28235	12·70620
	2 "	3·36493	8·56288	15·99780	22·53854	27·57259	30·73989	31·82052
	1 "	4·03214	16·59318	31·86885	45·03075	55·13592	61·48950	63·65674
1, 7	10 per cent.	1·89458	2·66006	3·60383	4·64137	5·53302	6·11373	6·31375
	5 "	2·36462	3·98044	6·54601	9·06502	11·03582	12·28086	12·70620
	2 "	2·99795	8·42578	15·97736	22·53172	27·57008	30·73931	31·82052
	1 "	3·49948	16·55705	31·86146	45·02768	55·13470	61·48922	63·65674
3, 3	10 per cent.	2·35336	2·38618	2·44512	2·47143	2·44512	2·38618	2·35336
	5 "	3·18245	3·19135	3·22536	3·24395	3·22536	3·19135	3·18245
	2 "	4·54070	4·50128	4·46705	4·45965	4·46705	4·50128	4·54070
	1 "	5·84091	5·75394	5·63969	5·59790	5·63969	5·75394	5·84091
3, 5	10 per cent.	2·01505	2·07070	2·17274	2·25780	2·30968	2·33931	2·35336
	5 "	2·57058	2·62621	2·75634	2·89717	3·02568	3·13386	3·18245
	2 "	3·36493	3·41493	3·59516	3·86108	4·16752	4·43370	4·54070
	1 "	4·03214	4·07640	4·31559	4·73914	5·25584	5·68061	5·84091
3, 7	10 per cent.	1·89458	1·95810	2·07525	2·18345	2·26712	2·32781	2·35336
	5 "	2·36462	2·43719	2·60042	2·78743	2·97188	3·12248	3·18245
	2 "	2·99795	3·08203	3·33586	3·70143	4·10751	4·42422	4·54070
	1 "	3·49948	3·59533	3·95817	4·54802	5·19922	5·67297	5·84091
5, 5	10 per cent.	2·01505	2·02391	2·04600	2·05747	2·04600	2·02391	2·01505
	5 "	2·57058	2·56353	2·56237	2·56518	2·56237	2·56353	2·57058
	2 "	3·36493	3·32820	3·27402	3·25431	3·27402	3·32820	3·36493
	1 "	4·03214	3·96770	3·85575	3·80866	3·85575	3·96770	4·03214
5, 7	10 per cent.	1·89458	1·91182	1·95200	1·98705	2·00449	2·01136	2·01505
	5 "	2·36462	2·37363	2·40968	2·45802	2·50521	2·54902	2·57058
	2 "	2·99795	2·98964	3·01057	3·08317	3·19488	3·31251	3·36493
	1 "	3·49948	3·47400	3·47645	3·57550	3·76019	3·95209	4·03214
7, 7	10 per cent.	1·89458	1·89902	1·91113	1·91788	1·91113	1·89902	1·89458
	5 "	2·36462	2·35807	2·35215	2·35161	2·35215	2·35807	2·36462
	2 "	2·99795	2·97119	2·92662	2·90869	2·92662	2·97119	2·99795
	1 "	3·49948	3·45397	3·36875	3·33071	3·36875	3·45397	3·49948

Table VI2. Significance of Difference between Two Means

One component of error distributed normally, the other in Student's distribution

P	n_2	0°(t)	10°	20°	30°	40°	50°	60°	70°	80°	90°(x)
	10	1·812	1·808	1·794	1·774	1·749	1·721	1·693	1·668	1·651	1·645
	12	1·782	1·778	1·767	1·751	1·730	1·707	1·684	1·664	1·650	1·645
	15	1·753	1·750	1·741	1·728	1·711	1·693	1·675	1·659	1·649	1·645
10 per cent.	20	1·725	1·722	1·716	1·706	1·694	1·680	1·667	1·656	1·648	1·645
	30	1·697	1·696	1·692	1·685	1·677	1·668	1·659	1·652	1·647	1·645
	60	1·671	1·670	1·668	1·665	1·661	1·656	1·652	1·648	1·646	1·645
	∞	1·645	1·645	1·645	1·645	1·645	1·645	1·645	1·645	1·645	1·645
	10	2·228	2·219	2·194	2·157	2·112	2·066	2·024	1·989	1·967	1·960
	12	2·179	2·171	2·151	2·120	2·083	2·046	2·011	1·984	1·966	1·960
	15	2·131	2·126	2·109	2·085	2·056	2·026	1·999	1·978	1·965	1·960
5 per cent.	20	2·086	2·082	2·069	2·051	2·030	2·008	1·989	1·973	1·963	1·960
	30	2·042	2·039	2·031	2·019	2·005	1·991	1·978	1·968	1·962	1·960
	60	2·000	1·999	1·995	1·989	1·982	1·975	1·969	1·964	1·961	1·960
	∞	1·960	1·960	1·960	1·960	1·960	1·960	1·960	1·960	1·960	1·960
	10	2·764	2·748	2·704	2·637	2·559	2·481	2·414	2·364	2·335	2·326
	12	2·681	2·668	2·631	2·576	2·513	2·450	2·396	2·356	2·334	2·326
	15	2·602	2·592	2·563	2·520	2·470	2·421	2·379	2·349	2·332	2·326
2 per cent.	20	2·528	2·520	2·498	2·466	2·430	2·394	2·364	2·343	2·330	2·326
	30	2·457	2·452	2·438	2·417	2·393	2·370	2·351	2·337	2·329	2·326
	60	2·390	2·388	2·380	2·370	2·358	2·347	2·338	2·331	2·328	2·326
	∞	2·326	2·326	2·326	2·326	2·326	2·326	2·326	2·326	2·326	2·326
	10	3·169	3·148	3·086	2·993	2·883	2·775	2·684	2·620	2·586	2·576
	12	3·055	3·037	2·985	2·909	2·820	2·733	2·661	2·611	2·584	2·576
	15	2·947	2·932	2·892	2·831	2·762	2·695	2·640	2·603	2·582	2·576
1 per cent.	20	2·845	2·835	2·804	2·760	2·709	2·661	2·622	2·595	2·580	2·576
	30	2·750	2·743	2·723	2·693	2·661	2·630	2·605	2·588	2·579	2·576
	60	2·660	2·657	2·647	2·632	2·616	2·601	2·590	2·582	2·577	2·576
	∞	2·576	2·576	2·576	2·576	2·576	2·576	2·576	2·576	2·576	2·576
	10	3·581	3·553	3·473	3·350	3·203	3·058	2·939	2·859	2·818	2·807
	12	3·429	3·405	3·338	3·237	3·119	3·003	2·910	2·848	2·816	2·807
	15	3·286	3·267	3·214	3·134	3·042	2·954	2·884	2·838	2·814	2·807
0·5 per cent.	20	3·153	3·139	3·099	3·040	2·974	2·911	2·861	2·829	2·812	2·807
	30	3·030	3·020	2·994	2·955	2·912	2·872	2·841	2·821	2·810	2·807
	60	2·915	2·910	2·897	2·878	2·857	2·838	2·823	2·814	2·809	2·807
	∞	2·807	2·807	2·807	2·807	2·807	2·807	2·807	2·807	2·807	2·807
	10	4·144	4·106	3·999	3·832	3·630	3·425	3·259	3·152	3·103	3·090
	12	3·930	3·898	3·809	3·671	3·508	3·347	3·219	3·138	3·100	3·090
	15	3·733	3·708	3·636	3·528	3·401	3·280	3·185	3·126	3·098	3·090
0·2 per cent.	20	3·552	3·533	3·479	3·399	3·308	3·222	3·156	3·116	3·096	3·090
	30	3·386	3·372	3·336	3·284	3·226	3·172	3·131	3·106	3·094	3·090
	60	3·232	3·225	3·207	3·181	3·153	3·128	3·110	3·098	3·092	3·090
	∞	3·090	3·090	3·090	3·090	3·090	3·090	3·090	3·090	3·090	3·090

TABLE VII. Values of the Correlation Coefficient for Different Levels of Significance

n	·1	·05	·02	·01	·001	n	·1	·05	·02	·01	·001
1	·98769	·99692	·999507	·999877	·9999988	16	·4000	·4683	·5425	·5897	·7084
2	·90000	·95000	·98000	·990000	·99900	17	·3887	·4555	·5285	·5751	·6932
3	·8054	·8783	·93433	·95873	·99116	18	·3783	·4438	·5155	·5614	·6787
4	·7293	·8114	·8822	·91720	·97406	19	·3687	·4329	·5034	·5487	·6652
5	·6694	·7545	·8329	·8745	·95074	20	·3598	·4227	·4921	·5368	·6524
6	·6215	·7067	·7887	·8343	·92493	25	·3233	·3809	·4451	·4869	·5974
7	·5822	·6664	·7498	·7977	·8982	30	·2960	·3494	·4093	·4487	·5541
8	·5494	·6319	·7155	·7646	·8721	35	·2746	·3246	·3810	·4182	·5189
9	·5214	·6021	·6851	·7348	·8471	40	·2573	·3044	·3578	·3932	·4896
10	·4973	·5760	·6581	·7079	·8233	45	·2428	·2875	·3384	·3721	·4648
11	·4762	·5529	·6339	·6835	·8010	50	·2306	·2732	·3218	·3541	·4433
12	·4575	·5324	·6120	·6614	·7800	60	·2108	·2500	·2948	·3248	·4078
13	·4409	·5139	·5923	·6411	·7603	70	·1954	·2319	·2737	·3017	·3799
14	·4259	·4973	·5742	·6226	·7420	80	·1829	·2172	·2565	·2830	·3568
15	·4124	·4821	·5577	·6055	·7246	90	·1726	·2050	·2422	·2673	·3375
						100	·1638	·1946	·2301	·2540	·3211

TABLE VIII. Transformation of r to z

z	·00	·01	·02	·03	·04	·05	·06	·07	·08	·09	Mean Diff.
·0	·0000	·0100	·0200	·0300	·0400	·0500	·0599	·0699	·0798	·0898	100
·1	·0997	·1096	·1194	·1293	·1391	·1489	·1586	·1684	·1781	·1877	98
·2	·1974	·2070	·2165	·2260	·2355	·2449	·2543	·2636	·2729	·2821	94
·3	·2913	·3004	·3095	·3185	·3275	·3364	·3452	·3540	·3627	·3714	89
·4	·3800	·3885	·3969	·4053	·4136	·4219	·4301	·4382	·4462	·4542	82
·5	·4621	·4699	·4777	·4854	·4930	·5005	·5080	·5154	·5227	·5299	75
·6	·5370	·5441	·5511	·5580	·5649	·5717	·5784	·5850	·5915	·5980	68
·7	·6044	·6107	·6169	·6231	·6291	·6351	·6411	·6469	·6527	·6584	60
·8	·6640	·6696	·6751	·6805	·6858	·6911	·6963	·7014	·7064	·7114	53
·9	·7163	·7211	·7259	·7306	·7352	·7398	·7443	·7487	·7531	·7574	46
1·0	·7616	·7658	·7699	·7739	·7779	·7818	·7857	·7895	·7932	·7969	39
1·1	·8005	·8041	·8076	·8110	·8144	·8178	·8210	·8243	·8275	·8306	33
1·2	·8337	·8367	·8397	·8426	·8455	·8483	·8511	·8538	·8565	·8591	28
1·3	·8617	·8643	·8668	·8692	·8717	·8741	·8764	·8787	·8810	·8832	24
1·4	·8854	·8875	·8896	·8917	·8937	·8957	·8977	·8996	·9015	·9033	20
1·5	·9051	·9069	·9087	·9104	·9121	·9138	·9154	·9170	·9186	·9201	17
1·6	·9217	·9232	·9246	·9261	·9275	·9289	·9302	·9316	·9329	·9341	14
1·7	·9354	·9366	·9379	·9391	·9402	·9414	·9425	·9436	·9447	·9458	12
1·8	·94681	·94783	·94884	·94983	·95080	·95175	·95268	·95359	·95449	·95537	95
1·9	·95624	·95709	·95792	·95873	·95953	·96032	·96109	·96185	·96259	·96331	79
2·0	·96403	·96473	·96541	·96609	·96675	·96739	·96803	·96865	·96926	·96986	65
2·1	·97045	·97103	·97159	·97215	·97269	·97323	·97375	·97426	·97477	·97526	53
2·2	·97574	·97622	·97668	·97714	·97759	·97803	·97846	·97888	·97929	·97970	44
2·3	·98010	·98049	·98087	·98124	·98161	·98197	·98233	·98267	·98301	·98335	36
2·4	·98367	·98399	·98431	·98462	·98492	·98522	·98551	·98579	·98607	·98635	30
2·5	·98661	·98688	·98714	·98739	·98764	·98788	·98812	·98835	·98858	·98881	24
2·6	·98903	·98924	·98945	·98966	·98987	·99007	·99026	·99045	·99064	·99083	20
2·7	·99101	·99118	·99136	·99153	·99170	·99186	·99202	·99218	·99233	·99248	16
2·8	·99263	·99278	·99292	·99306	·99320	·99333	·99346	·99359	·99372	·99384	13
2·9	·99396	·99408	·99420	·99431	·99443	·99454	·99464	·99475	·99485	·99495	11

z	·0	·1	·2	·3	·4	·5	·6	·7	·8	·9	
3	·99505	·99595	·99668	·99728	·99777	·99818	·99851	·99878	·99900	·99918	—
4	·99933	·99945	·99955	·99963	·99970	·99975	·99980	·99983	·99986	·99989	—

For notes see foot of page 64 overleaf.

TABLE VIII. TESTS OF SIGNIFICANCE FOR 2×2 CONTINGENCY TABLES

Calculate χ_c, the square root of χ^2 corrected for continuity (Fisher, *Statistical Methods*, 21.01). Determine m = the smallest expectation of any class, and $p = \dfrac{\text{the smallest expectation}}{\text{the smallest marginal total}}$.

The table gives the 2·5 and 0·5 per cent. points of χ_c for each tail separately, these being different for the two tails except when $p = 0·5$. The observed set of values will lie on the longer tail if the observed number in the cell with smallest expectation is greater than expectation.

Since three of the four marginal totals are independent their values are not completely determined by m and p. For a given m and p there is a whole set of contingency distributions, and a corresponding set of values of χ_c for each level of significance. The variation of these χ_c is, however, small. The table gives the greatest and least values, *i.e.* those for the binomial distribution (Roman type), and the limiting contingency distribution (black type). When $p = 0$ there is only one value, this being derived from the Poisson distribution.

The table should always be used in place of the ordinary χ^2 table when m is less than 10. Even with values of m as great as 100 the gain in precision is appreciable when p is small and the interest centres in the probability of a single tail (*i.e.* deviations in one direction only). The correction for continuity should be used when m is less than 500.

In regions of the table where no values are given, and to determine the exact probability in doubtful cases, the exact solution (on which this table is based) must be used (Fisher, *Statistical Methods*, 21.02).

The table can be used for testing the deviation of a sample of a binomial distribution with known p. Remember to include the contributions to χ^2 from both classes when calculating χ_c.

	p \ m	1	2	3	4	5	6	8	12	24	48	96
P = ·025	0	...	...	...	1·68	1·71	1·74	1·77	1·81	1·86	1·89	1·91
Shorter tail	0·25	...	...	1·73	1·77	1·80	1·82	1·84	1·87	1·90	1·91	1·93
		...	...	**1·83**	**1·85**	**1·86**	**1·87**	**1·89**	**1·90**	**1·92**	**1·94**	**1·94**
	0·5	...	...	1·88	1·90	1·91	1·92	1·93	1·94	1·95	1·96	1·96
		...	**1·91**	**1·93**	**1·94**	**1·94**	**1·94**	**1·95**	**1·95**	**1·95**	**1·96**	**1·96**
Longer tail	0·25	2·10	2·08	2·07	2·06	2·06	2·05	2·04	2·02	2·01	2·00	1·99
		2·04	**2·03**	**2·02**	**2·02**	**2·01**	**2·01**	**2·00**	**1·99**	**1·99**	**1·98**	**1·97**
	0	2·32	2·24	2·19	2·16	2·14	2·13	2·11	2·08	2·05	2·03	2·01
P = ·005	0	...	...	...	...	2·06	2·13	2·19	2·27	2·37	2·43	2·48
Shorter tail	0·25	...	...	...	2·18	2·23	2·27	2·32	2·38	2·45	2·49	2·52
		...	...	...	**2·32**	**2·36**	**2·38**	**2·42**	**2·46**	**2·50**	**2·52**	**2·54**
	0·5	...	...	...	2·41	2·44	2·47	2·50	2·52	2·55	2·56	2·57
		...	...	**2·48**	**2·50**	**2·52**	**2·53**	**2·54**	**2·55**	**2·56**	**2·57**	**2·57**
Longer tail	0·25	2·79	2·79	2·78	2·76	2·75	2·73	2·72	2·70	2·67	2·64	2·63
		2·67	**2·67**	**2·67**	**2·67**	**2·66**	**2·66**	**2·65**	**2·64**	**2·63**	**2·62**	**2·60**
	0	3·33	3·13	3·05	2·97	2·95	2·92	2·88	2·83	2·76	2·70	2·67

Notes on Tables VII and VIII1

For a total correlation, n is 2 less than the number of pairs in the sample; for a partial correlation, the number of eliminated variates also should be subtracted. The probability given at the head of each column of Table VII represents the chance that r will be greater than the values given or less than *minus* these values.

Table VIII1 gives the transformation $r = (e^{2z}-1)/(e^{2z}+1)$ or $z = \frac{1}{2}\{\log_e(1+r)-\log_e(1-r)\}$. With n defined as above z is distributed approximately normally with variance $1/(n-1)$. For exact work correct for bias in z by subtracting $r/2(n+1)$ from z. (See also Table XI, which is the inverse of Table VIII1, with $p = \frac{1}{2}(r+1)$.)

TABLE VIII1. BINOMIAL AND POISSON DISTRIBUTIONS: LIMITS OF THE EXPECTATION
(Based on W. L. Stevens)

a	N	$p = a/N$	Probability P of a or more ·005	·025	·1	Probability P of a or fewer ·1	·025	·005
0	5	(·2)	For $a = 0$, $1-P^{1/N}$ is the greatest probability allowable.			1·84	2·61	3·27
		(·15)				1·95	2·83	3·66
	10	(·1)				2·06	3·09	4·11
	20	(·05)				2·17	3·37	4·65
	∞	0				2·30	3·69	5·30
1	2		·0050	·0252	·103	1·897	1·9748	1·9950
	3		·0050	·0252	·104	2·413	2·7171	2·8758
	4		·0050	·0252	·104	2·718	3·224	3·556
	5	·2	·0050	·0253	·104	2·92	3·58	4·07
		·15	·0050	·0253	·105	3·14	3·99	4·69
	10	·1	·0050	·0253	·105	3·37	4·45	5·44
	20	·05	·0050	·0253	·105	3·62	4·97	6·34
	∞	0	·0050	·0253	·105	3·89	5·57	7·43
2	4		·118	·270	·570	3·430	3·730	3·882
	5		·114	·264	·561	3·77	4·267	4·586
	6		·112	·260	·556	4·00	4·67	5·138
	7		·111	·257	·552	4·17	4·97	5·54
	8	·25	·110	·255	·549	4·31	5·21	5·94
	10	·2	·108	·252	·545	4·50	5·56	6·48
		·15	·107	·249	·542	4·69	5·94	7·09
	20	·1	·106	·247	·538	4·90	6·34	7·74
	40	·05	·105	·245	·535	5·11	6·77	8·44
	∞	0	·103	·242	·532	5·32	7·22	9·27
3	6		·398	·709	1·21	4·79	5·291	5·602
	7		·387	·693	1·19	5·05	5·71	6·181
	8		·380	·682	1·17	5·24	6·04	6·642
	9		·374	·674	1·17	5·39	6·31	7·03
	10	·3	·370	·667	1·16	5·52	6·52	7·35
	15	·2	·358	·650	1·14	5·89	7·21	8·41
	30	·1	·348	·634	1·12	6·28	7·96	9·61
	∞	0	·338	·619	1·10	6·68	8·77	10·98
4	8	·5	·799	1·26	1·92	6·08	6·74	7·201
	10	·4	·768	1·22	1·88	6·46	7·38	8·091
		·3	·741	1·18	1·84	6·83	8·04	9·06
	20	·2	·716	1·15	1·80	7·21	8·73	10·13
	40	·1	·694	1·12	1·77	7·60	9·47	11·31
	∞	0	·672	1·09	1·74	7·99	10·24	12·59
5	10	·5	1·28	1·87	2·67	7·33	8·13	8·72
		·4	1·23	1·81	2·61	7·72	8·79	9·66
		·3	1·19	1·76	2·56	8·10	9·47	10·67
	25	·2	1·15	1·71	2·52	8·49	10·18	11·75
	50	·1	1·11	1·66	2·47	8·88	10·91	12·90
	∞	0	1·08	1·62	2·43	9·27	11·67	14·15
6	12	·5	1·83	2·53	3·46	8·54	9·47	10·17
	15	·4	1·75	2·45	3·38	8·95	10·16	11·16
	20	·3	1·69	2·38	3·32	9·35	10·86	12·19
	30	·2	1·63	2·31	3·26	9·74	11·57	13·28
	60	·1	1·58	2·26	3·20	10·14	12·30	14·44
	∞	0	1·54	2·20	3·15	10·53	13·06	15·66
7	14	·5	2·41	3·23	4·26	9·74	10·77	11·59
		·4	2·32	3·12	4·17	10·15	11·48	12·61
		·3	2·24	3·03	4·09	10·56	12·20	13·67
	35	·2	2·16	2·95	4·02	10·97	12·93	14·77
	70	·1	2·10	2·88	3·96	11·37	13·67	15·92
	∞	0	2·04	2·81	3·89	11·77	14·42	17·13

a	$p = a/N$	Probability P of a or more ·005	·025	·1	Probability P of a or fewer ·1	·025	·005
8	·5	3·04	3·94	5·09	10·91	12·06	12·96
	·4	2·92	3·82	4·98	11·35	12·79	14·02
	·3	2·82	3·72	4·89	11·77	13·52	15·10
	·2	2·73	3·62	4·80	12·18	14·26	16·22
	·1	2·65	3·53	4·73	12·59	15·01	17·37
	0	2·57	3·45	4·66	12·99	15·76	18·58
9	·5	3·68	4·68	5·92	12·08	13·32	14·32
	·4	3·55	4·54	5·80	12·52	14·07	15·40
	·3	3·43	4·42	5·70	12·96	14·82	16·50
	·2	3·32	4·31	5·60	13·38	15·57	17·63
	·1	3·22	4·21	5·51	13·79	16·32	18·80
	0	3·13	4·12	5·43	14·21	17·08	20·00
10	·5	4·35	5·44	6·76	13·24	14·56	15·65
	·4	4·20	5·28	6·63	13·69	15·33	16·76
	·3	4·06	5·14	6·51	14·13	16·10	17·88
	·2	3·93	5·02	6·41	14·56	16·86	19·02
	·1	3·82	4·90	6·31	14·99	17·62	20·20
	0	3·72	4·80	6·22	15·41	18·39	21·40
9	·5	+0·14	−0·16	−0·36	+0·36	+0·16	−0·14
	·4	+0·53	+0·10	−0·22	+0·55	+0·52	+0·41
	·3	+0·89	+0·34	−0·09	+0·74	+0·90	+1·04
	·2	+1·23	+0·57	+0·04	+0·94	+1·31	+1·72
	·1	+1·55	+0·79	+0·16	+1·15	+1·74	+2·47
	0	+1·86	+1·00	+0·28	+1·36	+2·20	+3·27
16	·5	−0·02	−0·25	−0·40	+0·40	+0·25	+0·02
	·4	+0·40	+0·02	−0·26	+0·58	+0·60	+0·57
	·3	+0·79	+0·28	−0·12	+0·76	+0·96	+1·16
	·2	+1·17	+0·52	+0·01	+0·94	+1·34	+1·80
	·1	+1·52	+0·76	+0·14	+1·13	+1·73	+2·47
	0	+1·87	+0·99	+0·26	+1·33	+2·14	+3·18
36	·5	−0·18	−0·34	−0·44	+0·44	+0·34	+0·18
	·4	+0·26	−0·06	−0·29	+0·60	+0·67	+0·72
	·3	+0·68	+0·21	−0·15	+0·77	+1·01	+1·28
	·2	+1·09	+0·47	−0·02	+0·94	+1·36	+1·86
	·1	+1·49	+0·73	+0·12	+1·11	+1·71	+2·46
	0	+1·88	+0·97	+0·25	+1·29	+2·08	+3·08
144	·5	−0·35	−0·43	−0·47	+0·47	+0·43	+0·35
	·4	+0·12	−0·14	−0·33	+0·63	+0·73	+0·86
	·3	+0·57	+0·14	−0·18	+0·78	+1·05	+1·37
	·2	+1·01	+0·42	−0·04	+0·94	+1·37	+1·90
	·1	+1·45	+0·69	+0·09	+1·09	+1·69	+2·44
	0	+1·88	+0·96	+0·23	+1·25	+2·01	+2·98
∞	·5	−0·50	−0·50	−0·50	+0·50	+0·50	+0·50
	·4	−0·02	−0·21	−0·36	+0·64	+0·79	+0·98
	·3	+0·45	+0·08	−0·21	+0·79	+1·08	+1·45
	·2	+0·93	+0·37	−0·07	+0·93	+1·37	+1·93
	·1	+1·40	+0·66	+0·07	+1·07	+1·66	+2·40
	0	+1·88	+0·95	+0·21	+1·21	+1·95	+2·88

To obtain the limits of the probability of an event (observed to occur a times out of N) corresponding to a given probability level P, divide the tabulated values by N, first interpolating if necessary by linear interpolation with reference to $p = a/N$. The limits of the expectation of Poisson distributions are given directly, taking $p = 0$. For $a > \frac{1}{2}N$ enter the table with $a' = N - a$. For $a > 10$ add the tabulated corrections to the limits of the expectation calculated from the standard error $\sqrt{\{a(1-p)\}}$, using asymptotic interpolation on $12/\sqrt{a}$.

Table VIII2. Densities of Organisms Estimated by the Dilution Method

(W. L. Stevens)

Two-fold

x	Number of Levels (s). 4	5	6	7	8	9	10	11 or more
0·4	·757	·773	·781	·785	·787	·788	·789	·789
0·6	·622	·640	·649	·653	·655	·656	·657	·657
0·8	·537	·556	·566	·571	·573	·574	·575	·575
1·0	·479	·500	·511	·516	·518	·520	·520	·521
1·2	·437	·461	·472	·478	·480	·482	·482	·483
1·4	·406	·432	·444	·450	·453	·455	·456	·456
1·6	·381	·411	·424	·431	·435	·436	·437	·438
1·8	·361	·394	·410	·417	·421	·423	·424	·425
2·0	·344	·382	·399	·408	·412	·414	·415	·416
2·5		·358	·382	·394	·399	·402	·403	·405
3·0			·370	·386	·394	·398	·400	·402
3·5				·379	·390	·396	·399	·401
4·0					·386	·394	·397	·401
4·5						·390	·396	·401
5·0							·394	·401
								·401*
y								
7·0								·399
6·0								·397
5·0							·394	·394
4·5						·390	·390	·390
4·0					·386	·386	·386	·386
3·5				·379	·379	·379	·379	·379
3·0			·370	·370	·370	·370	·370	·370
2·5		·358	·356	·356	·356	·356	·356	·356
2·0	·344	·334	·334	·334	·334	·334	·334	·334
1·8	·327	·323	·323	·323	·323	·323	·323	·323
1·6	·311	·309	·309	·309	·309	·309	·309	·309
1·4	·293	·292	·292	·292	·292	·292	·292	·292
1·2	·271	·271	·271	·271	·271	·271	·271	·271
1·0	·245	·245	·245	·245	·245	·245	·245	·245
0·8	·212	·212	·212	·212	·212	·212	·212	·212
0·6	·167	·167	·167	·167	·167	·167	·167	·167
0·4	·101	·101	·101	·101	·101	·101	·101	·101

Four-fold

x	Number of Levels 4	5	6 or more
0·4	·704	·706	·707
0·6	·615	·617	·618
0·8	·573	·576	·577
1·0	·555	·558	·559
1·5	·545	·551	·553
2·0	·537	·548	·551
2·5		·545	·552
			·552*
y			
3·5			·550
3·0			·548
2·5		·545	·545
2·0	·537	·537	·537
1·5	·522	·522	·522
1·0	·488	·488	·488
0·8	·464	·464	·464
0·6	·431	·431	·431
0·4	·375	·375	·375

Ten-fold (Three or more levels)

$x \leq 1$		$x>1, y>2$			$y \leq 2$			
x	K	x	K	y	K	y	K	y
		·0	·763	·0	·761	2·0	·744	1·0
		·1	·768	·9	·766	1·9	·744	0·9
		·2	·768	·8	·764	1·8	·734	0·8
		·3	·760	·7	·755	1·7	·712	0·7
0·4	·761	·4	·747	·6	·741	1·6	·684	0·6
0·5	·740	·5	·736	·5	·729	1·5	·658	0·5
0·6	·733	·6	·733	·4	·724	1·4	·638	0·4
0·7	·736	·7	·736	·3	·726	1·3		
0·8	·744	·8	·744	·2	·732	1·2		
0·9	·753	·9	·753	·1	·739	1·1		
1·0	·763	·0	·763	·0	·744	1·0		

When $x>1$ and $y>2$ enter the table with the decimal part of x or y only.

Calculate the mean fertile level x, and the mean sterile level y, where x is the number of fertile plates/number of cultures at each level (n), and $x+y$ is the number of levels. Enter the table with x or y, as indicated, and determine the corresponding value of the tabular entry K. When x and y fall outside the tabulated range use the value marked with an asterisk (*). The estimate of the number, λ, of organisms in the quantity of the medium used for one culture at the highest concentration is then given by $\log \lambda = x \log a - K$, where a is the dilution factor. The average value of the variance of the mean fertile level is $\frac{1}{n}\frac{\log 2}{\log a}$, and the average value of the variance of $\log \lambda$ is $\frac{1}{n} \log 2 \log a$ (see Introduction). Thus:

Two-fold

$$\log \lambda = 0{\cdot}30103\, x - K$$
$$V(x) = 1/n$$
$$V(\log \lambda) = 0{\cdot}091/n$$

Four-fold

$$\log \lambda = 0{\cdot}60206\, x - K$$
$$V(x) = 1/2n$$
$$V(\log \lambda) = 0{\cdot}201/n$$

Ten-fold

$$\log \lambda = x - K$$
$$V(x) = 0{\cdot}301/n$$

Table VIII3. Significance of Leading Periodic Components

n	5 per cent.	1 per cent.	n	5 per cent.	1 per cent.	n	5 per cent.	1 per cent.
5	·68377	·78874	20	·27040	·32971	35	·17513	·21338
6	·61615	·72179	21	·26060	·31783	36	·17124	·20860
7	·56115	·66440	22	·25155	·30683	37	·16754	·20405
8	·51569	·61517	23	·24315	·29661	38	·16400	·19970
9	·47749	·57271	24	·23534	·28709	39	·16062	·19554
10	·44495	·53584	25	·22805	·27819	40	·15738	·19156
11	·41688	·50357	26	·22123	·26986	41	·15429	·18776
12	·39240	·47510	27	·21483	·26205	42	·15132	·18411
13	·37085	·44982	28	·20883	·25470	43	·14847	·18060
14	·35172	·42722	29	·20317	·24778	44	·14573	·17724
15	·33461	·40689	30	·19784	·24124	45	·14310	·17401
16	·31922	·38851	31	·19280	·23506	46	·14057	·17089
17	·30529	·37180	32	·18803	·22921	47	·13814	·16789
18	·29262	·35655	33	·18351	·22366	48	·13579	·16501
19	·28104	·34257	34	·17921	·21839	49	·13353	·16222
20	·27040	·32971	35	·17513	·21338	50	·13135	·15954

Table of g; for testing the significance of the leading periodic component of $2n+1$ or $2n+2$ consecutive values. Each of n periods contributes a certain fraction to the sum of squares for all n periods, and g is taken to be the largest of these fractions. If this exceeds the corresponding tabulated value, significant evidence of periodicity is indicated.

Table IX. Probits

Transformation of the Sigmoid Dosage Mortality Curve to a Straight Line. (C. I. Bliss.)

	0·0	0·1	0·2	0·3	0·4	0·5	0·6	0·7	0·8	0·9	1	2	3	4	5
0	...	1·9098	2·1218	2·2522	2·3479	2·4242	2·4879	2·5427	2·5911	2·6344	For more detail see values for 95-100.				
1	2·6737	2·7096	2·7429	2·7738	2·8027	2·8299	2·8556	2·8799	2·9031	2·9251					
2	2·9463	2·9665	2·9859	3·0046	3·0226	3·0400	3·0569	3·0732	3·0890	3·1043					
3	3·1192	3·1337	3·1478	3·1616	3·1750	3·1881	3·2009	3·2134	3·2256	3·2376					
4	3·2493	3·2608	3·2721	3·2831	3·2940	3·3046	3·3151	3·3253	3·3354	3·3454					
5	3·3551	3·3648	3·3742	3·3836	3·3928	3·4018	3·4107	3·4195	3·4282	3·4368	9	18	27	36	45
6	3·4452	3·4536	3·4618	3·4699	3·4780	3·4859	3·4937	3·5015	3·5091	3·5167	8	16	24	32	40
7	3·5242	3·5316	3·5389	3·5462	3·5534	3·5605	3·5675	3·5745	3·5813	3·5882	7	14	21	28	36
8	3·5949	3·6016	3·6083	3·6148	3·6213	3·6278	3·6342	3·6405	3·6468	3·6531	6	13	19	26	32
9	3·6592	3·6654	3·6715	3·6775	3·6835	3·6894	3·6953	3·7012	3·7070	3·7127	6	12	18	24	30
10	3·7184	3·7241	3·7298	3·7354	3·7409	3·7464	3·7519	3·7574	3·7628	3·7681	6	11	17	22	28
11	3·7735	3·7788	3·7840	3·7893	3·7945	3·7996	3·8048	3·8099	3·8150	3·8200	5	10	16	21	26
12	3·8250	3·8300	3·8350	3·8399	3·8448	3·8497	3·8545	3·8593	3·8641	3·8689	5	10	15	20	24
13	3·8736	3·8783	3·8830	3·8877	3·8923	3·8969	3·9015	3·9061	3·9107	3·9152	5	9	14	18	23
14	3·9197	3·9242	3·9286	3·9331	3·9375	3·9419	3·9463	3·9506	3·9550	3·9593	4	9	13	18	22
15	3·9636	3·9678	3·9721	3·9763	3·9806	3·9848	3·9890	3·9931	3·9973	4·0014	4	8	13	17	21
16	4·0055	4·0096	4·0137	4·0178	4·0218	4·0259	4·0299	4·0339	4·0379	4·0419	4	8	12	16	20
17	4·0458	4·0498	4·0537	4·0576	4·0615	4·0654	4·0693	4·0731	4·0770	4·0808	4	8	12	16	19
18	4·0846	4·0884	4·0922	4·0960	4·0998	4·1035	4·1073	4·1110	4·1147	4·1184	4	8	11	15	19
19	4·1221	4·1258	4·1295	4·1331	4·1367	4·1404	4·1440	4·1476	4·1512	4·1548	4	7	11	15	18
20	4·1584	4·1619	4·1655	4·1690	4·1726	4·1761	4·1796	4·1831	4·1866	4·1901	4	7	11	14	18
21	4·1936	4·1970	4·2005	4·2039	4·2074	4·2108	4·2142	4·2176	4·2210	4·2244	3	7	10	14	17
22	4·2278	4·2312	4·2345	4·2379	4·2412	4·2446	4·2479	4·2512	4·2546	4·2579	3	7	10	13	17
23	4·2612	4·2644	4·2677	4·2710	4·2743	4·2775	4·2808	4·2840	4·2872	4·2905	3	7	10	13	16
24	4·2937	4·2969	4·3001	4·3033	4·3065	4·3097	4·3129	4·3160	4·3192	4·3224	3	6	10	13	16
25	4·3255	4·3287	4·3318	4·3349	4·3380	4·3412	4·3443	4·3474	4·3505	4·3536	3	6	9	12	16
26	4·3567	4·3597	4·3628	4·3659	4·3689	4·3720	4·3750	4·3781	4·3811	4·3842	3	6	9	12	15
27	4·3872	4·3902	4·3932	4·3962	4·3992	4·4022	4·4052	4·4082	4·4112	4·4142	3	6	9	12	15
28	4·4172	4·4201	4·4231	4·4260	4·4290	4·4319	4·4349	4·4378	4·4408	4·4437	3	6	9	12	15
29	4·4466	4·4495	4·4524	4·4554	4·4583	4·4612	4·4641	4·4670	4·4698	4·4727	3	6	9	12	14
30	4·4756	4·4785	4·4813	4·4842	4·4871	4·4899	4·4928	4·4956	4·4985	4·5013	3	6	9	11	14
31	4·5041	4·5070	4·5098	4·5126	4·5155	4·5183	4·5211	4·5239	4·5267	4·5295	3	6	8	11	14
32	4·5323	4·5351	4·5379	4·5407	4·5435	4·5462	4·5490	4·5518	4·5546	4·5573	3	6	8	11	14
33	4·5601	4·5628	4·5656	4·5684	4·5711	4·5739	4·5766	4·5793	4·5821	4·5848	3	5	8	11	14
34	4·5875	4·5903	4·5930	4·5957	4·5984	4·6011	4·6039	4·6066	4·6093	4·6120	3	5	8	11	14
35	4·6147	4·6174	4·6201	4·6228	4·6255	4·6281	4·6308	4·6335	4·6362	4·6389	3	5	8	11	13
36	4·6415	4·6442	4·6469	4·6495	4·6522	4·6549	4·6575	4·6602	4·6628	4·6655	3	5	8	11	13
37	4·6681	4·6708	4·6734	4·6761	4·6787	4·6814	4·6840	4·6866	4·6893	4·6919	3	5	8	11	13
38	4·6945	4·6971	4·6998	4·7024	4·7050	4·7076	4·7102	4·7129	4·7155	4·7181	3	5	8	10	13
39	4·7207	4·7233	4·7259	4·7285	4·7311	4·7337	4·7363	4·7389	4·7415	4·7441	3	5	8	10	13
40	4·7467	4·7492	4·7518	4·7544	4·7570	4·7596	4·7622	4·7647	4·7673	4·7699	3	5	8	10	13
41	4·7725	4·7750	4·7776	4·7802	4·7827	4·7853	4·7879	4·7904	4·7930	4·7955	3	5	8	10	13
42	4·7981	4·8007	4·8032	4·8058	4·8083	4·8109	4·8134	4·8160	4·8185	4·8211	3	5	8	10	13
43	4·8236	4·8262	4·8287	4·8313	4·8338	4·8363	4·8389	4·8414	4·8440	4·8465	3	5	8	10	13
44	4·8490	4·8516	4·8541	4·8566	4·8592	4·8617	4·8642	4·8668	4·8693	4·8718	3	5	8	10	13
45	4·8743	4·8769	4·8794	4·8819	4·8844	4·8870	4·8895	4·8920	4·8945	4·8970	3	5	8	10	13
46	4·8996	4·9021	4·9046	4·9071	4·9096	4·9122	4·9147	4·9172	4·9197	4·9222	3	5	8	10	13
47	4·9247	4·9272	4·9298	4·9323	4·9348	4·9373	4·9398	4·9423	4·9448	4·9473	3	5	8	10	13
48	4·9498	4·9524	4·9549	4·9574	4·9599	4·9624	4·9649	4·9674	4·9699	4·9724	3	5	8	10	13
49	4·9749	4·9774	4·9799	4·9825	4·9850	4·9875	4·9900	4·9925	4·9950	4·9975	3	5	8	10	13

The probit corresponding to a given percentage is the normal deviate (increased by 5 to avoid negative values) for which the probability (single tail) equals this percentage (see Tables I and II1).

TABLE IX. PROBITS—*continued*

	0·0	0·1	0·2	0·3	0·4	0·5	0·6	0·7	0·8	0·9	1	2	3	4	5
50	5·0000	5·0025	5·0050	5·0075	5·0100	5·0125	5·0150	5·0175	5·0201	5·0226	3	5	8	10	13
51	5·0251	5·0276	5·0301	5·0326	5·0351	5·0376	5·0401	5·0426	5·0451	5·0476	3	5	8	10	13
52	5·0502	5·0527	5·0552	5·0577	5·0602	5·0627	5·0652	5·0677	5·0702	5·0728	3	5	8	10	13
53	5·0753	5·0778	5·0803	5·0828	5·0853	5·0878	5·0904	5·0929	5·0954	5·0979	3	5	8	10	13
54	5·1004	5·1030	5·1055	5·1080	5·1105	5·1130	5·1156	5·1181	5·1206	5·1231	3	5	8	10	13
55	5·1257	5·1282	5·1307	5·1332	5·1358	5·1383	5·1408	5·1434	5·1459	5·1484	3	5	8	10	13
56	5·1510	5·1535	5·1560	5·1586	5·1611	5·1637	5·1662	5·1687	5·1713	5·1738	3	5	8	10	13
57	5·1764	5·1789	5·1815	5·1840	5·1866	5·1891	5·1917	5·1942	5·1968	5·1993	3	5	8	10	13
58	5·2019	5·2045	5·2070	5·2096	5·2121	5·2147	5·2173	5·2198	5·2224	5·2250	3	5	8	10	13
59	5·2275	5·2301	5·2327	5·2353	5·2378	5·2404	5·2430	5·2456	5·2482	5·2508	3	5	8	10	13
60	5·2533	5·2559	5·2585	5·2611	5·2637	5·2663	5·2689	5·2715	5·2741	5·2767	3	5	8	10	13
61	5·2793	5·2819	5·2845	5·2871	5·2898	5·2924	5·2950	5·2976	5·3002	5·3029	3	5	8	10	13
62	5·3055	5·3081	5·3107	5·3134	5·3160	5·3186	5·3213	5·3239	5·3266	5·3292	3	5	8	11	13
63	5·3319	5·3345	5·3372	5·3398	5·3425	5·3451	5·3478	5·3505	5·3531	5·3558	3	5	8	11	13
64	5·3585	5·3611	5·3638	5·3665	5·3692	5·3719	5·3745	5·3772	5·3799	5·3826	3	5	8	11	13
65	5·3853	5·3880	5·3907	5·3934	5·3961	5·3989	5·4016	5·4043	5·4070	5·4097	3	5	8	11	14
66	5·4125	5·4152	5·4179	5·4207	5·4234	5·4261	5·4289	5·4316	5·4344	5·4372	3	5	8	11	14
67	5·4399	5·4427	5·4454	5·4482	5·4510	5·4538	5·4565	5·4593	5·4621	5·4649	3	6	8	11	14
68	5·4677	5·4705	5·4733	5·4761	5·4789	5·4817	5·4845	5·4874	5·4902	5·4930	3	6	8	11	14
69	5·4959	5·4987	5·5015	5·5044	5·5072	5·5101	5·5129	5·5158	5·5187	5·5215	3	6	9	11	14
70	5·5244	5·5273	5·5302	5·5330	5·5359	5·5388	5·5417	5·5446	5·5476	5·5505	3	6	9	12	14
71	5·5534	5·5563	5·5592	5·5622	5·5651	5·5681	5·5710	5·5740	5·5769	5·5799	3	6	9	12	15
72	5·5828	5·5858	5·5888	5·5918	5·5948	5·5978	5·6008	5·6038	5·6068	5·6098	3	6	9	12	15
73	5·6128	5·6158	5·6189	5·6219	5·6250	5·6280	5·6311	5·6341	5·6372	5·6403	3	6	9	12	15
74	5·6433	5·6464	5·6495	5·6526	5·6557	5·6588	5·6620	5·6651	5·6682	5·6713	3	6	9	12	16
75	5·6745	5·6776	5·6808	5·6840	5·6871	5·6903	5·6935	5·6967	5·6999	5·7031	3	6	10	13	16
76	5·7063	5·7095	5·7128	5·7160	5·7192	5·7225	5·7257	5·7290	5·7323	5·7356	3	7	10	13	16
77	5·7388	5·7421	5·7454	5·7488	5·7521	5·7554	5·7588	5·7621	5·7655	5·7688	3	7	10	13	17
78	5·7722	5·7756	5·7790	5·7824	5·7858	5·7892	5·7926	5·7961	5·7995	5·8030	3	7	10	14	17
79	5·8064	5·8099	5·8134	5·8169	5·8204	5·8239	5·8274	5·8310	5·8345	5·8381	4	7	11	14	18
80	5·8416	5·8452	5·8488	5·8524	5·8560	5·8596	5·8633	5·8669	5·8705	5·8742	4	7	11	14	18
81	5·8779	5·8816	5·8853	5·8890	5·8927	5·8965	5·9002	5·9040	5·9078	5·9116	4	7	11	15	19
82	5·9154	5·9192	5·9230	5·9269	5·9307	5·9346	5·9385	5·9424	5·9463	5·9502	4	8	12	15	19
83	5·9542	5·9581	5·9621	5·9661	5·9701	5·9741	5·9782	5·9822	5·9863	5·9904	4	8	12	16	20
84	5·9945	5·9986	6·0027	6·0069	6·0110	6·0152	6·0194	6·0237	6·0279	6·0322	4	8	13	17	21
85	6·0364	6·0407	6·0450	6·0494	6·0537	6·0581	6·0625	6·0669	6·0714	6·0758	4	9	13	18	22
86	6·0803	6·0848	6·0893	6·0939	6·0985	6·1031	6·1077	6·1123	6·1170	6·1217	5	9	14	18	23
87	6·1264	6·1311	6·1359	6·1407	6·1455	6·1503	6·1552	6·1601	6·1650	6·1700	5	10	15	19	24
88	6·1750	6·1800	6·1850	6·1901	6·1952	6·2004	6·2055	6·2107	6·2160	6·2212	5	10	15	21	26
89	6·2265	6·2319	6·2372	6·2426	6·2481	6·2536	6·2591	6·2646	6·2702	6·2759	5	11	16	22	27
90	6·2816	6·2873	6·2930	6·2988	6·3047	6·3106	6·3165	6·3225	6·3285	6·3346	6	12	18	24	29
91	6·3408	6·3469	6·3532	6·3595	6·3658	6·3722	6·3787	6·3852	6·3917	6·3984	6	13	19	26	32
92	6·4051	6·4118	6·4187	6·4255	6·4325	6·4395	6·4466	6·4538	6·4611	6·4684	7	14	21	28	35
93	6·4758	6·4833	6·4909	6·4985	6·5063	6·5141	6·5220	6·5301	6·5382	6·5464	8	16	24	31	39
94	6·5548	6·5632	6·5718	6·5805	6·5893	6·5982	6·6072	6·6164	6·6258	6·6352	9	18	27	36	45
95	6·6449	6·6546	6·6646	6·6747	6·6849	6·6954	6·7060	6·7169	6·7279	6·7392					
	97	100	101	102	105	106	109	110	113	115					
96	6·7507	6·7624	6·7744	6·7866	6·7991	6·8119	6·8250	6·8384	6·8522	6·8663					
	117	120	122	125	128	131	134	138	141	145					
97	6·8808	6·8957	6·9110	6·9268	6·9431	6·9600	6·9774	6·9954	7·0141	7·0335					
	149	153	158	163	169	174	180	187	194	202					

Continued on next page.

Table IX. Probits—*continued*

	0·00	0·01	0·02	0·03	0·04	0·05	0·06	0·07	0·08	0·09	1	2	3	4	5
98·0	7·0537	7·0558	7·0579	7·0600	7·0621	7·0642	7·0663	7·0684	7·0706	7·0727	2	4	6	8	11
98·1	7·0749	7·0770	7·0792	7·0814	7·0836	7·0858	7·0880	7·0902	7·0924	7·0947	2	4	7	9	11
98·2	7·0969	7·0992	7·1015	7·1038	7·1061	7·1084	7·1107	7·1130	7·1154	7·1177	2	5	7	9	12
98·3	7·1201	7·1224	7·1248	7·1272	7·1297	7·1321	7·1345	7·1370	7·1394	7·1419	2	5	7	10	12
98·4	7·1444	7·1469	7·1494	7·1520	7·1545	7·1571	7·1596	7·1622	7·1648	7·1675	3	5	8	10	13
98·5	7·1701	7·1727	7·1754	7·1781	7·1808	7·1835	7·1862	7·1890	7·1917	7·1945	3	5	8	11	14
98·6	7·1973	7·2001	7·2029	7·2058	7·2086	7·2115	7·2144	7·2173	7·2203	7·2232	3	6	9	12	14
98·7	7·2262	7·2292	7·2322	7·2353	7·2383	7·2414	7·2445	7·2476	7·2508	7·2539	3	6	9	12	15
98·8	7·2571	7·2603	7·2636	7·2668	7·2701	7·2734	7·2768	7·2801	7·2835	7·2869	3	7	10	13	17
98·9	7·2904	7·2938	7·2973	7·3009	7·3044	7·3080	7·3116	7·3152	7·3189	7·3226	4	7	11	14	18
99·0	7·3263	7·3301	7·3339	7·3378	7·3416	7·3455	7·3495	7·3535	7·3575	7·3615	4	8	12	16	20
99·1	7·3656	7·3698	7·3739	7·3781	7·3824	7·3867	7·3911	7·3954	7·3999	7·4044	4	9	13	17	22
99·2	7·4089	7·4135	7·4181	7·4228	7·4276	7·4324	7·4372	7·4422	7·4471	7·4522	5	10	14	19	24
99·3	7·4573	7·4624	7·4677	7·4730	7·4783	7·4838	7·4893	7·4949	7·5006	7·5063	5	11	16	22	27
99·4	7·5121	7·5181	7·5241	7·5302	7·5364	7·5427	7·5491	7·5556	7·5622	7·5690	6	13	19	25	32
99·5	7·5758	7·5828	7·5899	7·5972	7·6045	7·6121	7·6197	7·6276	7·6356	7·6437					
99·6	7·6521	7·6606	7·6693	7·6783	7·6874	7·6968	7·7065	7·7164	7·7266	7·7370					
99·7	7·7478	7·7589	7·7703	7·7822	7·7944	7·8070	7·8202	7·8338	7·8480	7·8627					
99·8	7·8782	7·8943	7·9112	7·9290	7·9478	7·9677	7·9889	8·0115	8·0357	8·0618					
99·9	8·0902	8·1214	8·1559	8·1947	8·2389	8·2905	8·3528	8·4316	8·5401	8·7190					

Table IXi. Simple Quantiles of the Normal Distribution

Proper Fractions, with Denominators up to 30, and the corresponding Normal Deviates

*1/30	1·8339	*1/10	1·2816	5/27	0·8958	8/29	0·5952	*11/30	0·3407	9/20	0·1257
1/29	1·8187	3/29	1·2621	3/16	0·8871	5/18	0·5895	7/19	0·3360	5/11	0·1142
1/28	1·8028	2/19	1·2521	4/21	0·8761	7/25	0·5828	10/27	0·3309	11/24	0·1046
1/27	1·7861	3/28	1·2419	5/26	0·8694	2/7	0·5661	3/8	0·3186	6/13	0·0966
1/26	1·7688	1/9	1·2206	*1/5	0·8416	7/24	0·5485	11/29	0·3073	13/28	0·0896
1/25	1·7507	3/26	1·1984	6/29	0·8172	5/17	0·5414	8/21	0·3030	*7/15	0·0836
1/24	1·7317	2/17	1·1868	5/24	0·8122	8/27	0·5351	5/13	0·2934	8/17	0·0738
1/23	1·7117	3/25	1·1750	4/19	0·8046	*3/10	0·5244	7/18	0·2822	9/19	0·0660
1/22	1·6906	1/8	1·1503	3/14	0·7916	7/23	0·5119	9/23	0·2759	10/21	0·0597
1/21	1·6684	3/23	1·1244	5/23	0·7810	4/13	0·5024	11/28	0·2719	11/23	0·0545
1/20	1·6449	*2/15	1·1108	2/9	0·7647	9/29	0·4949	*2/5	0·2533	12/25	0.0502
1/19	1·6199	3/22	1·0968	5/22	0·7479	5/16	0·4888	11/27	0·2342	13/27	0·0464
1/18	1·5932	4/29	1·0897	3/13	0·7363	6/19	0·4795	9/22	0·2299	14/29	0·0432
1/17	1·5648	1/7	1·0676	*7/30	0·7279	7/22	0·4728	7/17	0·2230	*1/2	0
1/16	1·5341	4/27	1·0444	4/17	0·7215	8/25	0·4677	12/29	0·2178		
*1/15	1·5011	3/20	1·0364	5/21	0·7124	9/28	0·4637	5/12	0·2104		
2/29	1·4835	2/13	1·0201	6/25	0·7063	*1/3	0·4307	8/19	0·1992		
1/14	1·4652	3/19	1·0031	7/29	0·7019	10/29	0·3993	11/26	0·1940		
2/27	1·4461	4/25	0·9945	1/4	0·6745	9/26	0·3957	3/7	0·1800		
1/13	1·4261	*1/6	0·9674	7/27	0·6456	8/23	0·3912	*13/30	0·1679		
2/25	1·4051	5/29	0·9447	6/23	0·6407	7/20	0·3853	10/23	0·1642		
1/12	1·3829	4/23	0·9388	5/19	0·6336	6/17	0·3774	7/16	0·1573		
2/23	1·3597	3/17	0·9289	*4/15	0·6229	5/14	0·3661	11/25	0·1510		
1/11	1·3352	5/28	0·9208	7/26	0·6151	9/25	0·3585	4/9	0·1397		
2/21	1·3092	2/11	0·9085	3/11	0·6046	4/11	0·3488	13/29	0·1300		

Many fractions with higher denominators may be quickly obtained by simple interpolation between adjacent values, *e.g.* from 5/13 and 7/18, because $\frac{5+7}{13+18} = \frac{12}{31}$, the deviate for 12/31 is judged to be $0{\cdot}2822 + \frac{13}{31}({\cdot}0112) = 0{\cdot}2868{,}97$. True value 0·2868,94. Each thirtieth is marked by an asterisk.

To convert to probit values subtract from 5, and for fractions λ greater than $\frac{1}{2}$ add the value for $1-\lambda$ to 5

TABLE IX2. PROBITS

Weighting Coefficients and Probit Values to be used for Final Adjustments (Adapted from Bliss, 1935)

Expected Probit Y.	Minimum Working Probit $Y-P/Z$.	Range $1/Z$	Maximum working Probit $Y+Q/Z$.	Weighting Coefficient Z^2/PQ.	Expected Probit Y.	Minimum Working Probit $Y-P/Z$.	Range $1/Z$	Maximum working Probit $Y+Q/Z$.	Weighting Coefficient Z^2/PQ.
1·1	0·8579	5034	5035	·00082	5·0	3·7467	2·5066	6·2533	·63662
1·2	0·9522	3425	3426	·00118	5·1	3·7401	2·5192	6·2593	·63431
1·3	1·0462	2354	2355	·00167	5·2	3·7186	2·5573	6·2759	·62742
1·4	1·1400	1634	1635	·00235	5·3	3·6798	2·6220	6·3018	·61609
1·5	1·2335	1146	1147	·00327	5·4	3·6203	2·7154	6·3357	·60052
1·6	1·3266	811·5	812·8	·00451	5·5	3·5360	2·8404	6·3764	·58099
1·7	1·4194	580·5	581·9	·00614	5·6	3·4220	3·0010	6·4230	·55788
1·8	1·5118	419·4	420·9	·00828	5·7	3·2724	3·2025	6·4749	·53159
1·9	1·6038	306·1	307·7	·01104	5·8	3·0794	3·4519	6·5313	·50260
2·0	1·6954	225·6	227·3	·01457	5·9	2·8335	3·7582	6·5917	·47144
2·1	1·7866	168·00	169·79	·01903	6·0	2·5230	4·1327	6·6557	·43863
2·2	1·8772	126·34	128·22	·02459	6·1	2·1324	4·5903	6·7227	·40474
2·3	1·9673	95·96	97·93	·03143	6·2	1·6429	5·1497	6·7926	·37031
2·4	2·0568	73·62	75·68	·03977	6·3	1·0295	5·8354	6·8649	·33589
2·5	2·1457	57·05	59·20	·04979	6·4	0·2606	6·6788	6·9394	·30199
2·6	2·2340	44·654	46·888	·06169	6·5	−0·705	7·721	7·0158	·26907
2·7	2·3214	35·302	37·623	·07563	6·6	−1·921	9·015	7·0940	·23753
2·8	2·4081	28·189	30·597	·09179	6·7	−3·459	10·633	7·1739	·20774
2·9	2·4938	22·736	25·230	·11026	6·8	−5·411	12·666	7·2551	·17994
3·0	2·5786	18·522	21·101	·13112	6·9	−7·902	15·240	7·3376	·15436
3·1	2·6624	15·240	17·902	·15436	7·0	−11·101	18·522	7·4214	·13112
3·2	2·7449	12·666	15·411	·17994	7·1	−15·230	22·736	7·5062	·11026
3·3	2·8261	10·633	13·459	·20774	7·2	−20·597	28·189	7·5919	·09179
3·4	2·9060	9·015	11·921	·23753	7·3	−27·623	35·302	7·6786	·07564
3·5	2·9842	7·721	10·705	·26907	7·4	−36·888	44·654	7·7661	·06168
3·6	3·0606	6·6788	9·7394	·30199	7·5	−49·20	57·05	7·8543	·04979
3·7	3·1351	5·8354	8·9705	·33589	7·6	−65·68	73·62	7·9432	·03977
3·8	3·2074	5·1497	8·3571	·37031	7·7	−87·93	95·96	8·0327	·03143
3·9	3·2773	4·5903	7·8676	·40474	7·8	−118·22	126·34	8·1228	·02458
4·0	3·3443	4·1327	7·4770	·43863	7·9	−159·79	168·00	8·2134	·01903
4·1	3·4083	3·7582	7·1665	·47144	8·0	−217·3	225·6	8·3046	·01457
4·2	3·4687	3·4519	6·9206	·50260	8·1	−297·7	306·1	8·3962	·01104
4·3	3·5251	3·2025	6·7276	·53159	8·2	−410·9	419·4	8·4882	·00828
4·4	3·5770	3·0010	6·5780	·55788	8·3	−571·9	580·5	8·5806	·00614
4·5	3·6236	2·8404	6·4640	·58099	8·4	−802·8	811·5	8·6734	·00451
4·6	3·6643	2·7154	6·3797	·60052	8·5	−1137	1146	8·7666	·00327
4·7	3·6982	2·6220	6·3202	·61609	8·6	−1625	1634	8·8600	·00235
4·8	3·7241	2·5573	6·2814	·62741	8·7	−2345	2354	8·9538	·00167
4·9	3·7407	2·5192	6·2599	·63431	8·8	−3416	3425	9·0478	·00118
5·0	3·7467	2·5066	6·2533	·63662	8·9	−5025	5034	9·1421	·00082

In toxicology the probit values used are found by adding 5 to a normal deviate. The weighting coefficient for any test Z^2/PQ is derived from the probit Y given by some provisional formula. The appropriate score for a test in which an observed proportion p die and q survive is y_c, where $y_c = (Y-P/Z)+p/Z = (Y+Q/Z)-q/Z = q(Y-P/Z)+p(Y+Q/Z)$.

Table IX3. Probits

Weighting Coefficients for Use when there is a Natural Mortality
(D. J. Finney)

Y	Q/Z	Natural Mortality, K ·00	·01	·02	·04	·06	·08	·10	·15	·20	·25	·30	·35	·40
1·1	5034	·001	·000											
1·2	3425	·001	·000											
1·3	2354	·002	·000											
1·4	1634	·002	·000											
1·5	1146	·003	·000											
1·6	811·2	·005	·000											
1·7	580·2	·006	·000											
1·8	419·1	·008	·001	·000										
1·9	305·8	·011	·001	·000										
2·0	225·3	·015	·002	·001	·000	·000								
2·1	167·69	·019	·003	·002	·001	·001	·000	·000						
2·2	126·02	·025	·005	·003	·001	·001	·001	·001	·000					
2·3	95·63	·031	·008	·005	·002	·002	·001	·001	·001	·000	·000			
2·4	73·28	·040	·013	·007	·004	·003	·002	·002	·001	·001	·001	·000	·000	
2·5	56·70	·050	·019	·012	·006	·004	·003	·003	·002	·001	·001	·001	·001	·000
2·6	44·288	·062	·028	·018	·010	·007	·005	·004	·003	·002	·001	·001	·001	·001
2·7	34·923	·076	·039	·026	·015	·011	·008	·007	·004	·003	·002	·002	·001	·001
2·8	27·797	·092	·053	·037	·023	·016	·013	·010	·007	·005	·004	·003	·002	·002
2·9	22·330	·110	·070	·051	·033	·024	·019	·015	·010	·007	·006	·004	·004	·003
3·0	18·101	·131	·091	·069	·046	·034	·027	·022	·015	·011	·008	·007	·005	·004
3·1	14·802	·154	·114	·090	·063	·048	·038	·032	·022	·016	·012	·010	·008	·006
3·2	12·211	·180	·140	·115	·083	·065	·053	·044	·030	·023	·018	·014	·011	·009
3·3	10·159	·208	·169	·142	·107	·085	·070	·059	·042	·031	·024	·020	·016	·013
3·4	8·521	·238	·201	·173	·135	·110	·092	·078	·056	·043	·034	·027	·022	·018
3·5	7·205	·269	·234	·206	·166	·138	·117	·101	·074	·057	·045	·036	·030	·025
3·6	6·1394	·302	·268	·241	·199	·169	·145	·127	·095	·074	·059	·048	·039	·033
3·7	5·2705	·336	·304	·277	·235	·202	·177	·156	·119	·094	·076	·062	·051	·043
3·8	4·5571	·370	·340	·315	·272	·238	·211	·188	·146	·117	·095	·078	·065	·055
3·9	3·9676	·405	·377	·352	·310	·275	·247	·222	·176	·142	·117	·097	·081	·068
4·0	3·4770	·439	·412	·389	·347	·313	·283	·258	·208	·170	·141	·119	·100	·084
4·1	3·0665	·471	·447	·424	·384	·350	·320	·294	·241	·200	·168	·142	·120	·102
4·2	2·7206	·503	·480	·458	·420	·386	·356	·330	·274	·231	·195	·166	·142	·121
4·3	2·4276	·532	·510	·490	·454	·421	·391	·364	·307	·261	·224	·192	·165	·142
4·4	2·1780	·558	·538	·519	·484	·453	·424	·397	·339	·292	·252	·218	·188	·163
4·5	1·9640	·581	·563	·545	·512	·481	·453	·427	·370	·321	·279	·243	·212	·184
4·6	1·7797	·601	·583	·567	·536	·507	·480	·454	·397	·348	·305	·268	·234	·205
4·7	1·6202	·616	·600	·585	·556	·528	·502	·477	·421	·372	·329	·290	·256	·224
4·8	1·4814	·627	·613	·598	·571	·545	·520	·496	·442	·394	·350	·311	·275	·243
4·9	1·3599	·634	·621	·607	·582	·557	·534	·511	·458	·411	·368	·328	·292	·259
5·0	1·2533	·637	·624	·612	·588	·565	·542	·521	·471	·424	·382	·343	·307	·273

TABLE IX3. PROBITS—*continued*

Weighting Coefficients for Use when there is Natural Mortality

Y	Q/Z	Natural Mortality, K ·00	·01	·02	·04	·06	·08	·10	·15	·20	·25	·30	·35	·40
5·0	1·2533	·637	·624	·612	·588	·565	·542	·521	·471	·424	·382	·343	·307	·273
5·1	1·1593	·634	·623	·611	·589	·567	·546	·526	·478	·434	·392	·354	·318	·284
5·2	1·0759	·627	·617	·606	·585	·565	·546	·526	·481	·438	·398	·361	·325	·292
5·3	1·0018	·616	·606	·596	·577	·558	·540	·522	·479	·439	·400	·364	·329	·296
5·4	0·9357	·601	·591	·582	·565	·547	·530	·513	·473	·435	·398	·363	·330	·298
5·5	0·8764	·581	·573	·564	·548	·532	·516	·501	·463	·427	·392	·359	·327	·296
5·6	0·8230	·558	·550	·543	·528	·513	·498	·484	·449	·415	·382	·351	·320	·291
5·7	0·7749	·532	·525	·518	·504	·490	·477	·464	·431	·400	·369	·340	·311	·283
5·8	0·7313	·503	·496	·490	·477	·465	·453	·440	·411	·382	·353	·326	·299	·272
5·9	0·6917	·471	·466	·460	·449	·437	·426	·415	·388	·361	·334	·309	·284	·259
6·0	0·6557	·439	·433	·428	·418	·408	·398	·387	·363	·338	·314	·291	·267	·245
6·1	0·6227	·405	·400	·395	·386	·377	·368	·359	·336	·314	·292	·271	·249	·228
6·2	0·5926	·370	·366	·362	·354	·345	·337	·329	·309	·289	·269	·249	·230	·211
6·3	0·5649	·336	·332	·328	·321	·314	·306	·299	·281	·263	·245	·228	·210	·193
6·4	0·5394	·302	·299	·295	·289	·282	·276	·269	·253	·237	·222	·206	·190	·175
6·5	0·5158	·269	·266	·263	·258	·252	·246	·240	·226	·212	·198	·184	·171	·157
6·6	0·4940	·238	·235	·233	·228	·223	·218	·213	·200	·188	·176	·163	·151	·139
6·7	0·4739	·208	·206	·203	·199	·195	·190	·186	·175	·165	·154	·143	·133	·122
6·8	0·4551	·180	·178	·176	·172	·169	·165	·161	·152	·143	·134	·125	·115	·106
6·9	0·4376	·154	·153	·151	·148	·145	·142	·139	·131	·123	·115	·107	·099	·092
7·0	0·4214	·131	·130	·128	·126	·123	·120	·118	·111	·104	·098	·091	·085	·078
7·1	0·4062	·110	·109	·108	·106	·104	·101	·099	·093	·088	·082	·077	·071	·066
7·2	0·3919	·092	·091	·090	·088	·086	·084	·082	·078	·073	·069	·064	·059	·055
7·3	0·3786	·076	·075	·074	·073	·071	·070	·068	·064	·060	·057	·053	·049	·045
7·4	0·3661	·062	·061	·060	·059	·058	·057	·055	·052	·049	·046	·043	·040	·037
7·5	0·3543	·050	·049	·049	·048	·047	·046	·045	·042	·040	·037	·035	·032	·030
7·6	0·3432	·040	·039	·039	·038	·037	·037	·036	·034	·032	·030	·028	·026	·024
7·7	0·3327	·031	·031	·031	·030	·030	·029	·028	·027	·025	·024	·022	·020	·019
7·8	0·3228	·025	·024	·024	·024	·023	·023	·022	·021	·020	·018	·017	·016	·015
7·9	0·3134	·019	·019	·019	·018	·018	·018	·017	·016	·015	·014	·013	·012	·011
8·0	0·3046	·015	·014	·014	·014	·014	·013	·013	·012	·012	·011	·010	·009	·009
8·1	0·2962	·011	·011	·011	·011	·010	·010	·010	·009	·009	·008	·008	·007	·007
8·2	0·2882	·008	·008	·008	·008	·008	·008	·007	·007	·007	·006	·006	·005	·005
8·3	0·2806	·006	·006	·006	·006	·006	·006	·006	·005	·005	·005	·004	·004	·004
8·4	0·2734	·005	·004	·004	·004	·004	·004	·004	·004	·004	·003	·003	·003	·003
8·5	0·2666	·003	·003	·003	·003	·003	·003	·003	·003	·003	·002	·002	·002	·002
8·6	0·2600	·002	·002	·002	·002	·002	·002	·002	·002	·002	·002	·002	·002	·001
8·7	0·2538	·002	·002	·002	·002	·002	·002	·002	·001	·001	·001	·001	·001	·001
8·8	0·2478	·001	·001	·001	·001	·001	·001	·001	·001	·001	·001	·001	·001	·001
8·9	0·2421	·001	·001	·001	·001	·001	·001	·001	·001	·001	·001	·001	·001	·000

Table X. Angular Transformation

p%	0·0	0·1	0·2	0·3	0·4	0·5	0·6	0·7	0·8	0·9
0	0·00	1·81	2·56	3·14	3·63	4·05	4·44	4·80	5·13	5·44
1	5·74	6·02	6·29	6·55	6·80	7·03	7·27	7·49	7·71	7·92
2	8·13	8·33	8·53	8·72	8·91	9·10	9·28	9·46	9·63	9·80
3	9·97	10·14	10·30	10·47	10·63	10·78	10·94	11·09	11·24	11·39
4	11·54	11·68	11·83	11·97	12·11	12·25	12·38	12·52	12·66	12·79
5	12·92	13·05	13·18	13·31	13·44	13·56	13·69	13·81	13·94	14·06
6	14·18	14·30	14·42	14·54	14·65	14·77	14·89	15·00	15·12	15·23
7	15·34	15·45	15·56	15·68	15·79	15·89	16·00	16·11	16·22	16·32
8	16·43	16·54	16·64	16·74	16·85	16·95	17·05	17·15	17·26	17·36
9	17·46	17·56	17·66	17·76	17·85	17·95	18·05	18·15	18·24	18·34
10	18·43	18·53	18·63	18·72	18·81	18·91	19·00	19·09	19·19	19·28
11	19·37	19·46	19·55	19·64	19·73	19·82	19·91	20·00	20·09	20·18
12	20·27	20·36	20·44	20·53	20·62	20·70	20·79	20·88	20·96	21·05
13	21·13	21·22	21·30	21·39	21·47	21·56	21·64	21·72	21·81	21·89
14	21·97	22·06	22·14	22·22	22·30	22·38	22·46	22·54	22·63	22·71
15	22·79	22·87	22·95	23·03	23·11	23·18	23·26	23·34	23·42	23·50
16	23·58	23·66	23·73	23·81	23·89	23·97	24·04	24·12	24·20	24·27
17	24·35	24·43	24·50	24·58	24·65	24·73	24·80	24·88	24·95	25·03
18	25·10	25·18	25·25	25·33	25·40	25·47	25·55	25·62	25·70	25·77
19	25·84	25·91	25·99	26·06	26·13	26·21	26·28	26·35	26·42	26·49
20	26·57	26·64	26·71	26·78	26·85	26·92	26·99	27·06	27·13	27·20
21	27·27	27·35	27·42	27·49	27·56	27·62	27·69	27·76	27·83	27·90
22	27·97	28·04	28·11	28·18	28·25	28·32	28·39	28·45	28·52	28·59
23	28·66	28·73	28·79	28·86	28·93	29·00	29·06	29·13	29·20	29·27
24	29·33	29·40	29·47	29·53	29·60	29·67	29·73	29·80	29·87	29·93
25	30·00	30·07	30·13	30·20	30·26	30·33	30·40	30·46	30·53	30·59
26	30·66	30·72	30·79	30·85	30·92	30·98	31·05	31·11	31·18	31·24
27	31·31	31·37	31·44	31·50	31·56	31·63	31·69	31·76	31·82	31·88
28	31·95	32·01	32·08	32·14	32·20	32·27	32·33	32·39	32·46	32·52
29	32·58	32·65	32·71	32·77	32·83	32·90	32·96	33·02	33·09	33·15
30	33·21	33·27	33·34	33·40	33·46	33·52	33·58	33·65	33·71	33·77
31	33·83	33·90	33·96	34·02	34·08	34·14	34·20	34·27	34·33	34·39
32	34·45	34·51	34·57	34·63	34·70	34·76	34·82	34·88	34·94	35·00
33	35·06	35·12	35·18	35·24	35·30	35·37	35·43	35·49	35·55	35·61
34	35·67	35·73	35·79	35·85	35·91	35·97	36·03	36·09	36·15	36·21
35	36·27	36·33	36·39	36·45	36·51	36·57	36·63	36·69	36·75	36·81
36	36·87	36·93	36·99	37·05	37·11	37·17	37·23	37·29	37·35	37·41
37	37·46	37·52	37·58	37·64	37·70	37·76	37·82	37·88	37·94	38·00
38	38·06	38·12	38·17	38·23	38·29	38·35	38·41	38·47	38·53	38·59
39	38·65	38·70	38·76	38·82	38·88	38·94	39·00	39·06	39·11	39·17
40	39·23	39·29	39·35	39·41	39·47	39·52	39·58	39·64	39·70	39·76
41	39·82	39·87	39·93	39·99	40·05	40·11	40·16	40·22	40·28	40·34
42	40·40	40·45	40·51	40·57	40·63	40·69	40·74	40·80	40·86	40·92
43	40·98	41·03	41·09	41·15	41·21	41·27	41·32	41·38	41·44	41·50
44	41·55	41·61	41·67	41·73	41·78	41·84	41·90	41·96	42·02	42·07
45	42·13	42·19	42·25	42·30	42·36	42·42	42·48	42·53	42·59	42·65
46	42·71	42·76	42·82	42·88	42·94	42·99	43·05	43·11	43·17	43·22
47	43·28	43·34	43·39	43·45	43·51	43·57	43·62	43·68	43·74	43·80
48	43·85	43·91	43·97	44·03	44·08	44·14	44·20	44·26	44·31	44·37
49	44·43	44·48	44·54	44·60	44·66	44·71	44·77	44·83	44·89	44·94

Table X. Angular Transformation—*continued*

p%	0·0	0·1	0·2	0·3	0·4	0·5	0·6	0·7	0·8	0·9
50	45·00	45·06	45·11	45·17	45·23	45·29	45·34	45·40	45·46	45·52
51	45·57	45·63	45·69	45·74	45·80	45·86	45·92	45·97	46·03	46·09
52	46·15	46·20	46·26	46·32	46·38	46·43	46·49	46·55	46·61	46·66
53	46·72	46·78	46·83	46·89	46·95	47·01	47·06	47·12	47·18	47·24
54	47·29	47·35	47·41	47·47	47·52	47·58	47·64	47·70	47·75	47·81
55	47·87	47·93	47·98	48·04	48·10	48·16	48·22	48·27	48·33	48·39
56	48·45	48·50	48·56	48·62	48·68	48·73	48·79	48·85	48·91	48·97
57	49·02	49·08	49·14	49·20	49·26	49·31	49·37	49·43	49·49	49·55
58	49·60	49·66	49·72	49·78	49·84	49·89	49·95	50·01	50·07	50·13
59	50·18	50·24	50·30	50·36	50·42	50·48	50·53	50·59	50·65	50·71
60	50·77	50·83	50·89	50·94	51·00	51·06	51·12	51·18	51·24	51·30
61	51·35	51·41	51·47	51·53	51·59	51·65	51·71	51·77	51·83	51·88
62	51·94	52·00	52·06	52·12	52·18	52·24	52·30	52·36	52·42	52·48
63	52·54	52·59	52·65	52·71	52·77	52·83	52·89	52·95	53·01	53·07
64	53·13	53·19	53·25	53·31	53·37	53·43	53·49	53·55	53·61	53·67
65	53·73	53·79	53·85	53·91	53·97	54·03	54·09	54·15	54·21	54·27
66	54·33	54·39	54·45	54·51	54·57	54·63	54·70	54·76	54·82	54·88
67	54·94	55·00	55·06	55·12	55·18	55·24	55·30	55·37	55·43	55·49
68	55·55	55·61	55·67	55·73	55·80	55·86	55·92	55·98	56·04	56·10
69	56·17	56·23	56·29	56·35	56·42	56·48	56·54	56·60	56·66	56·73
70	56·79	56·85	56·91	56·98	57·04	57·10	57·17	57·23	57·29	57·35
71	57·42	57·48	57·54	57·61	57·67	57·73	57·80	57·86	57·92	57·99
72	58·05	58·12	58·18	58·24	58·31	58·37	58·44	58·50	58·56	58·63
73	58·69	58·76	58·82	58·89	58·95	59·02	59·08	59·15	59·21	59·28
74	59·34	59·41	59·47	59·54	59·60	59·67	59·74	59·80	59·87	59·93
75	60·00	60·07	60·13	60·20	60·27	60·33	60·40	60·47	60·53	60·60
76	60·67	60·73	60·80	60·87	60·94	61·00	61·07	61·14	61·21	61·27
77	61·34	61·41	61·48	61·55	61·61	61·68	61·75	61·82	61·89	61·96
78	62·03	62·10	62·17	62·24	62·31	62·38	62·44	62·51	62·58	62·65
79	62·73	62·80	62·87	62·94	63·01	63·08	63·15	63·22	63·29	63·36
80	63·43	63·51	63·58	63·65	63·72	63·79	63·87	63·94	64·01	64·09
81	64·16	64·23	64·30	64·38	64·45	64·53	64·60	64·67	64·75	64·82
82	64·90	64·97	65·05	65·12	65·20	65·27	65·35	65·42	65·50	65·57
83	65·65	65·73	65·80	65·88	65·96	66·03	66·11	66·19	66·27	66·34
84	66·42	66·50	66·58	66·66	66·74	66·82	66·89	66·97	67·05	67·13
85	67·21	67·29	67·37	67·46	67·54	67·62	67·70	67·78	67·86	67·94
86	68·03	68·11	68·19	68·28	68·36	68·44	68·53	68·61	68·70	68·78
87	68·87	68·95	69·04	69·12	69·21	69·30	69·38	69·47	69·56	69·64
88	69·73	69·82	69·91	70·00	70·09	70·18	70·27	70·36	70·45	70·54
89	70·63	70·72	70·81	70·91	71·00	71·09	71·19	71·28	71·37	71·47
90	71·57	71·66	71·76	71·85	71·95	72·05	72·15	72·24	72·34	72·44
91	72·54	72·64	72·74	72·85	72·95	73·05	73·15	73·26	73·36	73·46
92	73·57	73·68	73·78	73·89	74·00	74·11	74·21	74·32	74·44	74·55
93	74·66	74·77	74·88	75·00	75·11	75·23	75·35	75·46	75·58	75·70
94	75·82	75·94	76·06	76·19	76·31	76·44	76·56	76·69	76·82	76·95
95	77·08	77·21	77·34	77·48	77·62	77·75	77·89	78·03	78·17	78·32
96	78·46	78·61	78·76	78·91	79·06	79·22	79·37	79·53	79·70	79·86
97	80·03	80·20	80·37	80·54	80·72	80·90	81·09	81·28	81·47	81·67
98	81·87	82·08	82·29	82·51	82·73	82·97	83·20	83·45	83·71	83·98
99	84·26	84·56	84·87	85·20	85·56	85·95	86·37	86·86	87·44	88·19

TABLE XI. TRANSFORMATION OF PROPER FRACTIONS TO DEGREES

*1/30	10·5	*1/10	18·4	5/27	25·5	8/29	31·7	*11/30	37·3	9/20	42·1
1/29	10·7	3/29	18·8	3/16	25·7	5/18	31·8	7/19	37·4	5/11	42·4
1/28	10·9	2/19	18·9	4/21	25·9	7/25	31·9	10/27	37·5	11/24	42·6
1/27	11·1	3/28	19·1	5/26	26·0	2/7	32·3	3/8	37·8	6/13	42·8
1/26	11·3	1/9	19·5	*1/5	26·6	7/24	32·7	11/29	38·0	13/28	43·0
1/25	11·5	3/26	19·9	6/29	27·1	5/17	32·8	8/21	38·1	*7/15	43·1
1/24	11·8	2/17	20·1	5/24	27·2	8/27	33·0	5/13	38·3	8/17	43·3
1/23	12·0	3/25	20·3	4/19	27·3	*3/10	33·2	7/18	38·6	9/19	43·5
1/22	12·3	1/8	20·7	3/14	27·6	7/23	33·5	9/23	38·7	10/21	43·6
1/21	12·6	3/23	21·2	5/23	27·8	4/13	33·7	11/28	38·8	11/23	43·8
1/20	12·9	*2/15	21·4	2/9	28·1	9/29	33·9	*2/5	39·2	12/25	43·9
1/19	13·3	3/22	21·7	5/22	28·5	5/16	34·0	11/27	39·7	13/27	43·9
1/18	13·6	4/29	21·8	3/13	28·7	6/19	34·2	9/22	39·8	14/29	44·0
1/17	14·0	1/7	22·2	*7/30	28·9	7/22	34·3	7/17	39·9	*1/2	45·0
1/16	14·5	4/27	22·6	4/17	29·0	8/25	34·4	12/29	40·0		
*1/15	15·0	3/20	22·8	5/21	29·2	9/28	34·5	5/12	40·2		
2/29	15·2	2/13	23·1	6/25	29·3	*1/3	35·3	8/19	40·5		
1/14	15·5	3/19	23·4	7/29	29·4	10/29	36·0	11/26	40·6		
2/27	15·8	4/25	23·6	1/4	30·0	9/26	36·0	3/7	40·9		
1/13	16·1	*1/6	24·1	7/27	30·6	8/23	36·1	*13/30	41·2		
2/25	16·4	5/29	24·5	6/23	30·7	7/20	36·3	10/23	41·3		
1/12	16·8	4/23	24·6	5/19	30·9	6/17	36·4	7/16	41·4		
2/23	17·2	3/17	24·8	*4/15	31·1	5/14	36·7	11/25	41·6		
1/11	17·5	5/28	25·0	7/26	31·3	9/25	36·9	4/9	41·8		
2/21	18·0	2/11	25·2	3/11	31·5	4/11	37·1	13/29	42·0		

For fractions exceeding $\frac{1}{2}$, subtract the fraction from 1, and the angle from 90°. Each thirtieth is marked by an asterisk; for interpolation see Table IX1.

In connection with an observed fraction p, it is often advisable to use the angular value ϕ, such that $p = \sin^2 \phi$. ϕ rises to 90° at 100 per cent. The amount of information supplied by n observations about ϕ, in circular measure, is $4n$, or in degrees, $\frac{4\pi^2 n}{180^2} = \frac{\pi^2 n}{8100}$. Consequently in large samples, the variance of ϕ, as estimated from a sample, is $\frac{8100}{\pi^2 n} = \frac{820 \cdot 7}{n}$ independently of the true value of ϕ.

Table X2. Angular Values for Final Adjustments

Expected value	Working values: Minimum	Range	Maximum	Expected value	Working values: Minimum	Range	Maximum
				45	16·352	57·296	73·648
1	0·500	1641·737	1642·237	46	16·334	57·331	73·665
2	1·000	821·368	822·368	47	16·279	57·436	73·715
3	1·499	548·135	549·634	48	16·183	57·612	73·795
4	1·997	411·687	413·684	49	16·044	57·859	73·903
5	2·494	329·953	332·447	50	15·859	58·179	74·038
6	2·989	275·577	278·566	51	15·623	58·576	74·199
7	3·482	236·836	240·318	52	15·332	59·050	74·382
8	3·974	207·866	211·840	53	14·983	59·605	74·588
9	4·463	185·413	189·876	54	14·570	60·244	74·814
10	4·949	167·521	172·470	55	14·087	60·972	75·059
11	5·431	152·950	158·381	56	13·528	61·795	75·323
12	5·911	140·867	146·778	57	12·886	62·718	75·604
13	6·386	130·702	137·088	58	12·154	63·747	75·901
14	6·857	122·043	128·900	59	11·322	64·891	76·213
15	7·324	114·591	121·915	60	10·380	66·160	76·540
16	7·785	108·122	115·907	61	9·318	67·562	76·880
17	8·241	102·462	110·703	62	8·121	69·111	77·232
18	8·692	97·477	106·169	63	6·775	70·822	77·597
19	9·136	93·064	102·200	64	5·263	72·710	77·973
20	9·573	89·136	98·709	65	3·564	74·795	78·359
21	10·003	85·627	95·630	66	1·656	77·099	78·755
22	10·426	82·480	92·906	67	−0·490	79·650	79·160
23	10·840	79·650	90·490	68	−2·906	82·480	79·574
24	11·245	77·099	88·344	69	−5·630	85·627	79·997
25	11·641	74·795	86·436	70	−8·709	89·136	80·427
26	12·027	72·710	84·737	71	−12·200	93·064	80·864
27	12·403	70·822	83·225	72	−16·169	97·477	81·308
28	12·768	69·111	81·879	73	−20·703	102·462	81·759
29	13·120	67·562	80·682	74	−25·907	108·122	82·215
30	13·460	66·160	79·620	75	−31·915	114·591	82·676
31	13·787	64·891	78·678	76	−38·900	122·043	83·143
32	14·099	63·747	77·846	77	−47·088	130·702	83·614
33	14·396	62·718	77·114	78	−56·778	140·867	84·089
34	14·677	61·795	76·472	79	−68·381	152·950	84·569
35	14·941	60·972	75·913	80	−82·470	167·521	85·051
36	15·186	60·244	75·430	81	−99·876	185·413	85·537
37	15·412	59·605	75·017	82	−121·840	207·866	86·026
38	15·618	59·050	74·668	83	−150·318	236·836	86·518
39	15·801	58·576	74·377	84	−188·566	275·577	87·011
40	15·962	58·179	74·141	85	−242·447	329·953	87·506
41	16·097	57·859	73·956	86	−323·684	411·687	88·003
42	16·205	57·612	73·817	87	−459·634	548·135	88·501
43	16·285	57·436	73·721	88	−732·368	821·368	89·000
44	16·335	57·331	73·666	89	−1552·237	1641·737	89·500

This table supplies for the angular transformation the same facilities as does Table IX2 for probits. The weighting coefficient, 1/820·7, is constant for all angular values, and is only needed in testing homogeneity. The adjusted score for *a* dying and *b* surviving out of *n* (where $a/n = p$) is the minimum working value multiplied by b/n plus the maximum working value multiplied by a/n.

For angles near 0° and 90° it is always to be questioned whether the angular transformation is appropriate.

TABLE XI. LOGITS

The Logit or r, z Transformation

p%	0·0	0·1	0·2	0·3	0·4	0·5	0·6	0·7	0·8	0·9
50	·0000	·0020	·0040	·0060	·0080	·0100	·0120	·0140	·0160	·0180
51	·0200	·0220	·0240	·0260	·0280	·0300	·0320	·0340	·0360	·0380
52	·0400	·0420	·0440	·0460	·0480	·0500	·0520	·0541	·0561	·0581
53	·0601	·0621	·0641	·0661	·0681	·0701	·0721	·0741	·0761	·0782
54	·0802	·0822	·0842	·0862	·0882	·0902	·0923	·0943	·0963	·0983
55	·1003	·1024	·1044	·1064	·1084	·1104	·1125	·1145	·1165	·1186
56	·1206	·1226	·1246	·1267	·1287	·1307	·1328	·1348	·1368	·1389
57	·1409	·1430	·1450	·1471	·1491	·1511	·1532	·1552	·1573	·1593
58	·1614	·1634	·1655	·1676	·1696	·1717	·1737	·1758	·1779	·1799
59	·1820	·1841	·1861	·1882	·1903	·1923	·1944	·1965	·1986	·2007
60	·2027	·2048	·2069	·2090	·2111	·2132	·2153	·2174	·2195	·2216
61	·2237	·2258	·2279	·2300	·2321	·2342	·2363	·2384	·2405	·2427
62	·2448	·2469	·2490	·2512	·2533	·2554	·2575	·2597	·2618	·2640
63	·2661	·2683	·2704	·2726	·2747	·2769	·2790	·2812	·2833	·2855
64	·2877	·2899	·2920	·2942	·2964	·2986	·3008	·3029	·3051	·3073
65	·3095	·3117	·3139	·3161	·3183	·3205	·3228	·3250	·3272	·3294
66	·3316	·3339	·3361	·3383	·3406	·3428	·3451	·3473	·3496	·3518
67	·3541	·3564	·3586	·3609	·3632	·3654	·3677	·3700	·3723	·3746
68	·3769	·3792	·3815	·3838	·3861	·3884	·3907	·3931	·3954	·3977
69	·4001	·4024	·4047	·4071	·4094	·4118	·4142	·4165	·4189	·4213
70	·4236	·4260	·4284	·4308	·4332	·4356	·4380	·4404	·4428	·4453
71	·4477	·4501	·4526	·4550	·4574	·4599	·4624	·4648	·4673	·4698
72	·4722	·4747	·4772	·4797	·4822	·4847	·4872	·4897	·4922	·4948
73	·4973	·4999	·5024	·5049	·5075	·5101	·5126	·5152	·5178	·5204
74	·5230	·5256	·5282	·5308	·5334	·5361	·5387	·5413	·5440	·5466
75	·5493	·5520	·5547	·5573	·5600	·5627	·5654	·5682	·5709	·5736
76	·5763	·5791	·5818	·5846	·5874	·5901	·5929	·5957	·5985	·6013
77	·6042	·6070	·6098	·6127	·6155	·6184	·6213	·6241	·6270	·6299
78	·6328	·6358	·6387	·6416	·6446	·6475	·6505	·6535	·6565	·6595
79	·6625	·6655	·6685	·6716	·6746	·6777	·6807	·6838	·6869	·6900
80	·6931	·6963	·6994	·7026	·7057	·7089	·7121	·7153	·7185	·7218
81	·7250	·7283	·7315	·7348	·7381	·7414	·7447	·7481	·7514	·7548
82	·7582	·7616	·7650	·7684	·7718	·7753	·7788	·7823	·7858	·7893
83	·7928	·7964	·7999	·8035	·8071	·8107	·8144	·8180	·8217	·8254
84	·8291	·8328	·8366	·8404	·8441	·8480	·8518	·8556	·8595	·8634
85	·8673	·8712	·8752	·8792	·8832	·8872	·8912	·8953	·8994	·9035
86	·9076	·9118	·9160	·9202	·9245	·9287	·9330	·9373	·9417	·9461
87	·9505	·9549	·9594	·9639	·9684	·9730	·9775	·9822	·9868	·9915
88	·9962	1·0010	1·0058	1·0106	1·0154	1·0203	1·0253	1·0302	1·0352	1·0403
89	1·0454	1·0505	1·0557	1·0609	1·0661	1·0714	1·0768	1·0822	1·0876	1·0931
90	1·0986	1·1042	1·1098	1·1155	1·1212	1·1270	1·1329	1·1388	1·1447	1·1507
91	1·1568	1·1630	1·1692	1·1754	1·1817	1·1881	1·1946	1·2011	1·2077	1·2144
92	1·2212	1·2280	1·2349	1·2419	1·2490	1·2562	1·2634	1·2707	1·2782	1·2857
93	1·2933	1·3011	1·3089	1·3169	1·3249	1·3331	1·3414	1·3498	1·3583	1·3670
94	1·3758	1·3847	1·3938	1·4030	1·4124	1·4219	1·4316	1·4415	1·4516	1·4618
95	1·4722	1·4828	1·4937	1·5047	1·5160	1·5275	1·5393	1·5513	1·5636	1·5762
96	1·5890	1·6022	1·6157	1·6296	1·6438	1·6584	1·6734	1·6888	1·7047	1·7211
97	1·7380	1·7555	1·7736	1·7923	1·8117	1·8318	1·8527	1·8745	1·8972	1·9210
98	1·9459	1·9721	1·9996	2·0287	2·0595	2·0923	2·1273	2·1649	2·2054	2·2494
99	2·2976	2·3507	2·4101	2·4774	2·5550	2·6467	2·7587	2·9031	3·1063	3·4534

The logit transformation, $z = \frac{1}{2}\log_e(p/q)$, is equivalent to the r, z transformation with $r = 2p-1$ (see Table VIII1). For values of $p < 0·5$ logits are *negative*, and numerically equal to the tabular values for $1-p$.

Table XI1. Logits

Weighting Coefficients and Logit Values to be used for Final Adjustments

Expected value	Working values Minimum	Range	Maximum	Weighting coefficient	Expected value	Working values Minimum	Range	Maximum	Weighting coefficient
0·00	−1·0000	2·0000	1·0000	1·00000	2·00	−25·799	28·308	2·5092	·070651
0·05	−1·0026	2·0050	1·0024	·99750	2·05	−28·620	31·178	2·5583	·064147
0·10	−1·0107	2·0201	1·0094	·99007	2·10	−31·743	34·351	2·6075	·058223
0·15	−1·0249	2·0453	1·0204	·97783	2·15	−35·200	37·857	2·6568	·052831
0·20	−1·0459	2·0811	1·0352	·96104	2·20	−39·025	41·732	2·7061	·047925
0·25	−1·0744	2·1277	1·0533	·94001	2·25	−43·259	46·014	2·7556	·043465
0·30	−1·1111	2·1855	1·0744	·91514	2·30	−47·942	50·747	2·8050	·039411
0·35	−1·1569	2·2552	1·0983	·88685	2·35	−53·124	55·978	2·8545	·035728
0·40	−1·2128	2·3375	1·1247	·85564	2·40	−58·855	61·759	2·9041	·032384
0·45	−1·2798	2·4331	1·1533	·82200	2·45	−65·195	68·149	2·9537	·029348
0·50	−1·3591	2·5430	1·1839	·78645	2·50	−72·207	75·210	3·0034	·026592
0·55	−1·4521	2·6685	1·2164	·74948	2·55	−79·961	83·014	3·0530	·024092
0·60	−1·5601	2·8107	1·2506	·71158	2·60	−88·536	91·639	3·1028	·021825
0·65	−1·6846	2·9709	1·2863	·67319	2·65	−98·018	101·17	3·1525	·019769
0·70	−1·8276	3·1509	1·3233	·63474	2·70	−108·50	111·71	3·2023	·017904
0·75	−1·9908	3·3524	1·3616	·59659	2·75	−120·10	123·35	3·2520	·016214
0·80	−2·1765	3·5774	1·4009	·55906	2·80	−132·91	136·22	3·3018	·014683
0·85	−2·3870	3·8283	1·4413	·52242	2·85	−147·08	150·44	3·3517	·013295
0·90	−2·6248	4·1074	1·4826	·48692	2·90	−162·75	166·15	3·4015	·012037
0·95	−2·8929	4·4177	1·5248	·45272	2·95	−180·07	183·52	3·4514	·010898
1·00	−3·1945	4·7622	1·5677	·41997	3·00	−199·21	202·72	3·5012	·0098660
1·05	−3·5331	5·1443	1·6112	·38878	3·05	−220·38	223·93	3·5511	·0089314
1·10	−3·9125	5·5679	1·6554	·35920	3·10	−243·77	247·38	3·6010	·0080849
1·15	−4·3371	6·0372	1·7001	·33128	3·15	−269·64	273·29	3·6509	·0073183
1·20	−4·8116	6·5570	1·7454	·30502	3·20	−298·22	301·92	3·7008	·0066242
1·25	−5·3412	7·1322	1·7910	·28041	3·25	−329·82	333·57	3·7508	·0059957
1·30	−5·9319	7·7690	1·8371	·25743	3·30	−364·75	368·55	3·8007	·0054267
1·35	−6·5899	8·4735	1·8836	·23603	3·35	−403·35	407·20	3·8506	·0049115
1·40	−7·3223	9·2527	1·9304	·21615	3·40	−446·02	449·92	3·9006	·0044452
1·45	−8·1371	10·115	1·9775	·19773	3·45	−493·19	497·14	3·9505	·0040230
1·50	−9·0428	11·068	2·0249	·18071	3·50	−545·32	549·32	4·0005	·0036409
1·55	−10·049	12·121	2·0725	·16500	3·55	−602·93	606·98	4·0504	·0032950
1·60	−11·166	13·287	2·1204	·15053	3·60	−666·62	670·72	4·1004	·0029819
1·65	−12·406	14·575	2·1684	·13722	3·65	−737·00	741·15	4·1503	·0026985
1·70	−13·782	15·999	2·2167	·12501	3·70	−814·79	818·99	4·2003	·0024420
1·75	−15·308	17·573	2·2651	·11381	3·75	−900·77	905·02	4·2503	·0022099
1·80	−16·999	19·313	2·3137	·10356	3·80	−995·80	1000·1	4·3003	·0019998
1·85	−18·874	21·236	2·3624	·094180	3·85	−1100·8	1105·2	4·3502	·0018097
1·90	−20·951	23·362	2·4112	·085610	3·90	−1216·9	1221·3	4·4002	·0016376
1·95	−23·251	25·711	2·4601	·077787	3·95	−1345·2	1349·6	4·4502	·0014819

Table XI1 is used in the same manner as the probit table for final adjustments (Table IX2), the working value being given by q Min. Value$+p$ Max. Value, Min. Value$+p$ Range, or Max. Value$-q$ Range. If the expected value is negative interchange the tabular minimum and maximum values with change of sign. Thus with an expected value of −0·65 and $p = 0·3$ the working value is 0·7(−1·2863)+0·3(1·6846) = −0·3950.

Table XII. Complementary Loglog Transformation

p%	0·0	0·2	0·4	0·6	0·8	p%	0·0	0·2	0·4	0·6	0·8
0		−6·2136	−5·5195	−5·1130	−4·8243	50	−·3665	−·3607	−·3550	−·3492	−·3435
1	−4·6001	−4·4168	−4·2617	−4·1271	−4·0083	51	−·3378	−·3321	−·3264	−·3207	−·3150
2	−3·9019	−3·8056	−3·7176	−3·6365	−3·5614	52	−·3093	−·3036	−·2980	−·2923	−·2866
3	−3·4914	−3·4258	−3·3641	−3·3060	−3·2509	53	−·2810	−·2754	−·2698	−·2641	−·2585
4	−3·1985	−3·1487	−3·1012	−3·0557	−3·0121	54	−·2529	−·2473	−·2417	−·2362	−·2306
5	−2·9702	−2·9299	−2·8911	−2·8537	−2·8176	55	−·2250	−·2194	−·2139	−·2083	−·2028
6	−2·7826	−2·7488	−2·7160	−2·6842	−2·6532	56	−·1973	−·1917	−·1862	−·1807	−·1752
7	−2·6232	−2·5940	−2·5655	−2·5378	−2·5107	57	−·1696	−·1641	−·1586	−·1531	−·1476
8	−2·4843	−2·4586	−2·4334	−2·4088	−2·3847	58	−·1421	−·1367	−·1312	−·1257	−·1202
9	−2·3612	−2·3381	−2·3155	−2·2934	−2·2717	59	−·1147	−·1093	−·1038	−·0983	−·0929
10	−2·2504	−2·2295	−2·2090	−2·1888	−2·1690	60	−·0874	−·0820	−·0765	−·0711	−·0656
11	−2·1496	−2·1305	−2·1116	−2·0931	−2·0749	61	−·0602	−·0547	−·0493	−·0438	−·0384
12	−2·0570	−2·0394	−2·0220	−2·0049	−1·9880	62	−·0330	−·0275	−·0221	−·0166	−·0112
13	−1·9714	−1·9550	−1·9388	−1.9229	−1·9072	63	−·0058	−·0003	·0051	·0105	·0160
14	−1·8916	−1·8763	−1·8612	−1·8463	−1·8315	64	·0214	·0269	·0323	·0377	·0432
15	−1·8170	−1·8026	−1·7883	−1·7743	−1·7604	65	·0486	·0541	·0595	·0650	·0704
16	−1·7467	−1·7331	−1·7197	−1·7064	−1·6932	66	·0759	·0813	·0868	·0922	·0977
17	−1·6802	−1·6674	−1·6546	−1·6420	−1·6296	67	·1032	·1086	·1141	·1196	·1250
18	−1·6172	−1·6050	−1·5929	−1·5809	−1·5690	68	·1305	·1360	·1415	·1470	·1525
19	−1·5572	−1·5456	−1·5340	−1·5225	−1·5112	69	·1580	·1635	·1690	·1746	·1801
20	−1·4999	−1·4888	−1·4777	−1·4668	−1·4559	70	·1856	·1912	·1967	·2023	·2078
21	−1·4451	−1·4344	−1·4238	−1·4133	−1·4028	71	·2134	·2190	·2246	·2301	·2357
22	−1·3925	−1·3822	−1·3720	−1·3619	−1·3518	72	·2413	·2470	·2526	·2582	·2639
23	−1·3418	−1·3319	−1·3221	−1·3123	−1·3027	73	·2695	·2752	·2809	·2865	·2922
24	−1·2930	−1·2835	−1·2740	−1·2646	−1·2552	74	·2979	·3037	·3094	·3151	·3209
25	−1·2459	−1·2367	−1·2275	−1·2184	−1·2093	75	·3266	·3324	·3382	·3440	·3498
26	−1·2003	−1·1913	−1·1825	−1·1736	−1·1648	76	·3557	·3615	·3674	·3732	·3791
27	−1·1561	−1·1474	−1·1388	−1·1302	−1·1217	77	·3850	·3910	·3969	·4029	·4088
28	−1·1132	−1·1048	−1·0964	−1·0881	−1·0798	78	·4148	·4209	·4269	·4329	·4390
29	−1·0715	−1·0633	−1·0552	−1·0470	−1·0390	79	·4451	·4512	·4573	·4635	·4697
30	−1·0309	−1·0229	−1·0150	−1·0071	−·9992	80	·4759	·4821	·4884	·4946	·5009
31	−·9914	−·9836	−·9758	−·9681	−·9604	81	·5073	·5136	·5200	·5264	·5328
32	−·9528	−·9452	−·9376	−·9301	−·9226	82	·5393	·5458	·5523	·5589	·5655
33	−·9151	−·9077	−·9003	−·8929	−·8855	83	·5721	·5787	·5854	·5922	·5989
34	−·8782	−·8710	−·8637	−·8565	−·8493	84	·6057	·6126	·6194	·6264	·6333
35	−·8422	−·8350	−·8279	−·8209	−·8138	85	·6403	·6474	·6545	·6616	·6688
36	−·8068	−·7998	−·7928	−·7859	−·7790	86	·6761	·6834	·6907	·6981	·7055
37	−·7721	−·7653	−·7584	−·7516	−·7448	87	·7131	·7206	·7283	·7360	·7437
38	−·7381	−·7313	−·7246	−·7179	−·7113	88	·7515	·7594	·7674	·7754	·7836
39	−·7046	−·6980	−·6914	−·6848	−·6783	89	·7918	·8000	·8084	·8169	·8254
40	−·6717	−·6652	−·6587	−·6522	−·6458	90	·8340	·8428	·8516	·8605	·8696
41	−·6394	−·6329	−·6265	−·6202	−·6138	91	·8788	·8881	·8975	·9070	·9167
42	−·6075	−·6011	−·5948	−·5886	−·5823	92	·9265	·9365	·9466	·9569	·9674
43	−·5760	−·5698	−·5636	−·5574	−·5512	93	·9780	·9889	·9999	1·0112	1·0227
44	−·5450	−·5389	−·5328	−·5266	−·5205	94	1·0344	1·0464	1·0586	1·0712	1·0840
45	−·5144	−·5084	−·5023	−·4963	−·4902	95	1·0972	1·1107	1·1246	1·1390	1·1538
46	−·4842	−·4782	−·4722	−·4662	−·4603	96	1·1690	1·1848	1·2012	1·2183	1·2361
47	−·4543	−·4484	−·4425	−·4365	−·4306	97	1·2546	1·2741	1·2946	1·3163	1·3394
48	−·4248	−·4189	−·4130	−·4072	−·4013	98	1·3641	1·3906	1·4195	1·4513	1·4868
49	−·3955	−·3897	−·3839	−·3781	−·3723	99	1·5272	1·5745	1·6324	1·7086	1·8269

The complementary loglog transformation is $v = \log_e\{-\log_e(1-p)\}$. If the direct loglog transformation $u = \log_e(-\log_e p)$ is required, enter the table with $(1-p)$. For small values of p, $v = \log_e p + \frac{1}{2}p$ approximately.

TABLE XII1. COMPLEMENTARY LOGLOG TRANSFORMATION: WORKING VALUES

Expected value	Working values			Weighting coefficient	Expected value	Working values			Weighting coefficient
	Minimum	Range	Maximum			Minimum	Range	Maximum	
−7·5	−8·5003	1809·0	1800·5	·0005529	−2·5	−3·5422	13·2247	9·6825	·07876
−7·4	−8·4003	1637·0	1628·6	·0006111	−2·4	−3·4468	12·0699	8·6231	·08667
−7·3	−8·3003	1481·3	1473·0	·0006753	−2·3	−3·3518	11·0260	7·6742	·09532
−7·2	−8·2004	1340·0	1331·8	·0007465	−2·2	−3·2575	10·0825	6·8250	·10478
−7·1	−8·1004	1213·0	1204·9	·0008248	−2·1	−3·1638	9·2300	6·0662	·11511
−7·0	−8·0005	1097·6	1089·6	·0009115	−2·0	−3·0708	8·4599	5·3891	·12638
−6·9	−7·9005	993·28	985·38	·001007	−1·9	−2·9787	7·7646	4·7859	·13866
−6·8	−7·8006	898·85	891·05	·001113	−1·8	−2·8874	7·1370	4·2496	·15201
−6·7	−7·7006	813·41	805·71	·001230	−1·7	−2·7972	6·5711	3·7739	·16650
−6·6	−7·6007	736·10	728·50	·001359	−1·6	−2·7081	6·0611	3·3530	·18220
−6·5	−7·5008	666·14	658·64	·001502	−1·5	−2·6203	5·6020	2·9817	·19916
−6·4	−7·4008	602·85	595·45	·001660	−1·4	−2·5341	5·1893	2·6552	·21744
−6·3	−7·3009	545·57	538·27	·001835	−1·3	−2·4495	4·8188	2·3693	·23708
−6·2	−7·2010	493·75	486·55	·002027	−1·2	−2·3669	4·4870	2·1201	·25811
−6·1	−7·1011	446·86	439·76	·002240	−1·1	−2·2865	4·1907	1·9042	·28054
−6·0	−7·0012	404·43	397·43	·002476	−1·0	−2·2087	3·9270	1·7183	·30435
−5·9	−6·9014	366·04	359·14	·002736	−0·9	−2·1339	3·6935	1·5596	·32951
−5·8	−6·8015	331·30	324·50	·003023	−0·8	−2·0625	3·4880	1·4255	·35592
−5·7	−6·7017	299·87	293·17	·003340	−0·7	−1·9950	3·3088	1·3138	·38345
−5·6	−6·6018	271·43	264·83	·003691	−0·6	−1·9323	3·1544	1·2221	·41192
−5·5	−6·5020	245·69	239·19	·004079	−0·5	−1·8751	3·0238	1·1487	·44107
−5·4	−6·4023	222·41	216·01	·004506	−0·4	−1·8425	2·9163	1·0918	·47057
−5·3	−6·3025	201·34	195·04	·004977	−0·3	−1·7817	2·8316	1·0499	·49999
−5·2	−6·2028	182·27	176·07	·005501	−0·2	−1·7483	2·7697	1·0214	·52880
−5·1	−6·1031	165·02	158·92	·006078	−0·1	−1·7263	2·7315	1·0052	·55638
−5·0	−6·0034	149·41	143·41	·006715	0·0	−1·7183	2·7183	1·0000	·58198
−4·9	−5·9037	135·29	129·39	·007419	0·1	−1·7275	2·7323	1·0048	·60473
−4·8	−5·8041	122·51	116·71	·008196	0·2	−1·7584	2·7771	1·0187	·62369
−4·7	−5·7046	110·95	105·25	·009053	0·3	−1·8164	2·8572	1·0408	·63780
−4·6	−5·6050	100·489	94·884	·01000	0·4	−1·9094	2·9797	1·0703	·64598
−4·5	−5·5056	91·023	85·517	·01105	0·5	−2·0476	3·1541	1·1065	·64716
−4·4	−5·4062	82·457	77·051	·01220	0·6	−2·2456	3·3944	1·1488	·64034
−4·3	−5·3068	74·707	69·400	·01348	0·7	−2·5235	3·7201	1·1966	·62471
−4·2	−5·2075	67·694	62·486	·01488	0·8	−2·9108	4·1601	1·2493	·59975
−4·1	−5·1083	61·348	56·240	·01644	0·9	−3·4097	4·7163	1·3066	·57071
−4·0	−5·0092	55·607	50·598	·01815	1·0	−4·2071	5·5750	1·3679	·52204
−3·9	−4·9102	50·412	45·502	·02004	1·1	−5·2809	6·7138	1·4329	·47080
−3·8	−4·8113	45·712	40·901	·02212	1·2	−6·8309	8·3321	1·5012	·41342
−3·7	−4·7125	41·459	36·747	·02442	1·3	−9·1174	10·6899	1·5725	·35223
−3·6	−4·6138	37·612	32·998	·02695	1·4	−12·581	14·228	1·6466	·29005
−3·5	−4·5153	34·130	29·615	·02974	1·5	−17·998	19·721	1·7231	·22985
−3·4	−4·4169	30·981	26·564	·03282	1·6	−26·787	28·589	1·8019	·17448
−3·3	−4·3187	28·132	23·813	·03621	1·7	−41·669	43·552	1·8827	·12622
−3·2	−4·2207	25·554	21·333	·03994	1·8	−68·115	70·080	1·9653	·08653
−3·1	−4·1229	23·221	19·098	·04404	1·9	−117·76	119·81	2·0496	·05587
−3·0	−4·0253	21·111	17·086	·04856	2·0	−216·86	219·00	2·1353	·03376
−2·9	−3·9280	19·202	15·274	·05352	2·1	−428·81	431·03	2·2225	·01895
−2·8	−3·8310	17·476	13·645	·05898	2·2	−918·28	920·59	2·3108	·009805
−2·7	−3·7344	15·914	12·180	·06497	2·3	−2149·7	2152·1	2·4003	·004635
−2·6	−3·6381	14·502	10·864	·07155	2·4	−5556·5	5559·0	2·4907	·001983

The working value is given by q Min. Value+p Max. Value, Min. Value+p Range, or Max. Value−q Range. For the direct loglog transformation interchange p and q in these formulæ.

Table XIII. Scores for Linkage Data from Intercrosses

Recombination fraction. y per cent.	Coupling			Repulsion		
	Double dominant +	Single recessives −	Double recessive +	Double dominant −	Single recessives +	Double recessive −
1	·6644072	99·4974874	2·0202020	·0099995	·0200020	200·0000000
2	·6620727	49·4949495	2·0408163	·0199960	·0400160	100·0000000
3	·6596620	32·8257191	2·0618557	·0299865	·0600540	66·6666667
4	·6571742	24·4897959	2·0833333	·0399680	·0801282	50·0000000
5	·6546081	19·4871795	2·1052632	·0499376	·1002506	40·0000000
6	·6519628	16·1512027	2·1276596	·0598922	·1204336	33·3333333
7	·6492373	13·7675796	2·1505376	·0698289	·1406894	28·5714286
8	·6464306	11·9791667	2·1739130	·0797448	·1610306	25·0000000
9	·6435416	10·5875509	2·1978022	·0896370	·1814699	22·2222222
10	·6405694	9·4736842	2·2222222	·0995025	·2020202	20·0000000
11	·6375130	8·5618086	2·2471910	·1093385	·2226946	18·1818182
12	·6343714	7·8014184	2·2727273	·1191422	·2435065	16·6666667
13	·6311437	7·1575483	2·2988506	·1289107	·2644695	15·3846154
14	·6278289	6·6052227	2·3255814	·1386413	·2855977	14·2857143
15	·6244261	6·1261261	2·3529412	·1483313	·3069054	13·3333333
16	·6209344	5·7065217	2·3809524	·1579779	·3284072	12·5000000
17	·6173528	5·3359049	2·4096386	·1675785	·3501184	11·7647059
18	·6136806	5·0061050	2·4390244	·1771305	·3720546	11·1111111
19	·6099168	4·7106717	2·4691358	·1866313	·3942318	10·5263158
20	·6060606	4·4444444	2·5000000	·1960784	·4166667	10·0000000
21	·6021112	4·2032455	2·5316456	·2054694	·4393765	9·5238095
22	·5980678	3·9836568	2·5641026	·2148018	·4623792	9·0909091
23	·5939296	3·7828543	2·5974026	·2240733	·4856932	8·6956522
24	·5896958	3·5984848	2·6315789	·2332815	·5093379	8·3333333
25	·5853659	3·4285714	2·6666667	·2424242	·5333333	8·0000000
26	·5809389	3·2714412	2·7027027	·2514993	·5577006	7·6923077
27	·5764144	3·1256690	2·7397260	·2605046	·5824614	7·4074074
28	·5717916	2·9900332	2·7777778	·2694380	·6076389	7·1428571
29	·5670700	2·8634805	2·8169014	·2782976	·6332569	6·8965517
30	·5622490	2·7450980	2·8571429	·2870813	·6593407	6·6666667
31	·5573281	2·6340905	2·8985507	·2957874	·6859166	6·4516129
32	·5523067	2·5297619	2·9411765	·3044140	·7130125	6·2500000
33	·5471845	2·4315006	2·9850746	·3129594	·7406576	6·0606061
34	·5419609	2·3387668	3·0303030	·3214218	·7688829	5·8823529
35	·5366357	2·2510823	3·0769231	·3297998	·7977208	5·7142857
36	·5312085	2·1680217	3·1250000	·3380917	·8272059	5·5555556
37	·5256790	2·0892058	3·1746032	·3462960	·8573746	5·4054054
38	·5200470	2·0142950	3·2258065	·3544115	·8882655	5·2631579
39	·5143122	1·9429846	3·2786885	·3624367	·9199198	5·1282051
40	·5084746	1·8750000	3·3333333	·3703704	·9523810	5·0000000
41	·5025340	1·8100936	3·3898305	·3782113	·9856954	4·8780488
42	·4964903	1·7480410	3·4482759	·3859585	1·0199126	4·7619048
43	·4903437	1·6886387	3·5087719	·3936107	1·0550853	4·6511628
44	·4840941	1·6317016	3·5714286	·4011670	1·0912698	4·5454545
45	·4777416	1·5770609	3·6363636	·4086266	1·1285266	4·4444444
46	·4712864	1·5245624	3·7037037	·4159884	1·1669203	4·3478261
47	·4647288	1·4740648	3·7735849	·4232518	1·2065203	4·2553191
48	·4580691	1·4254386	3·8461538	·4304161	1·2474012	4·1666667
49	·4513075	1·3785647	3·9215686	·4374805	1·2896434	4·0816327
50	·4444444	1·3333333	4·0000000	·4444444	1·3333333	4·0000000

TABLE XIII1. PRODUCT RATIOS FOR DIFFERENT RECOMBINATION FRACTIONS

Recombination per cent.	Coupling		Repulsion	
	bc/ad	$\sigma_p \sqrt{h}(\%)$	*ad/bc*	$\sigma_p \sqrt{h}(\%)$
1	·0^{3}13558	0·9917	·0^{3}20005	0·9998
2	·0^{3}55155	1·9667	·0^{3}80080	1·9980
3	·0^{2}12623	2·9251	·0^{2}18041	2·9933
4	·0^{2}22828	3·8670	·0^{2}32128	3·9840
5	·0^{2}36290	4·7922	·0^{2}50314	4·9689
6	·0^{2}53176	5·7009	·0^{2}72652	5·9463
7	·0^{2}73661	6·5931	·0^{2}99210	6·9150
8	·0^{2}97929	7·4686	·013007	7·8735
9	·012618	8·3276	·016532	8·8204
10	·015860	9·1700	·020508	9·7544
11	·019543	9·9958	·024946	10·6743
12	·023689	10·8050	·029861	11·5789
13	·028321	11·5975	·035268	12·4669
14	·033466	12·3733	·041183	13·3373
15	·039149	13·1323	·047625	14·1890
16	·045400	13·8745	·054616	15·0209
17	·052248	14·5998	·062177	15·8322
18	·059725	15·3082	·070334	16·6219
19	·067866	15·9995	·079112	17·3892
20	·076705	16·6737	·088542	18·1333
21	·086280	17·3307	·098654	18·8536
22	·096632	17·9703	·10948	19·5494
23	·10780	18·5925	·12107	20·2202
24	·11984	19·1971	·13345	20·8653
25	·13279	19·7840	·14667	21·4844
26	·14671	20·3531	·16077	22·0770
27	·16164	20·9041	·17581	22·6428
28	·17766	21·4371	·19185	23·1815
29	·19481	21·9516	·20894	23·6929
30	·21318	22·4477	·22715	24·1767
31	·23283	22·9251	·24654	24·6329
32	·25383	23·3836	·26721	25·0613
33	·27627	23·8230	·28922	25·4619
34	·30025	24·2430	·31268	25·8347
35	·32585	24·6435	·33767	26·1796
36	·35317	25·0242	·36431	26·4969
37	·38234	25·3849	·39270	26·7865
38	·41346	25·7253	·42299	27·0486
39	·44667	26·0452	·45530	27·2833
40	·48211	26·3442	·48980	27·4909
41	·51993	26·6221	·52663	27·6715
42	·56029	26·8787	·56598	27·8254
43	·60337	27·1135	·60806	27·9527
44	·64937	27·3264	·65307	28·0538
45	·69850	27·5170	·70126	28·1289
46	·75098	27·6851	·75289	28·1783
47	·80709	27·8302	·80824	28·2023
48	·86708	27·9521	·86763	28·2012
49	·93128	28·0505	·93142	28·1754
50	1·00000	28·1250	1·00000	28·1250

Table XIV. Segmental Functions

x	$\alpha(x)$	$\beta(x)$	$\gamma(x)$	$\delta(x)$	cosh x $\alpha(x)+\gamma(x)$	sinh x $\beta(x)+\delta(x)$
·00	1·00000 00	·00000 00	·00000 00	·00000 00	1·00000 00	·00000 00
·02	1·00000 00	·02000 00	·00020 00	·00000 13	1·00020 00	·02000 13
·04	1·00000 01	·04000 00	·00080 00	·00001 07	1·00080 01	·04001 07
·06	1·00000 05	·06000 00	·00180 00	·00003 60	1·00180 05	·06003 60
·08	1·00000 17	·08000 00	·00320 00	·00008 53	1·00320 17	·08008 53
·10	1·00000 42	·10000 01	·00500 00	·00016 67	1·00500 42	·10016 68
·12	1·00000 86	·12000 02	·00720 00	·00028 80	1·00720 86	·12028 82
·14	1·00001 60	·14000 04	·00980 00	·00045 73	1·00981 60	·14045 77
·16	1·00002 73	·16000 09	·01280 00	·00068 27	1·01282 73	·16068 36
·18	1·00004 37	·18000 16	·01620 00	·00097 20	1·01624 37	·18097 36
·20	1·00006 67	·20000 27	·02000 01	·00133 33	1·02006 68	·20133 60
·22	1·00009 76	·22000 43	·02420 02	·00177 47	1·02429 78	·22177 90
·24	1·00013 82	·24000 66	·02880 03	·00230 40	1·02893 85	·24231 06
·26	1·00019 04	·26000 99	·03380 04	·00292 93	1·03399 08	·26293 92
·28	1·00025 61	·28001 43	·03920 07	·00365 87	1·03945 68	·28367 30
·30	1·00033 75	·30002 03	·04500 10	·00450 00	1·04533 85	·30452 03
·32	1·00043 69	·32002 80	·05120 15	·00546 14	1·05163 84	·32548 94
·34	1·00055 68	·34003 79	·05780 21	·00655 08	1·05835 89	·34658 87
·36	1·00069 98	·36005 04	·06480 30	·00777 62	1·06550 28	·36782 66
·38	1·00086 88	·38006 60	·07220 42	·00914 56	1·07307 30	·38921 16
·40	1·00106 67	·40008 53	·08000 57	·01066 70	1·08107 24	·41075 23
·42	1·00129 66	·42010 89	·08820 76	·01234 85	1·08950 42	·43245 74
·44	1·00156 17	·44013 74	·09681 01	·01419 80	1·09837 18	·45433 54
·46	1·00186 57	·46017 16	·10581 32	·01622 35	1·10767 89	·47639 51
·48	1·00221 19	·48021 23	·11521 70	·01843 32	1·11742 89	·49864 55
·50	1·00260 43	·50026 04	·12502 17	·02083 49	1·12762 60	·52109 53
·52	1·00304 66	·52031 68	·13522 75	·02343 67	1·13827 41	·54375 35
·54	1·00354 31	·54038 26	·14583 44	·02624 67	1·14937 75	·56662 93
·56	1·00409 79	·56045 90	·15684 28	·02927 28	1·16094 07	·58973 18
·58	1·00471 55	·58054 70	·16825 29	·03252 30	1·17296 84	·61307 00
·60	1·00540 04	·60064 80	·18006 48	·03600 56	1·18546 52	·63665 36
·62	1·00615 73	·62076 35	·19227 89	·03972 83	1·19843 62	·66049 18
·64	1·00699 12	·64089 48	·20489 54	·04369 94	1·21188 66	·68459 42
·66	1·00790 70	·66104 37	·21791 48	·04792 68	1·22582 18	·70897 05
·68	1·00891 00	·68121 17	·23133 73	·05241 87	1·24024 73	·73363 04
·70	1·01000 56	·70140 07	·24516 34	·05718 30	1·25516 90	·75858 37
·72	1·01119 92	·72161 26	·25939 35	·06222 79	1·27059 27	·78384 05
·74	1·01249 66	·74184 94	·27402 81	·06756 14	1·28652 47	·80941 08
·76	1·01390 37	·76211 32	·28906 77	·07319 17	1·30297 14	·83530 49
·78	1·01542 63	·78240 63	·30451 28	·07912 69	1·31993 91	·86153 32
·80	1·01707 08	·80273 10	·32036 41	·08537 49	1·33743 49	·88810 59
·82	1·01884 35	·82309 00	·33662 23	·09194 41	1·35546 58	·91503 41
·84	1·02075 08	·84348 57	·35328 80	·09884 26	1·37403 88	·94232 83
·86	1·02279 94	·86392 09	·37036 20	·10607 84	1·39316 14	·96999 93
·88	1·02499 62	·88439 86	·38784 51	·11365 98	1·41284 13	·99805 84
·90	1·02734 82	·90492 18	·40573 82	·12159 49	1·43308 64	1·02651 67
·92	1·02986 24	·92549 36	·42404 23	·12989 20	1·45390 47	1·05538 56
·94	1·03254 63	·94611 74	·44275 83	·13855 93	1·47530 46	1·08467 67
·96	1·03540 73	·96679 67	·46188 73	·14760 51	1·49729 46	1·11440 18
·98	1·03845 31	·98753 50	·48143 06	·15703 76	1·51988 37	1·14457 26
1·00	1·04169 15	1·00833 61	·50138 92	·16686 51	1·54308 07	1·17520 12

x	$\alpha(x)$	$\beta(x)$	$\gamma(x)$	$\delta(x)$	$\cosh x$ $\alpha(x)+\gamma(x)$	$\sinh x$ $\beta(x)+\delta(x)$
1·00	1·04169 15	1·00833 61	·50138 92	·16686 51	1·54308 07	1·17520 12
1·02	1·04513 04	1·02920 40	·52176 45	·17709 59	1·56689 49	1·20629 99
1·04	1·04877 81	1·05014 27	·54255 78	·18773 85	1·59133 59	1·23788 12
1·06	1·05264 27	1·07115 65	·56377 07	·19880 11	1·61641 34	1·26995 76
1·08	1·05673 30	1·09224 99	·58540 46	·21029 21	1·64213 76	1·30254 20
1·10	1·06105 73	1·11342 74	·60746 12	·22222 01	1·66851 85	1·33564 75
1·12	1·06562 47	1·13469 38	·62994 23	·23459 34	1·69556 70	1·36928 72
1·14	1·07044 41	1·15605 41	·65284 96	·24742 06	1·72329 37	1·40347 47
1·16	1·07552 46	1·17751 33	·67618 51	·26071 02	1·75170 97	1·43822 35
1·18	1·08087 56	1·19907 69	·69995 08	·27447 09	1·78082 64	1·47354 78
1·20	1·08650 67	1·22075 02	·72414 89	·28871 11	1·81065 56	1·50946 13
1·22	1·09242 73	1·24253 91	·74878 16	·30343 97	1·84120 89	1·54597 88
1·24	1·09864 76	1·26444 93	·77385 13	·31866 53	1·87249 89	1·58311 46
1·26	1·10517 73	1·28648 70	·79936 04	·33439 67	1·90453 77	1·62088 37
1·28	1·11202 69	1·30865 85	·82531 17	·35064 27	1·93733 86	1·65930 12
1·30	1·11920 65	1·33097 03	·85170 77	·36741 21	1·97091 42	1·69838 24
1·32	1·12672 69	1·35342 91	·87855 14	·38471 40	2·00527 83	1·73814 31
1·34	1·13459 87	1·37604 17	·90584 59	·40255 72	2·04044 46	1·77859 89
1·36	1·14283 29	1·39881 54	·93359 42	·42095 08	2·07642 71	1·81976 62
1·38	1·15144 05	1·42175 75	·96179 96	·43990 40	2·11324 01	1·86166 15
1·40	1·16043 28	1·44487 56	·99046 57	·45942 59	2·15089 85	1·90430 15
1·42	1·16982 13	1·46817 75	1·01959 59	·47952 57	2·18941 72	1·94770 32
1·44	1·17961 77	1·49167 12	1·04919 40	·50021 28	2·22881 17	1·99188 40
1·46	1·18983 38	1·51536 50	1·07926 41	·52149 66	2·26909 79	2·03686 16
1·48	1·20048 17	1·53926 74	1·10981 00	·54338 66	2·31029 17	2·08265 40
1·50	1·21157 34	1·56338 72	1·14083 62	·56589 22	2·35240 96	2·12927 94
1·52	1·22312 15	1·58773 34	1·17234 71	·58902 33	2·39546 86	2·17675 67
1·54	1·23513 86	1·61231 52	1·20434 71	·61278 94	2·43948 57	2·22510 46
1·56	1·24763 74	1·63714 22	1·23684 13	·63720 04	2·48447 87	2·27434 26
1·58	1·26063 10	1·66222 40	1·26983 45	·66226 64	2·53046 55	2·32449 04
1·60	1·27413 25	1·68757 08	1·30333 20	·68799 72	2·57746 45	2·37556 80
1·62	1·28815 53	1·71319 28	1·33733 92	·71440 30	2·62549 45	2·42759 58
1·64	1·30271 32	1·73910 06	1·37186 16	·74149 42	2·67457 48	2·48059 48
1·66	1·31781 97	1·76530 50	1·40690 52	·76928 10	2·72472 49	2·53458 60
1·68	1·33348 91	1·79181 71	1·44247 59	·79777 39	2·77596 50	2·58959 10
1·70	1·34973 55	1·81864 84	1·47858 00	·82698 36	2·82831 55	2·64563 20
1·72	1·36657 33	1·84581 05	1·51522 40	·85692 07	2·88179 73	2·70273 12
1·74	1·38401 72	1·87331 53	1·55241 47	·88759 62	2·93643 19	2·76091 15
1·76	1·40208 21	1·90117 53	1·59015 90	·91902 10	2·99224 11	2·82019 63
1·78	1·42078 31	1·92940 29	1·62846 41	·95120 63	3·04924 72	2·88060 92
1·80	1·44013 55	1·95801 10	1·66733 76	·98416 33	3·10747 31	2·94217 43
1·82	1·46015 49	1·98701 27	1·70678 72	1·01790 36	3·16694 21	3·00491 63
1·84	1·48085 70	2·01642 17	1·74682 09	1·05243 87	3·22767 79	3·06886 04
1·86	1·50225 78	2·04625 17	1·78744 69	1·08778 04	3·28970 47	3·13403 21
1·88	1·52437 37	2·07651 68	1·82867 38	1·12394 06	3·35304 75	3·20045 74
1·90	1·54722 10	2·10723 15	1·87051 05	1·16093 14	3·41773 15	3·26816 29
1·92	1·57081 65	2·13841 06	1·91296 62	1·19876 51	3·48378 27	3·33717 57
1·94	1·59517 73	2·17006 93	1·95605 02	1·23745 42	3·55122 75	3·40752 35
1·96	1·62032 05	2·20222 29	1·99977 23	1·27701 14	3·62009 28	3·47923 43
1·98	1·64626 36	2·23488 74	2·04414 25	1·31744 95	3·69040 61	3·55233 69
2·00	1·67302 44	2·26807 89	2·08917 13	1·35878 15	3·76219 57	3·62686 04

Table XV. Latin Squares

The 4×4 Latin Squares

First Transformation Set: 3 Self-Conjugate Standard Squares			Second Transformation Set: 1 Self-Conjugate Standard Square
A B C D B A D C C D B A D C A B	A B C D B C D A C D A B D A B C	A B C D B D A C C A D B D C B A	A B C D B A D C C D A B D C B A
1	2	3	4

The 5×5 Latin Squares

First Transformation Set: 25 Standard Squares and their Conjugates

A B C D E B A E C D C D A E B D E B A C E C D B A	A B C D E B A D E C C E B A D D C E B A E D A C B	A B C D E B A E C D C E D A B D C B E A E D A B C	A B C D E B A D E C C D E A B D E B C A E C A B D	A B C D E B C D E A C E B A D D A E B C E D A C B
1, 2	3, 4	5, 6	7, 8	9, 10
A B C D E B C D E A C E A B D D A E C B E D B A C	A B C D E B C E A D C D B E A D E A C B E A D B C	A B C D E B C E A D C A D E B D E B C A E D A B C	A B C D E B C D E A C A E B D D E A C B E D B A C	A B C D E B C A E D C E D A B D A E B C E D B C A
11, 12	13, 14	15, 16	17, 18	19, 20
A B C D E B C A E D C D E B A D E B A C E A D C B	A B C D E B D E C A C A B E D D E A B C E C D A B	A B C D E B D A E C C E D B A D C E A B E A B C D	A B C D E B D E C A C A D E B D E A B C E C B A D	A B C D E B D A E C C E D B A D A E C B E C B A D
21, 22	23, 24	25, 26	27, 28	29, 30
A B C D E B D E A C C E D B A D C A E B E A B C D	A B C D E B D A E C C E B A D D A E C B E C D B A	A B C D E B D E A C C E A B D D C B E A E A D C B	A B C D E B E D A C C A B E D D C E B A E D A C B	A B C D E B E D A C C A E B D D C A E B E D B C A
31, 32	33, 34	35, 36	37, 38	39, 40
A B C D E B E A C D C D E B A D A B E C E C D A B	A B C D E B E D C A C A E B D D C A E B E D B A C	A B C D E B E D C A C D E A B D A B E C E C A B D	A B C D E B E A C D C D B E A D C E A B E A D B C	A B C D E B E D A C C D A E B D C E B A E A B C D
41, 42	43, 44	45, 46	47, 48	49, 50

Second Transformation Set: 6 Self-Conjugate Standard Squares

A B C D E B C E A D C E D B A D A B E C E D A C B	A B C D E B C D E A C D E A B D E A B C E A B C D	A B C D E B D E C A C E B A D D C A E B E A D B C	A B C D E B D A E C C A E B D D E B C A E C D A B	A B C D E B E D A C C D B E A D A E C B E C A B D	A B C D E B E A C D C A D E B D C E B A E D B A C
51	52	53	54	55	56

When constructing squares for use in experimental arrangements use the randomization procedure described in the Introduction.

The 6×6 Latin Squares

Five conjugate pairs of transformation sets, and twelve sets containing conjugates. The numbers above the squares indicate the number in each set, *s* standing for self-conjugate and *c* for conjugate. Each pair of conjugate sets can be turned into the following transformation set by interchanging rows or columns with letters.

I	II	III	IV	V
1080, 1080	180*s*+450+450*c*	540+540*c*	540+540*c*	540, 540
A B C D E F	A B C D E F	A B C D E F	A B C D E F	A B C D E F
B C F A D E	B C F E A D	B A F E C D	B A E F C D	B A E C F D
C F B E A D	C F B A D E	C F B A D E	C F B A D E	C F B A D E
D E A B F C	D E A B F C	D C E B F A	D E A B F C	D E F B C A
E A D F C B	E A D F C B	E D A F B C	E D F C B A	E D A F B C
F D E C B A	F D E C B A	F E D C A B	F C D E A B	F C D E A B
0001-1080	2161-3240	3241-4320	4321-5400	5401-5940
1081-2160				5941-6480

VI	VII	VIII	IX	X
90*s*+225+225*c*	270+270*c*	360, 360	180+180*c*	60*s*+60+60*c*
A B C D E F	A B C D E F	A B C D E F	A B C D E F	A B C D E F
B A F E C D	B C D E F A	B A E F C D	B A E F C D	B C F A D E
C F B A D E	C E A F B D	C F A E D B	C F A B D E	C F B E A D
D E A B F C	D F B A C E	D C B A F E	D E B A F C	D A E B F C
E C D F B A	E D F B A C	E D F C B A	E D F C B A	E D A F C B
F D E C A B	F A E C D B	F E D B A C	F C D E A B	F E D C B A
6481-7020	7021-7560	7561-7920	8281-8640	8641-8820
		7921-8280		

XI	XII	XIII	XIV	XV
120, 120	60*s*+30+30*c*	60*s*	20+20*c*	36, 36
A B C D E F	A B C D E F	A B C D E F	A B C D E F	A B C D E F
B C A F D E	B C A E F D	B C A F D E	B C A E F D	B A F E D C
C A B E F D	C A B F D E	C A B E F D	C A B F D E	C D A B F E
D F E B A C	D E F B A C	D F E B A C	D F E B A C	D F E A C B
E D F C B A	E F D A C B	E D F A C B	E D F C B A	E C B F A D
F E D A C B	F D E C B A	F E D C B A	F E D A C B	F E D C B A
8821-8940	9061-9180	9181-9240	9241-9280	9281-9316
8941-9060				9317-9352

XVI	XVII
6*s*+15+15*c*	10+10*c*
A B C D E F	A B C D E F
B A E C F D	B C A F D E
C E A F D B	C A B E F D
D C F A B E	D E F A B C
E F D B A C	E F D C A B
F D B E C A	F D E B C A
9353-9388	9389-9408

Examples of 7×7 Latin Squares

(*a*)	(*b*)	(*c*)	(*d*)
A B C D E F G	A B C D E F G	$A_1\ B_2\ C_3\ D_4\ E_5\ F_6\ G_7$	$A_1\ B_2\ C_3\ D_4\ E_5\ F_6\ G_7$
B E A G F D C	B E A G F D C	$B_3\ F_7\ E_6\ G_1\ C_4\ A_2\ D_5$	$B_3\ C_7\ D_6\ E_1\ F_4\ G_2\ A_5$
C F G B D A E	C F G B D A E	$C_7\ D_1\ A_5\ E_2\ B_6\ G_4\ F_3$	$C_6\ D_5\ E_2\ F_3\ G_1\ A_7\ B_4$
D G E F C B A	D G E F B C A	$D_6\ C_5\ G_2\ A_3\ F_1\ E_7\ B_4$	$D_2\ E_4\ F_7\ G_6\ A_3\ B_5\ C_1$
E D B C A G F	E D B C A G F	$E_4\ G_6\ B_1\ F_5\ A_7\ D_3\ C_2$	$E_7\ F_1\ G_5\ A_2\ B_6\ C_4\ D_3$
F C D A G E B	F C D A G E B	$F_2\ A_4\ D_7\ C_6\ G_3\ B_5\ E_1$	$F_5\ G_3\ A_4\ B_7\ C_2\ D_1\ E_6$
G A F E B C D	G A F E C B D	$G_5\ E_3\ F_4\ B_7\ D_2\ C_1\ A_6$	$G_4\ A_6\ B_1\ C_5\ D_7\ E_3\ F_2$

When constructing squares for use in experimental arrangements use the randomization procedure described in the Introduction.

Table XV. Latin Squares—*continued*

Examples of Squares from 8×8 to 12×12

8×8

A	B	C	D	E	F	G	H
B	C	A	E	F	D	H	G
C	A	D	G	H	E	F	B
D	F	G	C	A	H	B	E
E	H	B	F	G	C	A	D
F	D	H	A	B	G	E	C
G	E	F	H	C	B	D	A
H	G	E	B	D	A	C	F

9×9

A	B	C	D	E	F	G	H	I
B	C	E	G	D	I	F	A	H
C	D	F	A	H	G	I	E	B
D	H	A	B	F	E	C	I	G
E	G	B	I	C	H	D	F	A
F	I	H	E	B	D	A	G	C
G	F	I	C	A	B	H	D	E
H	E	G	F	I	A	B	C	D
I	A	D	H	G	C	E	B	F

10×10

A	B	C	D	E	F	G	H	I	J
B	G	A	E	H	C	F	I	J	D
C	H	J	G	F	B	E	A	D	I
D	A	G	I	J	E	C	B	F	H
E	F	H	J	I	G	A	D	B	C
F	E	B	C	D	I	J	G	H	A
G	I	F	B	A	D	H	J	C	E
H	C	I	F	G	J	D	E	A	B
I	J	D	A	C	H	B	F	E	G
J	D	E	H	B	A	I	C	G	F

11×11

A	B	C	D	E	F	G	H	I	J	K
B	A	J	I	D	C	F	K	H	G	E
C	K	H	A	B	I	J	F	D	E	G
D	C	G	J	I	K	E	B	F	A	H
E	J	B	G	K	H	D	C	A	I	F
F	E	I	C	G	A	K	J	B	H	D
G	F	D	B	H	J	A	I	E	K	C
H	I	K	F	A	D	B	E	G	C	J
I	D	E	H	J	B	C	G	K	F	A
J	G	A	K	F	E	H	D	C	B	I
K	H	F	E	C	G	I	A	J	D	B

12×12

A	B	C	D	E	F	G	H	I	J	K	L
B	L	G	C	D	J	K	E	H	A	F	I
C	K	A	B	F	L	I	D	G	H	J	E
D	F	I	A	L	E	C	G	J	B	H	K
E	D	F	G	J	K	A	L	C	I	B	H
F	H	K	E	G	C	D	B	A	L	I	J
G	I	D	F	K	H	J	A	L	C	E	B
H	E	L	J	C	A	B	I	K	D	G	F
I	J	B	L	H	G	F	K	D	E	A	C
J	C	E	K	A	I	H	F	B	G	L	D
K	G	J	H	I	B	L	C	E	F	D	A
L	A	H	I	B	D	E	J	F	K	C	G

Table XVI. Complete Sets of Orthogonal Latin Squares

3×3

I

1	2	3
2	3	1
3	1	2

II

1	2	3
3	1	2
2	3	1

4×4

I

1	2	3	4
2	1	4	3
3	4	1	2
4	3	2	1

AC, BD

II

1	2	3	4
3	4	1	2
4	3	2	1
2	1	4	3

ACD, BC

III

1	2	3	4
4	3	2	1
2	1	4	3
3	4	1	2

AD, BCD

5×5

I

1	2	3	4	5
2	3	4	5	1
3	4	5	1	2
4	5	1	2	3
5	1	2	3	4

II

1	2	3	4	5
3	4	5	1	2
5	1	2	3	4
2	3	4	5	1
4	5	1	2	3

III

1	2	3	4	5
4	5	1	2	3
2	3	4	5	1
5	1	2	3	4
3	4	5	1	2

IV

1	2	3	4	5
5	1	2	3	4
4	5	1	2	3
3	4	5	1	2
2	3	4	5	1

7×7

I

1	2	3	4	5	6	7
2	3	4	5	6	7	1
3	4	5	6	7	1	2
4	5	6	7	1	2	3
5	6	7	1	2	3	4
6	7	1	2	3	4	5
7	1	2	3	4	5	6

II

1	2	3	4	5	6	7
3	4	5	6	7	1	2
5	6	7	1	2	3	4
7	1	2	3	4	5	6
2	3	4	5	6	7	1
4	5	6	7	1	2	3
6	7	1	2	3	4	5

III

1	2	3	4	5	6	7
4	5	6	7	1	2	3
7	1	2	3	4	5	6
3	4	5	6	7	1	2
6	7	1	2	3	4	5
2	3	4	5	6	7	1
5	6	7	1	2	3	4

IV

1	2	3	4	5	6	7
5	6	7	1	2	3	4
2	3	4	5	6	7	1
6	7	1	2	3	4	5
3	4	5	6	7	1	2
7	1	2	3	4	5	6
4	5	6	7	1	2	3

V

1	2	3	4	5	6	7
6	7	1	2	3	4	5
4	5	6	7	1	2	3
2	3	4	5	6	7	1
7	1	2	3	4	5	6
5	6	7	1	2	3	4
3	4	5	6	7	1	2

VI

1	2	3	4	5	6	7
7	1	2	3	4	5	6
6	7	1	2	3	4	5
5	6	7	1	2	3	4
4	5	6	7	1	2	3
3	4	5	6	7	1	2
2	3	4	5	6	7	1

Table XVI. Complete Sets of Orthogonal Latin Squares—*continued*

8×8

I

1	2	3	4	5	6	7	8
2	1	4	3	6	5	8	7
3	4	1	2	7	8	5	6
4	3	2	1	8	7	6	5
5	6	7	8	1	2	3	4
6	5	8	7	2	1	4	3
7	8	5	6	3	4	1	2
8	7	6	5	4	3	2	1

AD, BE, CF

II

1	2	3	4	5	6	7	8
5	6	7	8	1	2	3	4
2	1	4	3	6	5	8	7
6	5	8	7	2	1	4	3
7	8	5	6	3	4	1	2
3	4	1	2	7	8	5	6
8	7	6	5	4	3	2	1
4	3	2	1	8	7	6	5

AEF, BD, CE

III

1	2	3	4	5	6	7	8
7	8	5	6	3	4	1	2
5	6	7	8	1	2	3	4
3	4	1	2	7	8	5	6
8	7	6	5	4	3	2	1
2	1	4	3	6	5	8	7
4	3	2	1	8	7	6	5
6	5	8	7	2	1	4	3

ADE, BEF, CD

IV

1	2	3	4	5	6	7	8
8	7	6	5	4	3	2	1
7	8	5	6	3	4	1	2
2	1	4	3	6	5	8	7
4	3	2	1	8	7	6	5
5	6	7	8	1	2	3	4
6	5	8	7	2	1	4	3
3	4	1	2	7	8	5	6

ADEF, BDE, CEF

V

1	2	3	4	5	6	7	8
4	3	2	1	8	7	6	5
8	7	6	5	4	3	2	1
5	6	7	8	1	2	3	4
6	5	8	7	2	1	4	3
7	8	5	6	3	4	1	2
3	4	1	2	7	8	5	6
2	1	4	3	6	5	8	7

ADF, BDEF, CDE

VI

1	2	3	4	5	6	7	8
6	5	8	7	2	1	4	3
4	3	2	1	8	7	6	5
7	8	5	6	3	4	1	2
3	4	1	2	7	8	5	6
8	7	6	5	4	3	2	1
2	1	4	3	6	5	8	7
5	6	7	8	1	2	3	4

AF, BDF, CDEF

VII

1	2	3	4	5	6	7	8
3	4	1	2	7	8	5	6
6	5	8	7	2	1	4	3
8	7	6	5	4	3	2	1
2	1	4	3	6	5	8	7
4	3	2	1	8	7	6	5
5	6	7	8	1	2	3	4
7	8	5	6	3	4	1	2

AE, BF, CDF

9×9

I

1	2	3	4	5	6	7	8	9
2	3	1	5	6	4	8	9	7
3	1	2	6	4	5	9	7	8
4	5	6	7	8	9	1	2	3
5	6	4	8	9	7	2	3	1
6	4	5	9	7	8	3	1	2
7	8	9	1	2	3	4	5	6
8	9	7	2	3	1	5	6	4
9	7	8	3	1	2	6	4	5

AC(J), BD(J),
AC(J)×BD(J).

II

1	2	3	4	5	6	7	8	9
7	8	9	1	2	3	4	5	6
4	5	6	7	8	9	1	2	3
2	3	1	5	6	4	8	9	7
8	9	7	2	3	1	5	6	4
5	6	4	8	9	7	2	3	1
3	1	2	6	4	5	9	7	8
9	7	8	3	1	2	6	4	5
6	4	5	9	7	8	3	1	2

BC(J), AD(I),
BC(J)×AD(I).

III

1	2	3	4	5	6	7	8	9
9	7	8	3	1	2	6	4	5
5	6	4	8	9	7	2	3	1
6	4	5	9	7	8	3	1	2
2	3	1	5	6	4	8	9	7
7	8	9	1	2	3	4	5	6
8	9	7	2	3	1	5	6	4
4	5	6	7	8	9	1	2	3
3	1	2	6	4	5	9	7	8

ABC(Y), ABD(W),
BCD(Y), ACD(Z).

IV

1	2	3	4	5	6	7	8	9
8	9	7	2	3	1	5	6	4
6	4	5	9	7	8	3	1	2
9	7	8	3	1	2	6	4	5
4	5	6	7	8	9	1	2	3
2	3	1	5	6	4	8	9	7
5	6	4	8	9	7	2	3	1
3	1	2	6	4	5	9	7	8
7	8	9	1	2	3	4	5	6

ABC(X), ABD(Y),
ACD(X), BCD(Z).

V

1	2	3	4	5	6	7	8	9
3	1	2	6	4	5	9	7	8
2	3	1	5	6	4	8	9	7
7	8	9	1	2	3	4	5	6
9	7	8	3	1	2	6	4	5
8	9	7	2	3	1	5	6	4
4	5	6	7	8	9	1	2	3
6	4	5	9	7	8	3	1	2
5	6	4	8	9	7	2	3	1

AC(I), BD(I),
AC(I)×BD(I).

VI

1	2	3	4	5	6	7	8	9
4	5	6	7	8	9	1	2	3
7	8	9	1	2	3	4	5	6
3	1	2	6	4	5	9	7	8
6	4	5	9	7	8	3	1	2
9	7	8	3	1	2	6	4	5
2	3	1	5	6	4	8	9	7
5	6	4	8	9	7	2	3	1
8	9	7	2	3	1	5	6	4

BC(I), AD(J),
BC(I)×AD(J).

VII

1	2	3	4	5	6	7	8	9
5	6	4	8	9	7	2	3	1
9	7	8	3	1	2	6	4	5
8	9	7	2	3	1	5	6	4
3	1	2	6	4	5	9	7	8
4	5	6	7	8	9	1	2	3
6	4	5	9	7	8	3	1	2
7	8	9	1	2	3	4	5	6
2	3	1	5	6	4	8	9	7

ABC(Z), ABD(X),
BCD(X), ACD(W).

VIII

1	2	3	4	5	6	7	8	9
6	4	5	9	7	8	3	1	2
8	9	7	2	3	1	5	6	4
5	6	4	8	9	7	2	3	1
7	8	9	1	2	3	4	5	6
3	1	2	6	4	5	9	7	8
9	7	8	3	1	2	6	4	5
2	3	1	5	6	4	8	9	7
4	5	6	7	8	9	1	2	3

ABC(W), ABD(Z),
ACD(Y), BCD(W).

TABLE XVII. Combinatorial Solutions

1 — r 5, k 3, v 6, b 10, λ 2
Use 2, deleting any set of five varieties occurring in the same block

3 — r 6, k 4, v 10, b 15, λ 2
Use 5, deleting any set of six varieties occurring in the same block

5 — rk 6, vb 16, λ 2
Dicyclic solution
$a_1\, a_2\, a_3\, b_1\, c_4\, d_1$

7† — r 7, v 15, k 3, b 35, λ 1

abc	*ade*	*afg*	*ahi*	*ajk*	*alm*
dhl	*bhj*	*bik*	*bmo*	*bln*	*beg*
ejo	*cmn*	*clo*	*cdg*	*cef*	*cij*
fkm	*fio*	*djn*	*ekn*	*dim*	*dko*
gin	*gkl*	*ehm*	*fjl*	*gho*	*fhn*
—	—	—	—	—	
ano	*bdf*	*chk*	*eil*	*gjm*	

11 — r 8, k 4, v 9, b 18, λ 3
Cyclic solution in two families
abce
adfi

15 — r 9, v 10, k 3, b 30, λ 2

$a_1\, d_1\, b_2$	$e_2\, a_2\, b_2$	Cyclic permutation $(a\,b\,c\,d\,e)$
$b_2\, c_1\, b_1$	$a_1\, e_2\, d_1$	
$e_2\, b_2\, e_1$	$b_1\, c_1\, e_2$	

18 — r 9, k 3, v 19, b 57, λ 1
Cyclic solution in three families
agk, *bcn*, *dfi*

20 — rk 9, vb 25, λ 3

a b c d e f g h i	*n x e q l k c g w*
b h j e s p n l u	*o u l f k i p r c*
c g o m j e v p y	*p i g o t s w x h*
d m x c h j u w r	*q a b p u c x y t*
e d y u w q s o f	*r c s a o w b v n*
f q t j n m i c s	*s k d h c v y t l*
g r m l d a q s p	*t j n g a d o u k*
h t r k q b m e o	*u n h i v y r q g*
i y p w b n k d m	*v w a t i u l m e*
j l w y g r t f b	*w p k v f h a j q*
k s u x m g f b v	*x f v r p t e n d*
l v q b x o d i j	*y e i s r x j k a*
m o f n y l h a x	

23 — rk 9, λ 2, vb 37
Cyclic solution
a b d h r y z D J

26 — r 10, k 7, λ 3, v 21, b 30
Use 27, deleting any set of ten varieties occurring in the same block

27 — rk 10, vb 31, λ 3

$a_1\, a_6\, b_2\, b_5\, c_3\, c_4\, d_3\, d_5\, d_6\, e$	Cyclic permutation (1234567)
$a_2\, a_5\, b_3\, b_4\, c_1\, c_6\, d_3\, d_5\, d_6\, f$	
$a_3\, a_4\, b_1\, b_6\, c_2\, c_5\, d_3\, d_5\, d_6\, g$	
$a_1\, a_2\, a_4\, b_1\, b_2\, b_4\, c_1\, c_2\, c_4\, d_7$	
$a_1\, a_2\, a_3\, a_4\, a_5\, a_6\, a_7\, e\, f\, g$	Permutation $(a\,b\,c)$

2 — rk 5, λ 2, vb 11
Cyclic solution
aefgi

4 — r 6, k 3, v 13, b 26, λ 1
Cyclic solution in two families
aci
bfe

6† — r 7, v 8, k 4, b 14, λ 3

abcd	*abgh*	*acfh*	*adfg*
efgh	*cdef*	*bdeg*	*bceh*
—	—	—	
abef	*aceg*	*adeh*	
cdgh	*bdfh*	*bcfg*	

9 — rk 7, λ 3, vb 15
Cyclic solution
abcefik

13 — r 8, k 4, v 25, b 50, λ 1
Dicyclic solution in two families
$a_1\, a_2\, b_1\, e_5$
$a_1\, a_3\, c_1\, d_4$

16 — r 9, k 5, v 10, b 18, λ 4
Use 19, deleting any set of nine varieties occurring in the same block

17 — r 9, k 6, v 16, b 24, λ 3
Use 20, deleting any set of nine varieties occurring in the same block

19 — rk 9, λ 4, vb 19
Cyclic solution
acefghkno

21† — r 9, v 28, k 4, b 63, λ 1
Invariant variety I, and 27 others $a_1^1 \ldots a_3^3$, $b_1^1 \ldots b_3^3$, $c_1^1 \ldots c_3^3$. Blocks $a_2^1\, a_3^1\, b_1^2\, b_1^3$, $a_2^3\, b_2^2\, a_3^2\, b_3^3$, I $a_1^1\, b_1^1\, c_1^1$. Cyclic permutation $(a\,b\,c)$ generates 7 blocks in a replication; (123) in suffixes and indices generates the 9 replications.

22 — r 9, k 7, v 28, b 36, λ 2
Use 23, deleting any set of nine varieties occurring in the same block

25† — r 10, v 21, k 3, b 70, λ 1

abc	*def*	*ghi*	*jkl*	*mno*	*pqr*	*stu*
ado	*bek*	*cip*	*fqt*	*gls*	*hmr*	*jnu*
aeq	*bdn*	*cgk*	*fjs*	*hpt*	*ior*	*lmu*
afi	*bgp*	*chu*	*dqs*	*ejm*	*kot*	*lnr*
agu	*bmq*	*cjr*	*dhk*	*eps*	*flo*	*int*
ahj	*brs*	*coq*	*dlp*	*eiu*	*fkn*	*gmt*
akr	*bjt*	*cel*	*dim*	*fpu*	*gnq*	*hos*
alt	*bfh*	*cns*	*dru*	*ego*	*ijq*	*kmp*
ams	*bil*	*cdt*	*ehn*	*fgr*	*jop*	*kqu*
anp	*bou*	*cfm*	*dgj*	*ert*	*hlq*	*iks*

29 — r 10, k 5, v 41, b 82, λ 1
Cyclic solution in two families
a j p r K
a s E F I

† Divisible into groups, each containing a complete replication, No. 25 by rows, Nos. 6, 7 and 21 as indicated. (See Introduction.)
$rv = bk,\ \lambda = r(k-1)/(v-1)$.

Table XVIII. Index by Number of Replications

Replications (*r*).	Varieties (*v*).	Blocks (*b*).	Units per block (*k*).	Reference.
3	4	6	2	*u*
	4	4	3	*u*
	7	7	3	*o.s.*
4	5	10	2	*u*
	5	5	4	*u*
	7	7	4	*o.s.*
	9	12	3	*o.s.*
	13	13	4	*o.s.*
5	6	15	2	*u*
	6	10	3	1
	6	6	5	*u*
	11	11	5	2
	16	20	4	*o.s.*
	21	21	5	*o.s.*
6	5	10	3	*u*
	7	21	2	*u*
	7	7	6	*u*
	10	15	4	3
	11	11	6	2
	13	26	3	4
	16	16	6	5
	25	30	5	*o.s.*
	31	31	6	*o.s.*

Replications (*r*).	Varieties (*v*).	Blocks (*b*).	Units per block (*k*).	Reference.
7	8	28	2	*u*
	8	14	4	6
	8	8	7	*u*
	15	35	3	7
	15	21	5	8*
	15	15	7	9
	22	22	7	10*
8	9	36	2	*u*
	9	18	4	11
	9	12	6	*o.s.*
	9	9	8	*u*
	15	15	8	9
	21	28	6	12*
	25	50	4	13
	29	29	8	14*
	49	56	7	*o.s.*
	57	57	8	*o.s.*
9	10	45	2	*u*
	10	30	3	15
	10	18	5	16
	10	15	6	3
	10	10	9	*u*
	13	13	9	*o.s.*
	16	24	6	17

Replications (*r*).	Varieties (*v*).	Blocks (*b*).	Units per block (*k*).	Reference.
9	19	57	3	18
	19	19	9	19
	25	25	9	20
	28	63	4	21
	28	36	7	22
	37	37	9	23
	46	69	6	24†
	64	72	8	*o.s.*
	73	73	9	*o.s.*
10	6	15	4	*u*
	9	18	5	11
	11	55	2	*u*
	11	11	10	*u*
	16	16	10	5
	19	19	10	19
	21	70	3	25
	21	30	7	26
	31	31	10	27
	36	45	8	28*
	41	82	5	29
	46	46	10	30*
	51	85	6	31†
	81	90	9	*o.s.*
	91	91	10	*o.s.*

Table XIX. Index by Number of Units in a Block

Units per block (*k*).	Blocks (*b*).	Varieties (*v*).	Replications (*r*).	Reference.
3	4	4	3	*u*
	7	7	3	*o.s.*
	10	5	6	*u*
	10	6	5	1
	12	9	4	*o.s.*
	26	13	6	4
	30	10	9	15
	35	15	7	7
	57	19	9	18
	70	21	10	25
4	5	5	4	*u*
	7	7	4	*o.s.*
	13	13	4	*o.s.*
	14	8	7	6
	15	6	10	*u*
	15	10	6	3
	18	9	8	11
	20	16	5	*o.s.*
	50	25	8	13
	63	28	9	21
5	6	6	5	*u*
	11	11	5	2

Units per block (*k*).	Blocks (*b*).	Varieties (*v*).	Replications (*r*).	Reference.
5	18	9	10	11
	18	10	9	16
	21	15	7	8*
	21	21	5	*o.s.*
	30	25	6	*o.s.*
	82	41	10	29
6	7	7	6	*u*
	11	11	6	2
	12	9	8	*o.s.*
	15	10	9	3
	16	16	6	5
	24	16	9	17
	28	21	8	12*
	31	31	6	*o.s.*
	69	46	9	24†
	85	51	10	31†
7	8	8	7	*u*
	15	15	7	9
	22	22	7	10*
	30	21	10	26
	36	28	9	22
	56	49	8	*o.s.*

Units per block (*k*).	Blocks (*b*).	Varieties (*v*).	Replications (*r*).	Reference.
8	9	9	8	*u*
	15	15	8	9
	29	29	8	14*
	45	36	10	28*
	57	57	8	*o.s.*
	72	64	9	*o.s.*
9	10	10	9	*u*
	13	13	9	*o.s.*
	19	19	9	19
	25	25	9	20
	37	37	9	23
	73	73	9	*o.s.*
	90	81	10	*o.s.*
10	11	11	10	*u*
	16	16	10	5
	19	19	10	19
	31	31	10	27
	46	46	10	30*
	91	91	10	*o.s.*

* Proved non-existent.   † No solution yet discovered.

TABLE XIXI. Cyclic solutions, $r=11-15$. (Based on C. R. Rao.)

32 (i)†	[(1 2 4), (5 6 10), (3 9 7), (**12** 8 11)] mod 11
(ii)	(1 2 4), (1 2 5), (1 3 7), (**12** 1 6) mod 11
33 †	[(1 2 4 8), (3 5 10 11), (**12** 6 7 9)] mod 11
34	(1 2 4 8 9 11), (**12** 1 6 7 9 11) mod 11
35	(1 2 3 4 6 8 9 12 13 16 18) mod 23
36	(121 131 213 313 112), (321 231 333 223 112) mod (3,3,5); (**111 112 113 114 115**) mod (3,3,–)
41	(1 2 4 7 8 12), (1 2 3 4 8 12) mod 13
42	(1 2 4 13), (1 2 6 14), (1 5 7 10) mod 19
43	(11 16 25 22 33 34), (11 12 14 21 22 24) mod (3,7)
45 (i)	(1 2 4), (1 5 14), (1 6 12), (1 8 18) mod 25
(ii)	(12 52 24), (21 44 23), (43 32 13), (22 35 31) mod (5,5)
46, 49	See Introduction
48	(1,1 1,2 2,3 2,6), (1,2 1,4 1,9 2,1), (1,**12** 1,8 2,6 3,2) mod (3,11); (I **1**,1 **2**,1 **3**,1) mod (–,11); (I **1,12 2,12 3,12**)
$E(2,11):1$	$PC(12)$ [(1 10 28 30 47 51 77 105 108 115 116)$S(12)$; (2 14 26 38 50 62 74 86 98 110 **121**)] mod 120
$P(2,11):1$	(1 2 4 13 21 35 39 82 89 95 105 110) mod 133
$E(3,3):1$	(1 2 23), (1 3 9), (1 4 15), (1 8 18) mod 26; $PC(13)$ (**27** 1 14) mod 26
$E(3,3):2$ †	$PC(13)$ [(1 2 3 9 12 19 21 23 24)$S(13)$; (4 5 7 13 17 18 20 26 **27**)] mod 26
53	(112 211 231 222 313 221 213 131 132 232 322 133 332) mod (3,3,3)
$P(3,3):1$	(1 2 27 33), (1 8 20 37), (1 4 17 39) mod 40; $PC(10)$ (1 11 21 31) mod 40
$P(3,3):2$	(1 2 3 6 13 19 23 25 27 28 30 33 34) mod 40
61	(1 2 5 10 12), (1 2 5 11 13), (**15** 1 2 3 8) mod 14
62	(I **11 21 22 23 25**), (I **21 11 17 16 14**), (**12 13 15 21 22 24**), (**13 14 16 21 22 24**), (**11 15 16 21 22 24**) mod (–,7)
63	(**1**,1 **1**,4 **1**,10 **1**,11), (**1**,1 **2**,1 **2**,3 **2**,8), (**1**,1 **2**,1 **2**,10 **2**,11), (**1**,1 **1**,3 **2**,6 **2**,9), (**1**,1 **1**,4 **2**,5 **2**,8), (**1**,1 **1**,5 **2**,4 **2**,10), (**1**,1 **1**,6 **2**,3 **2**,7) mod (–,11)
64	(1 2 3 7 13 16 21), (1 3 4 5 11 16 20) mod 22
65	(1 7 16 20 23 24 25), (3 21 19 2 11 14 17) mod 29
66 †	[(12 17 26 23 34 35) mod (5,–); (I **11 21 31 41 51**)] mod (–,7); [(12 17 46 43 24 25) mod (5,–); (I **11 21 31 41 51**)] mod (–,7)
67	(**61** 12 17 26 23 34 35), (**61** 12 17 46 43 24 25) mod (5,7); (I **11 21 31 41 51 61**), (I **11 21 31 41 51 61**) mod (–,7); (**61 62 63 64 65 66 67**), (**61 62 63 64 65 66 67**)
$E(2,13):1$ †	$PC(14)$ [(1 20 24 36 73 93 98 111 137 152 158 159 161)$S(14)$; (2 16 30 44 58 72 86 100 114 128 142 156 **169**)] mod 168
$P(2,13):1$	(1 2 4 25 42 53 58 67 71 97 103 150 165 177) mod 183
71	(1 2 4), (1 2 6), (1 3 8), (1 2 9), (1 4 6) mod 11
72	(1 2 3 5 9), (1 2 4 7 13), (1 3 6 7 11) mod 13
73 (i)	(1 2 4), (1 4 9), (1 3 13), (1 2 8), (1 5 10) mod 16
(ii)	(**21 22 23**), (**21 23** 26), (11 **18 21**), (**12 17 21**), (**13 16 21**), (**14 15 21**), (**12 18 21**), (**13 17 21**), (**14 16 21**), (11 **21 25**) mod (–,8)
74	(1 2 3 5 8), (1 2 6 9 11), (1 2 4 8 12) mod 16
75	(1 2 4 6 10 13), (1 2 3 4 7 13) mod 16; $PC(8)$ (1 2 4 9 10 12) mod 16
$E(4,2):3$ †	[(4 5 6 7 9 11 12 15), (1 2 3 8 10 13 14 **16**)] mod 15
76	(**11 12 13 15 21 22 23 25** 33), (**11 17 16 14** 37 35 34 33 **21**), (**21 27 26 24** 37 35 34 33 **11**), (**15 12 14 21 23 27** 35 32 33), (**11 13 17 25 22 24** 35 32 33) mod 7
77	(12 15 23 24 32 35), (12 15 33 34 43 44) mod (5,5); (I **11 24 32 45 53**), (I **11 23 35 42 54**), (I **11 21 31 41 51**) mod (–,5)
$P(4,2):1$	(1 2 19), (1 3 6), (1 5 11), (1 9 21), (1 10 17) mod 31
79	(1 2 4 8 16), (3 6 12 24 17), (9 18 5 10 20) mod 31
$P(4,2):3$ (i)	(1 2 3 4 6 7 9 12 13 19 20 21 24 28 30) mod 31
(ii)	(1 2 4 5 7 8 9 10 14 16 18 19 20 25 28) mod 31
(iii)	(1 2 3 4 6 8 12 15 16 17 23 24 27 29 30) mod 31
80	(11 12 14 16 21 24 26 35 41 42 46 53 61 62 64) mod (6,6)
84	(1 9 20 58 34), (4 36 19 49 14), (16 22 15 13 56) mod 61
87	(1,1 1,13 2,5 2,9 6,2 6,12 3,7), (4,4 4,10 5,3 5,11 7,6 7,8 3,7) mod (7,13); (**1**,1 **2**,1 **3**,1 **4**,1 **5**,1 **6**,1 **7**,1) mod (–,13)

† Divisible into complete replicates, indicated by [ ]. (i), (ii), (iii) indicate alternative solutions. Bold-face numbers are not cyclically permuted. I indicates an invariant variety. $PC(x)$ indicates a partial cycle, 1 to x; $S(x)$ indicates steps of x, all such steps being included.

TABLE XIX2. Index by Number of Replications, $r = 11 - 15$.

Replications (r).	Varieties (v).	Blocks (b).	Units per block (k).	Conjunctions (λ).	Reference
11	12	44	3	2	32
	12	33	4	3	33
	12	22	6	5	34
	23	23	11	5	35
	45	99	5	1	36
	45	55	9	2	37 †
	56	56	11	2	38 †
	100	110	10	1	39 ?
	111	111	11	1	40 ?
12	13	26	6	5	41
	19	57	4	2	42
	21	42	6	3	43
	22	33	8	4	44 †
	25	100	3	1	45
	33	44	9	3	46
	34	34	12	4	47 *
	37	111	4	1	48
	45	45	12	3	49
	55	66	10	2	50 *
	61	122	6	1	51 †
	67	67	12	2	52 *
	121	132	11	1	$E(2,11)$: 1
	133	133	12	1	$P(2,11)$: 1
13	27	117	3	1	$E(3,3)$: 1
	27	39	9	4	$E(3,3)$: 2
	27	27	13	6	53
	40	130	4	1	$P(3,3)$: 1
	40	52	10	3	54 †
	40	40	13	4	$P(3,3)$: 2
	53	53	13	3	55 *
	66	143	6	1	56 †
	66	78	11	2	57 †
	79	79	13	2	58 †
	144	156	12	1	59 ?
	157	157	13	1	60 ?
14	15	42	5	4	61(d)
	15	35	6	5	62
	22	77	4	2	63
	22	44	7	4	64(d)
	29	58	7	3	65
	36	84	6	2	66(d)
	43	86	7	2	67(d)
	78	91	12	2	68 *
	85	170	7	1	69 †
	92	92	14	2	70 *
	169	182	13	1	$E(2,13)$: 1
	183	183	14	1	$P(2,13)$: 1
15	11	55	3	3	71
	13	39	5	5	72
	16	80	3	2	73
	16	48	5	4	74
	16	40	6	5	75
	16	30	8	7	$E(4,2)$: 3
	21	35	9	6	76
	26	65	6	3	77
	28	42	10	5	78 †
	31	155	3	1	$P(4,2)$: 1
	31	93	5	2	79
	31	31	15	7	$P(4,2)$: 3
	36	36	15	6	80
	43	43	15	5	81 †
	46	69	10	3	82 †
	56	70	12	3	83 †
	61	183	5	1	84
	71	71	15	3	85 †
	76	190	6	1	86 †
	91	195	7	1	87
	91	105	13	2	88 *
	106	106	15	2	89 *
	136	204	10	1	90 †
	196	210	14	1	91 *
	211	211	15	1	92 *

* Proved non-existent. † No solution yet discovered. ? Presumably non-existent. (d) Double of insoluble type.

Table XX. Scores for Ordinal (or Ranked) Data

The mean deviations of the 1st, 2nd, 3rd . . . largest members of samples of different sizes; zero and negative values omitted.

Size of Sample

Ordinal number.	—	2	3	4	5	6	7	8	9	10
1		·56	·85	1·03	1·16	1·27	1·35	1·42	1·49	1·54
2				·30	·50	·64	·76	·85	·93	1·00
3						·20	·35	·47	·57	·66
4								·15	·27	·38
5										·12

Ordinal number.	11	12	13	14	15	16	17	18	19	20
1	1·59	1·63	1·67	1·70	1·74	1·76	1·79	1·82	1·84	1·87
2	1·06	1·12	1·16	1·21	1·25	1·28	1·32	1·35	1·38	1·41
3	·73	·79	·85	·90	·95	·99	1·03	1·07	1·10	1·13
4	·46	·54	·60	·66	·71	·76	·81	·85	·89	·92
5	·22	·31	·39	·46	·52	·57	·62	·67	·71	·75
6		·10	·19	·27	·34	·39	·45	·50	·55	·59
7				·09	·17	·23	·30	·35	·40	·45
8						·08	·15	·21	·26	·31
9								·07	·13	·19
10										·06

Ordinal number.	21	22	23	24	25	26	27	28	29	30
1	1·89	1·91	1·93	1·95	1·97	1·98	2·00	2·01	2·03	2·04
2	1·43	1·46	1·48	1·50	1·52	1·54	1·56	1·58	1·60	1·62
3	1·16	1·19	1·21	1·24	1·26	1·29	1·31	1·33	1·35	1·36
4	·95	·98	1·01	1·04	1·07	1·09	1·11	1·14	1·16	1·18
5	·78	·82	·85	·88	·91	·93	·96	·98	1·00	1·03
6	·63	·67	·70	·73	·76	·79	·82	·85	·87	·89
7	·49	·53	·57	·60	·64	·67	·70	·73	·75	·78
8	·36	·41	·45	·48	·52	·55	·58	·61	·64	·67
9	·24	·29	·33	·37	·41	·44	·48	·51	·54	·57
10	·12	·17	·22	·26	·30	·34	·38	·41	·44	·47
11		·06	·11	·16	·20	·24	·28	·32	·35	·38
12				·05	·10	·14	·19	·22	·26	·29
13						·05	·09	·13	·17	·21
14								·04	·09	·12
15										·04

Tests of psychological preference and some other experimental data suffice to place a series of magnitudes in order of preference, without supplying metrical values. Analyses of variance, correlations, etc., can be carried out on such data by using the normal scores, appropriate to each position in order, in a sample of the size observed. Ties may be scored with the means of the ordinal values involved, but in such cases the sums of squares given in Table XXI will require correction.

Table XXI. Sums of Squares of Mean Deviations Tabulated

n	1	2	3	4	5	6	7	8	9	10	n
1		0·6272	1·4450	2·3018	3·1912	4·1250	5·0452	5·9646	6·9656	7·9320	10
11	8·8892	9·8662	10·8104	11·7846	12·8232	13·6600	14·7258	15·7454	16·6864	17·7144	20
21	18·6242	19·6862	20·6176	21·6040	22·6352	23·5470	24·5992	25·5808	26·5806	27·5454	30
31	28·5730	29·5960	30·5562	31·5152	32·5618	33·5166	34·5346	35·4840	36·4414	37·4288	40
41	38·4660	39·4654	40·4366	41·5788	42·5194	43·4922	44·4548	45·3626	46·5110	47·3830	50

TABLE XX. SCORES FOR ORDINAL (OR RANKED) DATA—*continued*

Ordinal number.	Size of Sample 31	32	33	34	35	36	37	38	39	40
1	2·06	2·07	2·08	2·09	2·11	2·12	2·13	2·14	2·15	2·16
2	1·63	1·65	1·66	1·68	1·69	1·70	1·72	1·73	1·74	1·75
3	1·38	1·40	1·42	1·43	1·45	1·46	1·48	1·49	1·50	1·52
4	1·20	1·22	1·23	1·25	1·27	1·28	1·30	1·32	1·33	1·34
5	1·05	1·07	1·09	1·11	1·12	1·14	1·16	1·17	1·19	1·20
6	·92	·94	·96	·98	1·00	1·02	1·03	1·05	1·07	1·08
7	·80	·82	·85	·87	·89	·91	·92	·94	·96	·98
8	·69	·72	·74	·76	·79	·81	·83	·85	·86	·88
9	·60	·62	·65	·67	·69	·72	·73	·75	·77	·79
10	·50	·53	·56	·58	·60	·63	·65	·67	·69	·71
11	·41	·44	·47	·50	·52	·54	·57	·59	·61	·63
12	·33	·36	·39	·41	·44	·47	·49	·51	·54	·56
13	·24	·28	·31	·34	·36	·39	·42	·44	·46	·49
14	·16	·20	·23	·26	·29	·32	·34	·37	·39	·42
15	·08	·12	·15	·18	·22	·24	·27	·30	·33	·35
16		·04	·08	·11	·14	·17	·20	·23	·26	·28
17				·04	·07	·10	·14	·16	·19	·22
18						·03	·07	·10	·13	·16
19								·03	·06	·09
20										·03

Ordinal number.	41	42	43	44	45	46	47	48	49	50
1	2·17	2·18	2·19	2·20	2·21	2·22	2·22	2·23	2·24	2·25
2	1·76	1·78	1·79	1·80	1·81	1·82	1·83	1·84	1·85	1·85
3	1·53	1·54	1·55	1·57	1·58	1·59	1·60	1·61	1·62	1·63
4	1·36	1·37	1·38	1·40	1·41	1·42	1·43	1·44	1·45	1·46
5	1·22	1·23	1·25	1·26	1·27	1·28	1·30	1·31	1·32	1·33
6	1·10	1·11	1·13	1·14	1·16	1·17	1·18	1·19	1·21	1·22
7	·99	1·01	1·02	1·04	1·05	1·07	1·08	1·09	1·11	1·12
8	·90	·91	·93	·95	·96	·98	·99	1·00	1·02	1·03
9	·81	·83	·84	·86	·88	·89	·91	·92	·94	·95
10	·73	·75	·76	·78	·80	·81	·83	·84	·86	·87
11	·65	·67	·69	·71	·72	·74	·76	·77	·79	·80
12	·58	·60	·62	·64	·65	·67	·69	·70	·72	·74
13	·51	·53	·55	·57	·59	·60	·62	·64	·66	·67
14	·44	·46	·48	·50	·52	·54	·56	·58	·59	·61
15	·37	·40	·42	·44	·46	·48	·50	·52	·53	·55
16	·31	·33	·36	·38	·40	·42	·44	·46	·48	·49
17	·25	·27	·29	·32	·34	·36	·38	·40	·42	·44
18	·18	·21	·23	·26	·28	·30	·32	·34	·36	·38
19	·12	·15	·17	·20	·22	·25	·27	·29	·31	·33
20	·06	·09	·12	·14	·17	·19	·21	·24	·26	·28
21		·03	·06	·09	·11	·14	·16	·18	·21	·23
22				·03	·06	·08	·11	·13	·15	·18
23						·03	·05	·08	·10	·13
24								·03	·05	·08
25										·03

Table XXII. Initial Differences of Powers of Natural Numbers

These values, the so-called "differences of zero," enter into many combinatorial formulæ, as in Whitworth's *Choice and Chance*. They afford exact solutions of some important sampling problems (Stevens, (12), (13), 1937). In particular if s objects are distributed among n classes, the probabilities of any one object falling into each class being equal, the distribution of r, the number of occupied classes, is

$$\frac{1}{n^s}\ \frac{n!}{(n-r)!\,r!}\ \Delta^r 0^s$$

The values tabulated, $\Delta^r 0^s / r!$, are the rth differences of powers of natural numbers, advancing from 0^s, divided by $r!$, for values of s from 2 to 25.

r \ s	2	3	4	5	6	7	8
2	1	3	7	15	31	63	127
3		1	6	25	90	301	966
4			1	10	65	350	1701
5				1	15	140	1050
6					1	21	266
7						1	28
8							1

r \ s	9	10	11	12	13
2	255	511	1023	2047	4095
3	3025	9330	28501	86526	2 61625
4	7770	34105	1 45750	6 11501	25 32530
5	6951	42525	2 46730	13 79400	75 08501
6	2646	22827	1 79487	13 23652	93 21312
7	462	5880	63987	6 27396	57 15424
8	36	750	11880	1 59027	18 99612
9	1	45	1155	22275	3 59502
10		1	55	1705	39325
11			1	66	2431
12				1	78
13					1

r \ s	14	15	16	17	18
2	8191	16383	32767	65535	1 31071
3	7 88970	23 75101	71 41686	214 57825	644 39010
4	103 91745	423 55950	1717 98901	6943 37290	27988 06985
5	400 75035	2107 66920	10961 90550	56527 51651	2 89580 95545
6	634 36373	4206 93273	27349 26558	1 75057 49898	11 06872 51039
7	493 29280	4087 41333	32818 82604	2 57081 04786	19 74624 83400
8	209 12320	2166 27840	21417 64053	2 04159 95028	18 90360 65010
9	51 35130	671 28490	8207 84250	95288 22303	10 61753 95755
10	7 52752	126 62650	1937 54990	27583 34150	3 71121 63803
11	66066	14 79478	289 36908	5120 60978	83910 04908
12	3367	1 06470	27 57118	620 22324	12563 28866
13	91	4550	1 65620	49 10178	1258 54638
14	1	105	6020	2 49900	84 08778
15		1	120	7820	3 67200
16			1	136	9996
17				1	153
18					1

r \ s	19	20	21	22
2	2 62143	5 24287	10 48575	20 97151
3	1934 48101	5806 06446	17423 43625	52280 79450
4	1 12596 66950	4 52321 15901	18 15090 70050	72 77786 23825
5	14 75892 84710	74 92060 90500	379 12625 68401	1913 78219 12055
6	69 30816 01779	430 60788 95384	2658 56794 62804	16330 53393 45225
7	149 29246 34839	1114 35540 45652	8231 09572 14948	60276 23799 67440
8	170 97510 03480	1517 09326 62679	13251 10153 47084	1 14239 90799 91620
9	114 46146 26805	1201 12826 44725	12327 24764 65204	1 24196 33035 33920
10	47 72970 33785	591 75849 64655	7118 71322 91275	83514 37993 77954
11	12 94132 17791	190 08424 29486	2682 68516 89001	36628 25008 70286
12	2 34669 51300	41 10166 33391	683 30420 30178	10882 33560 51137
13	28924 39160	6 10686 60380	120 49092 18331	2249 68618 68481
14	2435 77530	63025 24580	14 93040 04500	329 51652 81331
15	139 16778	4523 29200	1 30874 62580	34 56159 43200
16	5 27136	223 50954	8099 44464	2 60465 74004
17	12597	7 41285	349 52799	14041 42047
18	171	15675	10 23435	533 74629
19	1	190	19285	13 89850
20		1	210	23485
21			1	231
22				1

r \ s	23	24	25
2	41 94303	83 88607	167 77215
3	1 56863 35501	4 70632 00806	14 11979 91025
4	291 63425 74750	1168 10566 34501	4677 12897 38810
5	9641 68881 84100	48500 07834 95250	2 43668 49741 10751
6	99896 98579 83405	6 09023 60360 84530	37 02641 70000 02430
7	4 38264 19991 17305	31 67746 38518 04540	227 83248 29987 16310
8	9 74195 50199 00400	82 31828 21583 20505	690 22372 11183 68580
9	12 32006 88117 96900	120 62257 43260 72500	1167 92145 10929 73005
10	9 59340 12973 13460	108 25408 17849 31500	1203 16339 21753 87500
11	4 86425 13089 51100	63 10016 56957 75560	802 35590 44384 62660
12	1 67216 27734 83930	24 93020 45907 58260	362 26262 07848 74680
13	40128 25603 41390	6 88883 60579 22000	114 48507 33437 44260
14	6862 91758 07115	1 36209 10216 41000	25 95811 03608 96000
15	847 94044 29331	19582 02422 47080	4 29939 46553 47200
16	76 23611 27264	2067 71824 65555	52665 51616 95960
17	4 99169 88803	161 09499 36915	4806 33313 93110
18	23648 85369	9 24849 25445	327 56785 94925
19	797 81779	38807 39170	16 62189 69675
20	18 59550	1169 72779	62201 94750
21	28336	24 54606	1685 19505
22	253	33902	32 00450
23	1	276	40250
24		1	300
25			1

TABLE XXIII. ORTHOGONAL POLYNOMIALS

(Values from $n' = 32$ to $n' = 75$ are due to V. Satakopan)

3

ξ'_1	ξ'_2
−1	+1
0	−2
+1	+1
2	6
1	3

4

ξ'_1	ξ'_2	ξ'_3
−3	+1	−1
−1	−1	+3
+1	−1	−3
+3	+1	+1
20	4	20
2	1	$\frac{10}{3}$

5

ξ'_1	ξ'_2	ξ'_3	ξ'_4
−2	+2	−1	+1
−1	−1	+2	−4
0	−2	0	+6
+1	−1	−2	−4
+2	+2	+1	+1
10	14	10	70
1	1	$\frac{5}{6}$	$\frac{35}{12}$

6

ξ'_1	ξ'_2	ξ'_3	ξ'_4	ξ'_5
−5	+5	−5	+1	−1
−3	−1	+7	−3	+5
−1	−4	+4	+2	−10
+1	−4	−4	+2	+10
+3	−1	−7	−3	−5
+5	+5	+5	+1	+1
70	84	180	28	252
2	$\frac{3}{2}$	$\frac{5}{3}$	$\frac{7}{12}$	$\frac{21}{10}$

7

ξ'_1	ξ'_2	ξ'_3	ξ'_4	ξ'_5
−3	+5	−1	+3	−1
−2	0	+1	−7	+4
−1	−3	+1	+1	−5
0	−4	0	+6	0
+1	−3	−1	+1	+5
+2	0	−1	−7	−4
+3	+5	+1	+3	+1
28	84	6	154	84
1	1	$\frac{1}{6}$	$\frac{7}{12}$	$\frac{7}{20}$

8

ξ'_1	ξ'_2	ξ'_3	ξ'_4	ξ'_5
−7	+7	−7	+7	−7
−5	+1	+5	−13	+23
−3	−3	+7	−3	−17
−1	−5	+3	+9	−15
+1	−5	−3	+9	+15
+3	−3	−7	−3	+17
+5	+1	−5	−13	−23
+7	+7	+7	+7	+7
168	168	264	616	2184
2	1	$\frac{2}{3}$	$\frac{7}{12}$	$\frac{7}{10}$

9

ξ'_1	ξ'_2	ξ'_3	ξ'_4	ξ'_5
0	−20	0	+18	0
+1	−17	−9	+9	+9
+2	−8	−13	−11	+4
+3	+7	−7	−21	−11
+4	+28	+14	+14	+4
60	2,772	990	2,002	468
1	3	$\frac{5}{6}$	$\frac{7}{12}$	$\frac{3}{20}$

10

ξ'_1	ξ'_2	ξ'_3	ξ'_4	ξ'_5
+1	−4	−12	+18	+6
+3	−3	−31	+3	+11
+5	−1	−35	−17	+1
+7	+2	−14	−22	−14
+9	+6	+42	+18	+6
330	132	8,580	2,860	780
2	$\frac{1}{2}$	$\frac{5}{3}$	$\frac{5}{12}$	$\frac{1}{10}$

11

ξ'_1	ξ'_2	ξ'_3	ξ'_4	ξ'_5
0	−10	0	+6	0
+1	−9	−14	+4	+4
+2	−6	−23	−1	+4
+3	−1	−22	−6	−1
+4	+6	−6	−6	−6
+5	+15	+30	+6	+3
110	858	4,290	286	156
1	1	$\frac{5}{6}$	$\frac{1}{12}$	$\frac{1}{40}$

12

ξ'_1	ξ'_2	ξ'_3	ξ'_4	ξ'_5
+1	−35	−7	+28	+20
+3	−29	−19	+12	+44
+5	−17	−25	−13	+29
+7	+1	−21	−33	−21
+9	+25	−3	−27	−57
+11	+55	+33	+33	+33
572	12,012	5,148	8,008	15,912
2	3	$\frac{2}{3}$	$\frac{7}{24}$	$\frac{3}{20}$

13

ξ'_1	ξ'_2	ξ'_3	ξ'_4	ξ'_5
0	−14	0	+84	0
+1	−13	−4	+64	+20
+2	−10	−7	+11	+26
+3	−5	−8	−54	+11
+4	+2	−6	−96	−18
+5	+11	0	−66	−33
+6	+22	+11	+99	+22
182	2,002	572	68,068	6,188
1	1	$\frac{1}{6}$	$\frac{7}{12}$	$\frac{7}{120}$

14

ξ'_1	ξ'_2	ξ'_3	ξ'_4	ξ'_5
+1	−8	−24	+108	+60
+3	−7	−67	+63	+145
+5	−5	−95	−13	+139
+7	−2	−98	−92	+28
+9	+2	−66	−132	−132
+11	+7	+11	−77	−187
+13	+13	+143	+143	+143
910	728	97,240	136,136	235,144
2	$\frac{1}{2}$	$\frac{5}{3}$	$\frac{7}{12}$	$\frac{7}{30}$

15

ξ'_1	ξ'_2	ξ'_3	ξ'_4	ξ'_5
0	−56	0	+756	0
+1	−53	−27	+621	+675
+2	−44	−49	+251	+1000
+3	−29	−61	−249	+751
+4	−8	−58	−704	−44
+5	+19	−35	−869	−979
+6	+52	+13	−429	−1144
+7	+91	+91	+1001	+1001
280	37,128	39,780	6,466,460	10,581,480
1	3	$\frac{5}{6}$	$\frac{35}{12}$	$\frac{21}{20}$

16

ξ'_1	ξ'_2	ξ'_3	ξ'_4	ξ'_5
+1	−21	−63	+189	+45
+3	−19	−179	+129	+115
+5	−15	−265	+23	+131
+7	−9	−301	−101	+77
+9	−1	−267	−201	−33
+11	+9	−143	−221	−143
+13	+21	+91	−91	−143
+15	+35	+455	+273	+143
1,360	5,712	1,007,760	470,288	201,552
2	1	$\frac{10}{3}$	$\frac{7}{12}$	$\frac{1}{10}$

17

ξ'_1	ξ'_2	ξ'_3	ξ'_4	ξ'_5
0	−24	0	+36	0
+1	−23	−7	+31	+55
+2	−20	−13	+17	+88
+3	−15	−17	−3	+83
+4	−8	−18	−24	+36
+5	+1	−15	−39	−39
+6	+12	−7	−39	−104
+7	+25	+7	−13	−91
+8	+40	+28	+52	+104
408	7,752	3,876	16,796	100,776
1	1	$\frac{1}{6}$	$\frac{1}{12}$	$\frac{1}{20}$

18

ξ'_1	ξ'_2	ξ'_3	ξ'_4	ξ'_5
+1	−40	−8	+44	+220
+3	−37	−23	+33	+583
+5	−31	−35	+13	+733
+7	−22	−42	−12	+588
+9	−10	−42	−36	+156
+11	+5	−33	−51	−429
+13	+23	−13	−47	−871
+15	+44	+20	−12	−676
+17	+68	+68	+68	+884
1,938	23,256	23,256	28,424	6,953,544
2	$\frac{3}{2}$	$\frac{1}{3}$	$\frac{1}{12}$	$\frac{3}{10}$

19

ξ'_1	ξ'_2	ξ'_3	ξ'_4	ξ'_5
0	−30	0	+396	0
+1	−29	−44	+352	+44
+2	−26	−83	+227	+74
+3	−21	−112	+42	+79
+4	−14	−126	−168	+54
+5	−5	−120	−354	+3
+6	+6	−89	−453	−58
+7	+19	−28	−388	−98
+8	+34	+68	−68	−68
+9	+51	+204	+612	+102
570	13,566	213,180	2,288,132	89,148
1	1	$\frac{5}{6}$	$\frac{7}{12}$	$\frac{1}{40}$

20

ξ'_1	ξ'_2	ξ'_3	ξ'_4	ξ'_5
+1	−33	−99	+1188	+396
+3	−31	−287	+948	+1076
+5	−27	−445	+503	+1441
+7	−21	−553	−77	+1351
+9	−13	−591	−687	+771
+11	−3	−539	−1187	−187
+13	+9	−377	−1402	−1222
+15	+23	−85	−1122	−1802
+17	+39	+357	−102	−1122
+19	+57	+969	+1938	+1938
2,660	17,556	4,903,140	22,881,320	31,201,800
2	1	$\frac{10}{3}$	$\frac{35}{24}$	$\frac{7}{20}$

21

ξ'_1	ξ'_2	ξ'_3	ξ'_4	ξ'_5
0	−110	0	+594	0
+1	−107	−54	+540	+1404
+2	−98	−103	+385	+2444
+3	−83	−142	+150	+2819
+4	−62	−166	−130	+2354
+5	−35	−170	−406	+1063
+6	−2	−149	−615	−788
+7	+37	−98	−680	−2618
+8	+82	−12	−510	−3468
+9	+133	+114	0	−1938
+10	+190	+285	+969	+3876
770	201,894	432,630	5,720,330	121,687,020
1	3	$\frac{5}{6}$	$\frac{7}{12}$	$\frac{21}{40}$

22

ξ'_1	ξ'_2	ξ'_3	ξ'_4	ξ'_5
+1	−20	−12	+702	+390
+3	−19	−35	+585	+1079
+5	−17	−55	+365	+1509
+7	−14	−70	+70	+1554
+9	−10	−78	−258	+1158
+11	−5	−77	−563	+363
+13	+1	−65	−775	−663
+15	+8	−40	−810	−1598
+17	+16	0	−570	−1938
+19	+25	+57	+57	−969
+21	+35	+133	+1197	+2261
3,542	7,084	96,140	8,748,740	40,562,340
2	$\frac{1}{2}$	$\frac{1}{3}$	$\frac{7}{12}$	$\frac{7}{30}$

23

ξ'_1	ξ'_2	ξ'_3	ξ'_4	ξ'_5
0	−44	0	+858	0
+1	−43	−13	+793	+65
+2	−40	−25	+605	+116
+3	−35	−35	+315	+141
+4	−28	−42	−42	+132
+5	−19	−45	−417	+87
+6	−8	−43	−747	+12
+7	+5	−35	−955	−77
+8	+20	−20	−950	−152
+9	+37	+3	−627	−171
+10	+56	+35	+133	−76
+11	+77	+77	+1463	+209
1,012	35,420	32,890	13,123,110	340,860
1	1	$\frac{1}{6}$	$\frac{7}{12}$	$\frac{1}{60}$

24

ξ'_1	ξ'_2	ξ'_3	ξ'_4	ξ'_5
+1	−143	−143	+143	+715
+3	−137	−419	+123	+2005
+5	−125	−665	+85	+2893
+7	−107	−861	+33	+3171
+9	−83	−987	−27	+2721
+11	−53	−1023	−87	+1551
+13	−17	−949	−137	−169
+15	+25	−745	−165	−2071
+17	+73	−391	−157	−3553
+19	+127	+133	−97	−3743
+21	+187	+847	+33	−1463
+23	+253	+1771	+253	+4807
4,600	394,680	17,760,600	394,680	177,928,920
2	3	$\frac{10}{3}$	$\frac{1}{12}$	$\frac{3}{10}$

25

ξ'_1	ξ'_2	ξ'_3	ξ'_4	ξ'_5
0	−52	0	+858	0
+1	−51	−77	+803	+275
+2	−48	−149	+643	+500
+3	−43	−211	+393	+631
+4	−36	−258	+78	+636
+5	−27	−285	−267	+501
+6	−16	−287	−597	+236
+7	−3	−259	−857	−119
+8	+12	−196	−982	−488
+9	+29	−93	−897	−753
+10	+48	+55	−517	−748
+11	+69	+253	+253	−253
+12	+92	+506	+1518	+1012
1,300	53,820	1,480,050	14,307,150	7,803,900
1	1	$\frac{5}{6}$	$\frac{5}{12}$	$\frac{1}{20}$

26

ξ'_1	ξ'_2	ξ'_3	ξ'_4	ξ'_5
+1	−28	−84	+1386	+330
+3	−27	−247	+1221	+935
+5	−25	−395	+905	+1381
+7	−22	−518	+466	+1582
+9	−18	−606	−54	+1482
+11	−13	−649	−599	+1067
+13	−7	−637	−1099	+377
+15	0	−560	−1470	−482
+17	+8	−408	−1614	−1326
+19	+17	−171	−1419	−1881
+21	+27	+161	−759	−1771
+23	+38	+598	+506	−506
+25	+50	+1150	+2530	+2530
5,850	16,380	7,803,900	40,060,020	48,384,180
2	$\frac{1}{2}$	$\frac{5}{3}$	$\frac{7}{12}$	$\frac{1}{10}$

27

ξ'_1	ξ'_2	ξ'_3	ξ'_4	ξ'_5
0	−182	0	+1638	0
+1	−179	−18	+1548	+3960
+2	−170	−35	+1285	+7304
+3	−155	−50	+870	+9479
+4	−134	−62	+338	+10058
+5	−107	−70	−262	+8803
+6	−74	−73	−867	+5728
+7	−35	−70	−1400	+1162
+8	+10	−60	−1770	−4188
+9	+61	−42	−1872	−9174
+10	+118	−15	−1587	−12144
+11	+181	+22	−782	−10879
+12	+250	+70	+690	−2530
+13	+325	+130	+2990	+16445
1,638	712,530	101,790	56,448,210	2,032,135,560
1	3	$\frac{1}{6}$	$\frac{7}{12}$	$\frac{21}{40}$

28

ξ'_1	ξ'_2	ξ'_3	ξ'_4	ξ'_5
+1	−65	−39	+936	+1560
+3	−63	−115	+840	+4456
+5	−59	−185	+655	+6701
+7	−53	−245	+395	+7931
+9	−45	−291	+81	+7887
+11	−35	−319	−259	+6457
+13	−23	−325	−590	+3718
+15	−9	−305	−870	−22
+17	+7	−255	−1050	−4182
+19	+25	−171	−1074	−7866
+21	+45	−49	−879	−9821
+23	+67	+115	−395	−8395
+25	+91	+325	+455	−1495
+27	+117	+585	+1755	+13455
7,308	95,004	2,103,660	19,634,160	1,354,757,040
2	1	$\frac{2}{3}$	$\frac{7}{24}$	$\frac{7}{20}$

29

ξ'_1	ξ'_2	ξ'_3	ξ'_4	ξ'_5
0	−70	0	+2184	0
+1	−69	−104	+2080	+1768
+2	−66	−203	+1775	+3298
+3	−61	−292	+1290	+4373
+4	−54	−366	+660	+4818
+5	−45	−420	−66	+4521
+6	−34	−449	−825	+3454
+7	−21	−448	−1540	+1694
+8	−6	−412	−2120	−556
+9	+11	−336	−2460	−2946
+10	+30	−215	−2441	−4958
+11	+51	−44	−1930	−5885
+12	+74	+182	−780	−4810
+13	+99	+468	+1170	−585
+14	+126	+819	+4095	+8190
2,030	113,274	4,207,320	107,987,880	500,671,080
1	1	$\frac{5}{6}$	$\frac{7}{12}$	$\frac{7}{40}$

30

ξ'_1	ξ'_2	ξ'_3	ξ'_4	ξ'_5
+1	−112	−112	+12376	+1768
+3	−109	−331	+11271	+5083
+5	−103	−535	+9131	+7753
+7	−94	−714	+6096	+9408
+9	−82	−858	+2376	+9768
+11	−67	−957	−1749	+8679
+13	−49	−1001	−5929	+6149
+15	−28	−980	−9744	+2384
+17	−4	−884	−12704	−2176
+19	+23	−703	−14249	−6821
+21	+53	−427	−13749	−10535
+23	+86	−46	−10504	−11960
+25	+122	+450	−3744	−9360
+27	+161	+1071	+7371	−585
+29	+203	+1827	+23751	+16965
8,990	302,064	21,360,240	3,671,587,920	2,145,733,200
2	$\frac{3}{2}$	$\frac{5}{3}$	$\frac{35}{12}$	$\frac{3}{10}$

31

ξ'_1	ξ'_2	ξ'_3	ξ'_4	ξ'_5
+0	−80	0	+408	0
+1	−79	−119	+391	+221
+2	−76	−233	+341	+416
+3	−71	−337	+261	+561
+4	−64	−426	+156	+636
+5	−55	−495	+33	+627
+6	−44	−539	−99	+528
+7	−31	−553	−229	+343
+8	−16	−532	−344	+88
+9	+1	−471	−429	−207
+10	+20	−365	−467	−496
+11	+41	−209	−439	−715
+12	+64	+2	−324	−780
+13	+89	+273	−99	−585
+14	+116	+609	+261	0
+15	+145	+1015	+783	+1131
2,480	158,224	6,724,520	4,034,712	9,536,592
1	1	$\frac{5}{6}$	$\frac{1}{12}$	$\frac{1}{60}$

32

ξ'_1	ξ'_2	ξ'_3	ξ'_4	ξ'_5
+1	−85	−51	+459	+255
+3	−83	−151	+423	+737
+5	−79	−245	+353	+1137
+7	−73	−329	+253	+1407
+9	−65	−399	+129	+1509
+11	−55	−451	−11	+1419
+13	−43	−481	−157	+1131
+15	−29	−485	−297	+661
+17	−13	−459	−417	+51
+19	+5	−399	−501	−627
+21	+25	−301	−531	−1267
+23	+47	−161	−487	−1725
+25	+71	+25	−347	−1815
+27	+97	+261	−87	−1305
+29	+125	+551	+319	+87
+31	+155	+899	+899	+2697
10,912	185,504	5,379,616	5,379,616	54,285,216
2	1	$\frac{2}{3}$	$\frac{1}{12}$	$\frac{1}{30}$

33

ξ'_1	ξ'_2	ξ'_3	ξ'_4	ξ'_5
0	−272	0	+3672	0
+1	−269	−27	+3537	+2565
+2	−260	−53	+3139	+4864
+3	−245	−77	+2499	+6649
+4	−224	−98	+1652	+7708
+5	−197	−115	+647	+7883
+6	−164	−127	−453	+7088
+7	−125	−133	−1571	+5327
+8	−80	−132	−2616	+2712
+9	−29	−123	−3483	−519
+10	+28	−105	−4053	−3984
+11	+91	−77	−4193	−7139
+12	+160	−38	−3756	−9260
+13	+235	+13	−2581	−9425
+14	+316	+77	−493	−6496
+15	+403	+155	+2697	+899
+16	+496	+248	+7192	+14384
2,992	1,947,792	417,384	348,330,136	1,547,128,656
1	3	$\frac{1}{6}$	$\frac{7}{12}$	$\frac{3}{20}$

34

ξ'_1	ξ'_2	ξ'_3	ξ'_4	ξ'_5
1	−48	−144	+4104	+6840
3	−47	−427	+3819	+19855
5	−45	−695	+3263	+30917
7	−42	−938	+2464	+38864
9	−38	−1146	+1464	+42744
11	−33	−1309	+319	+41899
13	−27	−1417	−901	+36049
15	−20	−1460	−2112	+25376
17	−12	−1428	−3216	+10608
19	−3	−1311	−4101	−6897
21	+7	−1099	−4641	−25067
23	+18	−782	−4696	−41032
25	+30	−350	−4112	−51040
27	+43	+207	−2721	−50373
29	+57	+899	−341	−33263
31	+72	+1736	+3224	+7192
33	+88	+2728	+8184	+79112
13,090	62,832	51,477,360	456,432,592	46,929,569,232
2	$\frac{1}{2}$	$\frac{5}{3}$	$\frac{7}{12}$	$\frac{1}{10}$

35

ξ'_1	ξ'_2	ξ'_3	ξ'_4	ξ'_5
0	−102	0	+23256	0
1	−101	−152	+22496	+3800
2	−98	−299	+20251	+7250
3	−93	−436	+16626	+10021
4	−86	−558	+11796	+11826
5	−77	−660	+6006	+12441
6	−66	−737	−429	+11726
7	−53	−784	−7124	+9646
8	−38	−796	−13624	+6292
9	−21	−768	−19404	+1902
10	−2	−695	−23869	−3118
11	+19	−572	−26354	−8173
12	+42	−394	−26124	−12458
13	+67	−156	−22374	−14937
14	+94	+147	−14229	−14322
15	+123	+520	−744	−9052
16	+154	+968	+19096	+2728
17	+187	+1496	+46376	+23188
3,570	290,598	15,775,320	14,834,059,240	4,045,652,520
1	1	$\frac{5}{6}$	$\frac{35}{12}$	$\frac{7}{40}$

36

ξ'_1	ξ'_2	ξ'_3	ξ'_4	ξ'_5
1	−323	−323	+2584	+12920
3	−317	−959	+2424	+37640
5	−305	−1565	+2111	+59063
7	−287	−2121	+1659	+75201
9	−263	−2607	+1089	+84381
11	−233	−3003	+429	+85371
13	−197	−3289	−286	+77506
15	−155	−3445	−1014	+60814
17	−107	−3451	−1706	+36142
19	−53	−3287	−2306	+5282
21	+7	−2933	−2751	−28903
23	+73	−2369	−2971	−62353
25	+145	−1575	−2889	−89685
27	+223	−531	−2421	−104067
29	+307	+783	−1476	−97092
31	+397	+2387	+44	−58652
33	+493	+4301	+2244	+23188
35	+595	+6545	+5236	+162316
15,540	3,011,652	307,618,740	191,407,216	199,046,103,984
2	3	$\frac{10}{3}$	$\frac{7}{24}$	$\frac{21}{20}$

37

ξ'_1	ξ'_2	ξ'_3	ξ'_4	ξ'_5
0	−114	0	+5814	0
1	−113	−34	+5644	+680
2	−110	−67	+5141	+1304
3	−105	−98	+4326	+1819
4	−98	−126	+3234	+2178
5	−89	−150	+1914	+2343
6	−78	−169	+429	+2288
7	−65	−182	−1144	+2002
8	−50	−188	−2714	+1492
9	−33	−186	−4176	+786
10	−14	−175	−5411	−64
11	+7	−154	−6286	−979
12	+30	−122	−6654	−1850
13	+55	−78	−6354	−2535
14	+82	−21	−5211	−2856
15	+111	+50	−3036	−2596
16	+142	+136	+374	−1496
17	+175	+238	+5236	+748
18	+210	+357	+11781	+4488
4,218	383,838	932,178	980,961,982	152,877,192
1	1	$\frac{1}{6}$	$\frac{7}{12}$	$\frac{1}{40}$

38

ξ'_1	ξ'_2	ξ'_3	ξ'_4	ξ'_5
1	−60	−36	+918	+1530
3	−59	−107	+867	+4471
5	−57	−175	+767	+7061
7	−54	−238	+622	+9086
9	−50	−294	+438	+10362
11	−45	−341	+223	+10747
13	−39	−377	−13	+10153
15	−32	−400	−258	+8558
17	−24	−408	−498	+6018
19	−15	−399	−717	+2679
21	−5	−371	−897	−1211
23	+6	−322	−1018	−5290
25	+18	−250	−1058	−9070
27	+31	−153	−993	−11925
29	+45	−29	−797	−13079
31	+60	+124	−442	−11594
33	+76	+308	+102	−6358
35	+93	+525	+867	+3927
37	+111	+777	+1887	+20757
18,278	109,668	4,496,388	25,479,532	3,286,859,628
2	$\frac{1}{2}$	$\frac{1}{3}$	$\frac{1}{12}$	$\frac{1}{10}$

39

ξ'_1	ξ'_2	ξ'_3	ξ'_4	ξ'_5
0	−380	0	+1026	0
1	−377	−189	+999	+5049
2	−368	−373	+919	+9724
3	−353	−547	+789	+13669
4	−332	−706	+614	+16564
5	−305	−845	+401	+18143
6	−272	−959	+159	+18212
7	−233	−1043	−101	+16667
8	−188	−1092	−366	+13512
9	−137	−1101	−621	+8877
10	−80	−1065	−849	+3036
11	−17	−979	−1031	−3575
12	+52	−838	−1146	−10340
13	+127	−637	−1171	−16445
14	+208	−371	−1081	−20860
15	+295	−35	−849	−22321
16	+388	+376	−446	−19312
17	+487	+867	+159	−10047
18	+592	+1443	+999	+7548
19	+703	+2109	+2109	+35853
4,940	4,496,388	33,722,910	32,224,114	9,860,578,884
1	3	$\frac{5}{6}$	$\frac{1}{12}$	$\frac{3}{20}$

40

ξ'_1	ξ'_2	ξ'_3	ξ'_4	ξ'_5
1	−133	−399	+39501	+627
3	−131	−1187	+37521	+1837
5	−127	−1945	+33631	+2917
7	−121	−2653	+27971	+3787
9	−113	−3291	+20751	+4377
11	−103	−3839	+12251	+4631
13	−91	−4277	+2821	+4511
15	−77	−4585	−7119	+4001
17	−61	−4743	−17079	+3111
19	−43	−4731	−26499	+1881
21	−23	−4529	−34749	+385
23	−1	−4117	−41129	−1265
25	+23	−3475	−44869	−2915
27	+49	−2583	−45129	−4365
29	+77	−1421	−40999	−5365
31	+107	+31	−31499	−5611
33	+139	+1793	−15579	−4741
35	+173	+3885	+7881	−2331
37	+209	+6327	+40071	+2109
39	+247	+9139	+82251	+9139
21,320	567,112	644,482,280	49,625,135,560	644,482,280
2	1	$\frac{10}{3}$	$\frac{35}{12}$	$\frac{1}{30}$

41

ξ'_1	ξ'_2	ξ'_3	ξ'_4	ξ'_5
0	−140	0	+8778	0
1	−139	−209	+8569	+4807
2	−136	−413	+7949	+9292
3	−131	−607	+6939	+13147
4	−124	−786	+5574	+16092
5	−115	−945	+3903	+17889
6	−104	−1079	+1989	+18356
7	−91	−1183	−91	+17381
8	−76	−1252	−2246	+14936
9	−59	−1281	−4371	+11091
10	−40	−1265	−6347	+6028
11	−19	−1199	−8041	+55
12	+4	−1078	−9306	−6380
13	+29	−897	−9981	−12675
14	+56	−651	−9891	−18060
15	+85	−335	−8847	−21583
16	+116	+56	−6646	−22096
17	+149	+527	−3071	−18241
18	+184	+1083	+2109	−8436
19	+221	+1729	+9139	+9139
20	+260	+2470	+18278	+36556
5,740	641,732	47,900,710	2,481,256,778	10,376,164,708
1	1	$\frac{5}{6}$	$\frac{7}{12}$	$\frac{7}{60}$

42

ξ'_1	ξ'_2	ξ'_3	ξ'_4	ξ'_5
1	−220	−44	+9614	+48070
3	−217	−131	+9177	+141151
5	−211	−215	+8317	+225181
7	−202	−294	+7062	+294546
9	−190	−366	+5454	+344262
11	−175	−429	+3549	+370227
13	−157	−481	+1417	+369473
15	−136	−520	−858	+340418
17	−112	−544	−3178	+283118
19	−85	−551	−5431	+199519
21	−55	−539	−7491	+93709
23	−22	−506	−9218	−27830
25	+14	−450	−10458	−155970
27	+53	−369	−11043	−278685
29	+95	−261	−10791	−380799
31	+140	−124	−9506	−443734
33	+188	+44	−6978	−445258
35	+239	+245	−2983	−359233
37	+293	+481	+2717	−155363
39	+350	+754	+10374	+201058
41	+410	+1066	+20254	+749398
24,682	1,629,012	9,075,924	3,084,805,724	4,389,117,671,484
2	$\frac{3}{2}$	$\frac{1}{3}$	$\frac{7}{12}$	$\frac{21}{10}$

43

ξ'_1	ξ'_2	ξ'_3	ξ'_4	ξ'_5
0	−154	0	+10626	0
1	−153	−46	+10396	+8740
2	−150	−91	+9713	+16948
3	−145	−134	+8598	+24113
4	−138	−174	+7086	+29766
5	−129	−210	+5226	+33501
6	−118	−241	+3081	+34996
7	−105	−266	+728	+34034
8	−90	−284	−1742	+30524
9	−73	−294	−4224	+24522
10	−54	−295	−6599	+16252
11	−33	−286	−8734	+6127
12	−10	−266	−10482	−5230
13	+15	−234	−11682	−16965
14	+42	−189	−12159	−27972
15	+71	−130	−11724	−36872
16	+102	−56	−10174	−41992
17	+135	+34	−7292	−41344
18	+170	+141	−2847	−32604
19	+207	+266	+3406	−13091
20	+246	+410	+11726	+20254
21	+287	+574	+22386	+70889
6,622	814,506	2,676,234	3,815,417,606	39,541,600,644
1	1	$\frac{1}{6}$	$\frac{7}{12}$	$\frac{7}{40}$

44

ξ'_1	ξ'_2	ξ'_3	ξ'_4	ξ'_5
1	−161	−483	+5796	+1380
3	−159	−1439	+5556	+4060
5	−155	−2365	+5083	+6503
7	−149	−3241	+4391	+8561
9	−141	−4047	+3501	+10101
11	−131	−4763	+2441	+11011
13	−119	−5369	+1246	+11206
15	−105	−5845	−42	+10634
17	−89	−6171	−1374	+9282
19	−71	−6327	−2694	+7182
21	−51	−6293	−3939	+4417
23	−29	−6049	−5039	+1127
25	−5	−5575	−5917	−2485
27	+21	−4851	−6489	−6147
29	+49	−3857	−6664	−9512
31	+79	−2573	−6344	−12152
33	+111	−979	−5424	−13552
35	+145	+945	−3792	−13104
37	+181	+3219	−1329	−10101
39	+219	+5863	+2091	−3731
41	+259	+8897	+6601	+6929
43	+301	+12341	+12341	+22919
28,380	913,836	1,257,829,980	1,173,974,648	4,162,273,752
2	1	$\frac{10}{3}$	$\frac{7}{24}$	$\frac{1}{20}$

45

ξ'_1	ξ'_2	ξ'_3	ξ'_4	ξ'_5
0	−506	0	+9108	0
1	−503	−252	+8928	+4500
2	−494	−499	+8393	+8750
3	−479	−736	+7518	+12509
4	−458	−958	+6328	+15554
5	−431	−1160	+4858	+17689
6	−398	−1337	+3153	+18754
7	−359	−1484	+1268	+18634
8	−314	−1596	−732	+17268
9	−263	−1668	−2772	+14658
10	−206	−1695	−4767	+10878
11	−143	−1672	−6622	+6083
12	−74	−1594	−8232	+518
13	+1	−1456	−9482	−5473
14	+82	−1253	−10247	−11438
15	+169	−980	−10392	−16808
16	+262	−632	−9772	−20888
17	+361	−204	−8232	−22848
18	+466	+309	−5607	−21714
19	+577	+912	−1722	−16359
20	+694	+1610	+3608	−5494
21	+817	+2408	+10578	+12341
22	+946	+3311	+19393	+38786
7,590	9,203,634	92,036,340	2,934,936,620	12,006,558,900
1	3	$\frac{5}{6}$	$\frac{5}{12}$	$\frac{3}{40}$

Table XXIII. Orthogonal Polynomials—*continued*

46				47				48				49				50			
ξ'_2	ξ'_3	ξ'_4	ξ'_5	ξ'_2	ξ'_3	ξ'_4	ξ'_5	ξ'_2	ξ'_3	ξ'_4	ξ'_5	ξ'_2	ξ'_3	ξ'_4	ξ'_5	ξ'_2	ξ'_3	ξ'_4	ξ'_5
-88	-264	+1980	+3300	-184	0	+15180	0	-575	-115	+16445	+82225	-200	0	+17940	0	-104	-312	+96876	+10764
-87	-787	+1905	+9725	-183	-55	+14905	+3575	-569	-343	+15873	+242671	-199	-299	+17641	+9867	-103	-931	+93771	+31809
-85	-1295	+1757	+15631	-180	-109	+14087	+6968	-557	-565	+14743	+391231	-196	-593	+16751	+19272	-101	-1535	+87631	+51419
-82	-1778	+1540	+20692	-175	-161	+12747	+10003	-539	-777	+13083	+520401	-191	-877	+15291	+27767	-98	-2114	+78596	+68684
-78	-2226	+1260	+24612	-168	-210	+10920	+12516	-515	-975	+10935	+623307	-184	-1146	+13296	+34932	-94	-2658	+66876	+82764
-73	-2629	+925	+27137	-159	-255	+8655	+14361	-485	-1155	+8355	+693957	-175	-1395	+10815	+40389	-89	-3157	+52751	+92917
-67	-2977	+545	+28067	-148	-295	+6015	+15416	-449	-1313	+5413	+727493	-164	-1619	+7911	+43816	-83	-3601	+36571	+98527
-60	-3260	+132	+27268	-135	-329	+3077	+15589	-407	-1445	+2193	+720443	-151	-1813	+4661	+44961	-76	-3980	+18756	+99132
-52	-3468	-300	+24684	-120	-356	-68	+14824	-359	-1547	-1207	+670973	-136	-1972	+1156	+43656	-68	-4284	-204	+94452
-43	-3591	-735	+20349	-103	-375	-3315	+13107	-305	-1615	-4675	+579139	-119	-2091	-2499	+39831	-59	-4503	-19749	+84417
-33	-3619	-1155	+14399	-84	-385	-6545	+10472	-245	-1645	-8085	+447139	-100	-2165	-6185	+33528	-49	-4627	-39249	+69195
-22	-3542	-1540	+7084	-63	-385	-9625	+7007	-179	-1633	-11297	+279565	-79	-2189	-9769	+24915	-38	-4646	-58004	+49220
-10	-3350	-1868	-1220	-40	-374	-12408	+2860	-107	-1575	-14157	+83655	-56	-2158	-13104	+14300	-26	-4550	-75244	+25220
+3	-3033	-2115	-9999	-15	-351	-14733	-1755	-29	-1467	-16497	-130455	-31	-2067	-16029	+2145	-13	-4329	-90129	-1755
+17	-2581	-2255	-18589	+12	-315	-16425	-6552	+55	-1305	-18135	-349479	-4	-1911	-18369	-10920	+1	-3973	-101749	-30305
+32	-1984	-2260	-26164	+41	-265	-17295	-11167	+145	-1085	-18875	-556729	+25	-1685	-19935	-24083	+16	-3472	-109124	-58652
+48	-1232	-2100	-31724	+72	-200	-17140	-15152	+241	-803	-18507	-731863	+56	-1384	-20524	-36336	+32	-2816	-111204	-84612
+65	-315	-1743	-34083	+105	-119	-15743	-17969	+343	-455	-16807	-850633	+89	-1003	-19919	-46461	+49	-1995	-106869	-105567
+83	+777	-1155	-31857	+140	-21	-12873	-18984	+451	-37	-13537	-884633	+124	-537	-17889	-53016	+67	-999	-94929	-118437
+102	+2054	-300	-23452	+177	+95	-8285	-17461	+565	+455	-8445	-801047	+161	+19	-14189	-54321	+86	+182	-74124	-119652
+122	+3526	+860	-7052	+216	+230	-1720	-12556	+685	+1025	-1265	-562397	+200	+670	-8560	-48444	+106	+1558	-43124	-105124
+143	+5203	+2365	+19393	+257	+385	+7095	-3311	+811	+1677	+8283	-126291	+241	+1421	-729	-33187	+127	+3139	-529	-70219
+165	+7095	+4257	+58179	+300	+561	+18447	+11352	+943	+2415	+20493	+554829	+284	+2277	+9591	-6072	+149	+4935	+55131	-9729
				+345	+759	+32637	+32637	+1081	+3243	+35673	+1533939	+329	+3243	+22701	+35673	+172	+6956	+125396	+82156
												+376	+4324	+38916	+95128	+196	+9212	+211876	+211876
285,384	429,502,920	143,167,640	27,214,866,840	1,271,256	4,994,220	8,518,474,580	8,629,104,120	12,712,560	92,620,080	10,301,411,120	19,208,385,771,120	1,566,040	167,230,700	12,408,517,940	74,451,107,640	433,160	770,715,400	372,255,538,200	372,255,538,200
$\frac{1}{2}$	$\frac{5}{3}$	$\frac{1}{12}$	$\frac{1}{10}$	1	$\frac{1}{6}$	$\frac{7}{12}$	$\frac{1}{20}$	3	$\frac{2}{3}$	$\frac{7}{12}$	$\frac{21}{10}$	1	$\frac{5}{6}$	$\frac{7}{12}$	$\frac{7}{60}$	$\frac{1}{2}$	$\frac{5}{3}$	$\frac{35}{12}$	$\frac{7}{30}$

From this point onwards ξ'_1 is omitted. The value of $S(\xi'^2_1)$ is $n'(n'^2-1)/3$ for even n', and $n'(n'^2-1)/12$ for odd n'.

Table XXIII. Orthogonal Polynomials—*continued*

51 ξ'_2	51 ξ'_3	51 ξ'_4	51 ξ'_5	52 ξ'_2	52 ξ'_3	52 ξ'_4	52 ξ'_5	53 ξ'_2	53 ξ'_3	53 ξ'_4	53 ξ'_5	54 ξ'_2	54 ξ'_3	54 ξ'_4	54 ξ'_5	55 ξ'_2	55 ξ'_3	55 ξ'_4	55 ξ'_5
-650	0	+21060	0	-225	-135	+1620	+2700	-234	0	+3510	0	-364	-364	+26390	+18850	-252	0	+142506	0
-647	-324	+20736	+7452	-223	-403	+1572	+7988	-233	-70	+3460	+2900	-361	-1087	+25665	+55825	-251	-377	+140621	+47125
-638	-643	+19771	+14582	-219	-665	+1477	+12943	-230	-139	+3311	+5684	-355	-1795	+24229	+90643	-248	-749	+135001	+92500
-623	-952	+18186	+21077	-213	-917	+1337	+17353	-225	-206	+3066	+8239	-346	-2478	+22110	+121926	-243	-1111	+125751	+134417
-602	-1246	+16016	+26642	-205	-1155	+1155	+21021	-218	-270	+2730	+10458	-334	-3126	+19350	+148386	-236	-1458	+113046	+171252
-575	-1520	+13310	+31009	-195	-1375	+935	+23771	-209	-330	+2310	+12243	-319	-3729	+16005	+168861	-227	-1785	+97131	+201507
-542	-1769	+10131	+33946	-183	-1573	+682	+25454	-198	-385	+1815	+13508	-301	-4277	+12145	+182351	-216	-2087	+78321	+223852
-503	-1988	+6556	+35266	-169	-1745	+402	+25954	-185	-434	+1256	+14182	-280	-4760	+7854	+188054	-203	-2359	+57001	+237167
-458	-2172	+2676	+34836	-153	-1887	+102	+25194	-170	-476	+646	+14212	-256	-5168	+3230	+185402	-188	-2596	+33626	+240584
-407	-2316	-1404	+32586	-135	-1995	-210	+23142	-153	-510	0	+13566	-229	-5491	-1615	+174097	-171	-2793	+8721	+233529
-350	-2415	-5565	+28518	-115	-2065	-525	+19817	-134	-535	-665	+12236	-199	-5719	-6555	+154147	-152	-2945	-17119	+215764
-287	-2464	-9674	+22715	-93	-2093	-833	+15295	-113	-550	-1330	+10241	-166	-5842	-11450	+125902	-131	-3047	-43229	+187429
-218	-2458	-13584	+15350	-69	-2075	-1123	+9715	-90	-554	-1974	+7630	-130	-5850	-16146	+90090	-108	-3094	-68874	+149084
-143	-2392	-17134	+6695	-43	-2007	-1383	+3285	-65	-546	-2574	+4485	-91	-5733	-20475	+47853	-83	-3081	-93249	+101751
-62	-2261	-20149	-2870	-15	-1885	-1600	-3712	-38	-525	-3105	+924	-49	-5481	-24255	+783	-56	-3003	-115479	+46956
+25	-2060	-22440	-12848	+15	-1705	-1760	-10912	-9	-490	-3540	-2896	-4	-5084	-27290	-49042	-27	-2855	-134619	-13229
+118	-1784	-23804	-22616	+47	-1463	-1848	-17864	+22	-440	-3850	-6776	+44	-4532	-29370	-99022	+4	-2632	-149654	-76144
+217	-1428	-24024	-31416	+81	-1155	-1848	-24024	+55	-374	-4004	-10472	+95	-3815	-30271	-145999	+37	-2329	-159499	-138499
+322	-987	-22869	-38346	+117	-777	-1743	-28749	+90	-291	-3969	-13692	+149	-2923	-29755	-186221	+72	-1941	-162999	-196332
+433	-456	-20094	-42351	+155	-325	-1515	-31291	+127	-190	-3710	-16093	+206	-1846	-27570	-215306	+109	-1463	-158929	-244967
+550	+170	-15440	-42214	+195	+205	-1145	-30791	+166	-70	-3190	-17278	+266	-574	-23450	-228206	+148	-890	-145994	-278972
+673	+896	-8634	-36547	+237	+817	-613	-26273	+207	+70	-2370	-16793	+329	+903	-17115	-219171	+189	-217	-122829	-292117
+802	+1727	+611	-23782	+281	+1515	+102	-16638	+250	+231	-1209	-14124	+395	+2595	-8271	-181713	+232	+561	-87999	-277332
+937	+2668	+12596	-2162	+327	+2303	+1022	-658	+295	+414	+336	-8694	+464	+4512	+3390	-108570	+277	+1449	-39999	-226665
+1078	+3724	+27636	+30268	+375	+3185	+2170	+23030	+342	+620	+2310	+140	+536	+6664	+18190	+8330	+324	+2452	+22746	-131240
+1225	+4900	+46060	+75670	+425	+4165	+3570	+55930	+391	+850	+4760	+13090	+611	+9061	+36465	+177905	+373	+3575	+101881	+18785
								+442	+1105	+7735	+30940	+689	+11713	+58565	+409955	+424	+4823	+199121	+234260
																+477	+6201	+316251	+527085
17,218,110	221,375,700	17,803,525,740	47,861,426,340	2,108,340	162,342,180	108,228,120	26,358,466,680	2,319,174	11,595,870	514,083,570	8,131,867,380	5,729,724	1,321,929,180	29.816,847,060	1,439,340,526,260	2.791,404	375,842,610	879,596,988,270	2,398,900,877,100
3	$\frac{5}{6}$	$\frac{7}{12}$	$\frac{3}{40}$	1	$\frac{2}{3}$	$\frac{1}{24}$	$\frac{1}{20}$	1	$\frac{1}{6}$	$\frac{1}{12}$	$\frac{1}{40}$	$\frac{3}{2}$	$\frac{5}{3}$	$\frac{7}{12}$	$\frac{3}{10}$	1	$\frac{5}{6}$	$\frac{35}{12}$	$\frac{7}{20}$

TABLE XXIII. ORTHOGONAL POLYNOMIALS—*continued*

56 ξ'_2	ξ'_3	ξ'_4	ξ'_5	57 ξ'_2	ξ'_3	ξ'_4	ξ'_5	58 ξ'_2	ξ'_3	ξ'_4	ξ'_5	59 ξ'_2	ξ'_3	ξ'_4	ξ'_5	60 ξ'_2	ξ'_3	ξ'_4	ξ'_5
-261	-783	+30537	+50895	-812	0	+32886	0	-140	-84	+35154	+2790	-290	0	+5394	0	-899	-899	+14384	+14384
-259	-2339	+29757	+150865	-809	-81	+32481	+163215	-139	-251	+34317	+8277	-289	-434	+5332	+1488	-893	-2687	+14064	+42704
-255	-3865	+28211	+245417	-800	-161	+31273	+320788	-137	-415	+32657	+13487	-286	-863	+5147	+2928	-881	-4445	+13429	+69689
-249	-5341	+25927	+331079	-785	-239	+29283	+467203	-134	-574	+30202	+18242	-281	-1282	+4842	+4273	-863	-6153	+12489	+94479
-241	-6747	+22947	+404589	-764	-314	+26546	+597196	-130	-726	+26994	+22374	-274	-1686	+4422	+5478	-839	-7791	+11259	+116259
-231	-8063	+19327	+462979	-737	-385	+23111	+705881	-125	-869	+23089	+25729	-265	-2070	+3894	+6501	-809	-9339	+9759	+134277
-219	-9269	+15137	+503659	-704	-451	+19041	+788876	-119	-1001	+18557	+28171	-254	-2429	+3267	+7304	-773	-10777	+8014	+147862
-205	-10345	+10461	+524501	-665	-511	+14413	+842429	-112	-1120	+13482	+29586	-241	-2758	+2552	+7854	-731	-12085	+6054	+156442
-189	-11271	+5397	+523923	-620	-564	+9318	+863544	-104	-1224	+7962	+29886	-226	-3052	+1762	+8124	-683	-13243	+3914	+159562
-171	-12027	+57	+500973	-569	-609	+3861	+850107	-95	-1311	+2109	+29013	-209	-3306	+912	+8094	-629	-14231	+1634	+156902
-151	-12593	-5433	+455413	-512	-645	-1839	+801012	-85	-1379	-3951	+26943	-190	-3515	+19	+7752	-569	-15029	-741	+148295
-129	-12949	-10933	+387803	-449	-671	-7649	+716287	-74	-1426	-10078	+23690	-169	-3674	-898	+7095	-503	-15617	-3161	+133745
-105	-13075	-16289	+299585	-380	-686	-13422	+597220	-62	-1450	-16118	+19310	-146	-3778	-1818	+6130	-431	-15975	-5571	+113445
-79	-12951	-21333	+193167	-305	-689	-18997	+446485	-49	-1449	-21903	+13905	-121	-3822	-2718	+4875	-353	-16083	-7911	+87795
-51	-12557	-25883	+72007	-224	-679	-24199	+268268	-35	-1421	-27251	+7627	-94	-3801	-3573	+3360	-269	-15921	-10116	+57420
-21	-11873	-29743	-59303	-137	-655	-28839	+68393	-20	-1364	-31966	+682	-65	-3710	-4356	+1628	-179	-15469	-12116	+23188
+11	-10879	-32703	-194953	-44	-616	-32714	-145552	-4	-1276	-35838	-6666	-34	-3544	-5038	-264	-83	-14707	-13836	-13772
+45	-9555	-34539	-327831	+55	-561	-35607	-364089	+13	-1155	-38643	-14091	-1	-3298	-5588	-2244	+19	-13615	-15196	-52052
+81	-7881	-35013	-449439	+160	-489	-37287	-575724	+31	-999	-40143	-21201	+34	-2967	-5973	-4224	+127	-12173	-16111	-89947
+119	-5837	-33873	-549809	+271	-399	-37509	-766821	+50	-806	-40086	-27534	+71	-2546	-6158	-6099	+241	-10361	-16491	-125437
+159	-3403	-30853	-617419	+388	-290	-36014	-921476	+70	-574	-38206	-32554	+110	-2030	-6106	-7746	+361	-8159	-16241	-156169
+201	-559	-25673	-639109	+511	-161	-32529	-1021391	+91	-301	-34223	-35647	+151	-1414	-5778	-9023	+487	-5547	-15261	-179439
+245	+2715	-18039	-599997	+640	-11	-26767	-1045748	+113	+15	-27843	-36117	+194	-693	-5133	-9768	+619	-2505	-13446	-192174
+291	+6439	-7643	-483395	+775	+161	-18427	-971083	+136	+376	-18758	-33182	+239	+138	-4128	-9798	+757	+987	-10686	-190914
+339	+10633	+5837	-270725	+916	+356	-7194	-771160	+160	+784	-6646	-25970	+286	+1084	-2718	-8908	+901	+4949	-6866	-171794
+389	+15317	+22737	+58565	+1063	+575	+7261	-416845	+185	+1241	+8829	-13515	+335	+2150	-856	-6870	+1051	+9401	-1866	-130526
+441	+20511	+43407	+527085	+1216	+819	+25281	+124020	+211	+1749	+28017	+5247	+386	+3341	+1507	-3432	+1207	+14363	+4439	-62381
+495	+26235	+68211	+1159587	+1375	+1089	+47223	+886743	+238	+2310	+51282	+31482	+439	+4662	+4422	+1683	+1369	+19855	+12179	+37829
				+1540	+1386	+73458	+1909908	+266	+2926	+79002	+66462	+494	+6118	+7942	+8778	+1537	+25897	+21489	+175769
												+551	+7714	+12122	+18183	+1711	+32509	+32509	+357599
3,054,744	6,822,988,920	41,392,799,448	11,706,636,280,248	30,038,316	19,310,346	48,556,937,814	32,021,093,354,796	910,252	87,254,156	56,802,455,556	39,097,794,084	3,966,098	614,745,190	1,352,439,418	2,950,776,912	38,826,012	11,065,413,420	9,835,923,040	1,150,802,995,680
1	$\frac{10}{3}$	$\frac{7}{12}$	$\frac{7}{10}$	3	$\frac{1}{6}$	$\frac{7}{12}$	$\frac{21}{20}$	$\frac{1}{2}$	$\frac{\ }{3}$	$\frac{7}{12}$	$\frac{1}{30}$	1	$\frac{5}{6}$	$\frac{1}{12}$	$\frac{1}{120}$	3	$\frac{10}{3}$	$\frac{5}{24}$	$\frac{3}{20}$

Table XXIII. Orthogonal Polynomials—*continued*

61 ξ'_2	61 ξ'_3	61 ξ'_4	61 ξ'_5	62 ξ'_2	62 ξ'_3	62 ξ'_4	62 ξ'_5	63 ξ'_2	63 ξ'_3	63 ξ'_4	63 ξ'_5	64 ξ'_2	64 ξ'_3	64 ξ'_4	64 ξ'_5	65 ξ'_2	65 ξ'_3	65 ξ'_4	65 ξ'_5
−310	0	+43152	0	−160	−96	+45936	+76560	−992	0	+49104	0	−341	−1023	+52173	+86955	−352	0	+278256	0
−309	−464	+42688	+5104	−159	−287	+44979	+227447	−989	−99	+48609	+244035	−339	−3059	+51153	+258485	−351	−527	+275621	+13175
−306	−923	+41303	+10054	−157	−475	+43079	+371677	−980	−197	+47131	+481168	−335	−5065	+49127	+422917	−348	−1049	+267751	+26000
−301	−1372	+39018	+14699	−154	−658	+40264	+504952	−965	−293	+44691	+704623	−329	−7021	+46123	+575659	−343	−1561	+254751	+38131
−294	−1806	+35868	+18894	−150	−834	+36576	+623184	−944	−386	+41324	+907876	−321	−8907	+42183	+712329	−336	−2058	+236796	+49236
−285	−2220	+31902	+22503	−145	−1001	+32071	+722579	−917	−475	+37079	+1084781	−311	−10703	+37363	+828839	−327	−2535	+214131	+59001
−274	−2609	+27183	+25402	−139	−1157	+26819	+799721	−884	−559	+32019	+1229696	−299	−12389	+31733	+921479	−316	−2987	+187071	+67136
−261	−2968	+21788	+27482	−132	−1300	+20904	+851656	−845	−637	+26221	+1337609	−285	−13945	+25377	+987001	−303	−3409	+156001	+73381
−246	−3292	+15808	+28652	−124	−1428	+14424	+875976	−800	−708	+19776	+1404264	−269	−15351	+18393	+1022703	−288	−3796	+121376	+77512
−229	−3576	+9348	+28842	−115	−1539	+7491	+870903	−749	−771	+12789	+1426287	−251	−16587	+10893	+1026513	−271	−4143	+83721	+79347
−210	−3815	+2527	+28006	−105	−1631	+231	+835373	−692	−825	+5379	+1401312	−231	−17633	+3003	+997073	−252	−4445	+43631	+78752
−189	−4004	−4522	+26125	−94	−1702	−7216	+769120	−629	−869	−2321	+1328107	−209	−18469	−5137	+933823	−231	−4697	+1771	+75647
−166	−4138	−11652	+23210	−82	−1750	−14696	+672760	−560	−902	−10164	+1206700	−185	−19075	−13373	+837085	−208	−4894	−41124	+70012
−141	−4212	−18702	+19305	−69	−1773	−22041	+547875	−485	−923	−17989	+1038505	−159	−19431	−21537	+708147	−183	−5031	−84249	+61893
−114	−4221	−25497	+14490	−55	−1769	−29069	+397097	−404	−931	−25621	+826448	−131	−19517	−29447	+549347	−156	−5103	−126729	+51408
−85	−4160	−31848	+8884	−40	−1736	−35584	+224192	−317	−925	−32871	+575093	−101	−19313	−36907	+364157	−127	−5105	−167619	+38753
−54	−4024	−37552	+2648	−24	−1672	−41376	+34144	−224	−904	−39536	+290768	−69	−18799	−43707	+157267	−96	−5032	−205904	+24208
−21	−3808	−42392	−4012	−7	−1575	−46221	−166761	−125	−867	−45399	−18309	−35	−17955	−49623	−65331	−63	−4879	−240499	+8143
+14	−3507	−46137	−10842	+11	−1443	−49881	−370851	−20	−813	−50229	−341904	+1	−16761	−54417	−296259	−28	−4641	−270249	−8976
+51	−3116	−48542	−17537	+30	−1274	−52104	−568984	+91	−741	−53781	−667641	+39	−15197	−57837	−526669	+9	−4313	−293929	−26581
+90	−2630	−49348	−23738	+50	−1066	−52624	−750464	+208	−650	−55796	−980876	+79	−13243	−59617	−746159	+48	−3890	−310244	−43996
+131	−2044	−48282	−29029	+71	−817	−51161	−902957	+331	−539	−56001	−1264571	+121	−10879	−59477	−942689	+89	−3367	−317829	−60431
+174	−1353	−45057	−32934	+93	−525	−47421	−1012407	+460	−407	−54109	−1499168	+165	−8085	−57123	−1102497	+132	−2739	−315249	−74976
+219	−552	−39372	−34914	+116	−188	−41096	−1062952	+595	−253	−49819	−1662463	+211	−4841	−52247	−1210015	+177	−2001	−300999	−86595
+266	+364	−30912	−34364	+140	+196	−31864	−1036840	+736	−76	−42816	−1729480	+259	−1127	−44527	−1247785	+224	−1148	−273504	−94120
+315	+1400	−19348	−30610	+165	+629	−19389	−914345	+883	+125	−32771	−1672345	+309	+3077	−33627	−1196375	+273	−175	−231119	−96245
+366	+2561	−4337	−22906	+191	+1113	−3321	−673683	+1036	+351	−19341	−1460160	+361	+7791	−19197	−1034295	+324	+923	−172129	−91520
+419	+3852	+14478	−10431	+218	+1650	+16704	−290928	+1195	+603	−2169	−1058877	+415	+13035	−873	−737913	+377	+2151	−94749	−78345
+474	+5278	+37468	+7714	+246	+2242	+41064	+260072	+1360	+882	+19116	−431172	+471	+18829	+21723	−281371	+432	+3514	+2876	−54964
+531	+6844	+65018	+32509	+275	+2891	+70151	+1007779	+1531	+1189	+44899	+463681	+529	+25193	+48983	+363499	+489	+5017	+122671	−19459
+590	+8555	+97527	+65018	+305	+3599	+104371	+1983049	+1708	+1525	+75579	+1669936	+589	+32147	+81313	+1227259	+548	+6665	+266631	+30256
								+1891	+1891	+111569	+3235501	+651	+39711	+119133	+2342949	+609	+8463	+436821	+96441
																+672	+10416	+635376	+181536
4,685,898	776,520,240	89,506,899,664	38,360,099,856	1,270,752	139,238,112	103,639,568,032	35,981,773,664,928	49,559,328	38,939,472	119,721,569,968	96,582,655,626,912	5,957,952	17,392,964,160	137,984,182,336	51,066,691,480,896	6,438,432	1,211,804,880	3,967,045,242,160	309,120,408,480
1	$\frac{5}{6}$	$\frac{7}{12}$	$\frac{1}{40}$	$\frac{1}{2}$	$\frac{1}{3}$	$\frac{7}{12}$	$\frac{7}{10}$	3	$\frac{1}{6}$	$\frac{7}{12}$	$\frac{21}{20}$	1	$\frac{10}{3}$	$\frac{7}{12}$	$\frac{7}{10}$	1	$\frac{5}{6}$	$\frac{35}{12}$	$\frac{1}{20}$

Table XXIII. Orthogonal Polynomials—*continued* (*also overleaf*)

n = 66

ξ'_2	ξ'_3	ξ'_4	ξ'_5
−544	−544	+8432	+42160
−541	−1627	+8277	+125395
−535	−2695	+7969	+205393
−526	−3738	+7512	+280056
−514	−4746	+6912	+347376
−499	−5709	+6177	+405471
−481	−6617	+5317	+452621
−460	−7460	+4344	+487304
−436	−8228	+3272	+508232
−409	−8911	+2117	+514387
−379	−9499	+897	+505057
−346	−9982	−368	+479872
−310	−10350	−1656	+438840
−271	−10593	−2943	+382383
−229	−10701	−4203	+311373
−184	−10664	−5408	+227168
−136	−10472	−6528	+131648
−85	−10115	−7531	+27251
−31	−9583	−8383	−82991
+26	−8866	−9048	−195416
+86	−7954	−9488	−305696
+149	−6837	−9663	−408801
+215	−5505	−9531	−498963
+284	−3948	−9048	−569640
+356	−2156	−8168	−613480
+431	−119	−6843	−622285
+509	+2173	−5023	−586975
+590	+4730	−2656	−497552
+674	+7562	+312	−343064
+761	+10679	+3937	−111569
+851	+14091	+8277	+209901
+944	+17808	+13392	+635376
+1040	+21840	+19344	+1179984
15,636,192	5,394,486,240	3,716,201,632	13,168,529,401,248
$\frac{3}{2}$	$\frac{5}{3}$	$\frac{1}{12}$	$\frac{3}{10}$

n = 67

ξ'_2	ξ'_3	ξ'_4	ξ'_5
−374	0	+8976	0
−373	−112	+8896	+2480
−370	−223	+8657	+4898
−365	−332	+8262	+7193
−358	−438	+7716	+9306
−349	−540	+7026	+11181
−338	−637	+6201	+12766
−325	−728	+5252	+14014
−310	−812	+4192	+14884
−293	−888	+3036	+15342
−274	−955	+1801	+15362
−253	−1012	+506	+14927
−230	−1058	−828	+14030
−205	−1092	−2178	+12675
−178	−1113	−3519	+10878
−149	−1120	−4824	+8668
−118	−1112	−6064	+6088
−85	−1088	−7208	+3196
−50	−1047	−8223	+66
−13	−988	−9074	−3211
+26	−910	−9724	−6526
+67	−812	−10134	−9751
+110	−693	−10263	−12738
+155	−552	−10068	−15318
+202	−388	−9504	−17300
+251	−200	−8524	−18470
+302	+13	−7079	−18590
+355	+252	−5118	−17397
+410	+518	−2588	−14602
+467	+812	+566	−9889
+526	+1135	+4401	−2914
+587	+1488	+8976	+6696
+650	+1872	+14352	+19344
+715	+2288	+20592	+35464
7,492,342	59,938,736	4,255,650,256	11,993,196,176
1	$\frac{1}{6}$	$\frac{1}{12}$	$\frac{1}{120}$

n = 68

ξ'_2	ξ'_3	ξ'_4	ξ'_5
−385	−231	+33264	+2640
−383	−691	+32688	+7856
−379	−1145	+31543	+12881
−373	−1589	+29843	+17591
−365	−2019	+27609	+21867
−355	−2431	+24869	+25597
−343	−2821	+21658	+28678
−329	−3185	+18018	+31018
−313	−3519	+13998	+32538
−295	−3819	+9654	+33174
−275	−4081	+5049	+32879
−253	−4301	+253	+31625
−229	−4475	−4657	+29405
−203	−4599	−9597	+26235
−175	−4669	−14476	+22156
−145	−4681	−19196	+17236
−113	−4631	−23652	+11572
−79	−4515	−27732	+5292
−43	−4329	−31317	−1443
−5	−4069	−34281	−8437
+35	−3731	−36491	−15457
+77	−3311	−37807	−22231
+121	−2805	−38082	−28446
+167	−2209	−37162	−33746
+215	−1519	−34886	−37730
+265	−731	−31086	−39950
+317	+159	−25587	−39909
+371	+1155	−18207	−37059
+427	+2261	−8757	−30799
+485	+3481	+2959	−20473
+545	+4819	+17144	−5368
+607	+6279	+34008	+15288
+671	+7865	+53768	+42328
+737	+9581	+76648	+76648
8,068,676	1,063,912,564	59,579,103,584	56,484,085,216
1	$\frac{2}{3}$	$\frac{7}{24}$	$\frac{1}{60}$

n = 69

ξ'_2	ξ'_3	ξ'_4	ξ'_5
−1190	0	+70686	0
−1187	−594	+70092	+175824
−1178	−1183	+68317	+347504
−1163	−1762	+65382	+510959
−1142	−2326	+61322	+662234
−1115	−2870	+56186	+797563
−1082	−3389	+50037	+913432
−1043	−3878	+42952	+1006642
−998	−4332	+35022	+1074372
−947	−4746	+26352	+1114242
−890	−5115	+17061	+1124376
−827	−5434	+7282	+1103465
−758	−5698	−2838	+1050830
−683	−5902	−13138	+966485
−602	−6041	−23443	+851200
−515	−6110	−33564	+706564
−422	−6104	−43298	+535048
−323	−6018	−52428	+340068
−218	−5847	−60723	+126048
−107	−5586	−67938	−101517
+10	−5230	−73814	−335998
+133	−4774	−78078	−569569
+262	−4213	−80443	−793144
+397	−3542	−80608	−996314
+538	−2756	−78258	−1167284
+685	−1850	−73064	−1292810
+838	−819	−64683	−1358136
+997	+342	−52758	−1346931
+1162	+1638	−36918	−1241226
+1333	+3074	−16778	−1021351
+1510	+4655	+8061	−665872
+1693	+6386	+38012	−151528
+1882	+8272	+73502	+546832
+2077	+10318	+114972	+1456312
+2278	+12529	+162877	+2606032
78,119,454	1,841,387,130	271,829,660,102	65,832,201,319,248
3	$\frac{5}{6}$	$\frac{7}{12}$	$\frac{21}{40}$

n = 70

ξ'_2	ξ'_3	ξ'_4	ξ'_5
−204	−612	+373626	+124542
−203	−1831	+367521	+370777
−201	−3035	+355381	+608507
−198	−4214	+337346	+832202
−194	−5358	+313626	+1036542
−189	−6457	+284501	+1216501
−183	−7501	+250321	+1367431
−176	−8480	+211506	+1485146
−168	−9384	+168546	+1566006
−159	−10203	+122001	+1607001
−149	−10927	+72501	+1605835
−138	−11546	+20746	+1561010
−126	−12050	−32494	+1471910
−113	−12429	−86379	+1338885
−99	−12673	−139999	+1163335
−84	−12772	−192374	+947794
−68	−12716	−242454	+696014
−51	−12495	−289119	+413049
−33	−12099	−331179	+105339
−14	−11518	−367374	−219206
+6	−10742	−396374	−551122
+27	−9761	−416779	−879307
+49	−8565	−427119	−1190937
+72	−7144	−425854	−1471382
+96	−5488	−411374	−1704122
+121	−3587	−381999	−1870663
+147	−1431	−335979	−1950453
+174	+990	−271494	−1920798
+202	+3686	−186654	−1756778
+231	+6667	−79499	−1431163
+261	+9943	+52001	−914329
+292	+13524	+209946	−174174
+324	+17420	+396506	+823966
+357	+21641	+613921	+2117401
+391	+26197	+864501	+3746171
2,331,924	8,146,743,060	7,736,690,325,980	137,150,419,415,100
$\frac{1}{2}$	$\frac{5}{3}$	$\frac{35}{12}$	$\frac{7}{10}$

71				72				73				74				75			
ξ'_2	ξ'_3	ξ'_4	ξ'_5	ξ'_2	ξ'_3	ξ'_4	ξ'_5	ξ'_2	ξ'_3	ξ'_4	ξ'_5	ξ'_2	ξ'_3	ξ'_4	ξ'_5	ξ'_2	ξ'_3	ξ'_4	ξ'_5
−420	0	+79254	0	−1295	−259	+83657	+59755	−444	0	+12654	0	−228	−684	+13338	+22230	−1406	0	+493506	0
−419	−629	+78625	+131461	−1289	−775	+82365	+177973	−443	−133	+12559	+20995	−227	−2047	+13143	+66235	−1403	−702	+489996	+35100
−416	−1253	+76745	+259996	−1277	−1285	+79795	+292333	−440	−265	+12275	+41548	−225	−3395	+12755	+108881	−1394	−1399	+479501	+69500
−411	−1867	+73635	+382721	−1259	−1785	+75975	+400323	−435	−395	+11805	+61223	−222	−4718	+12178	+149282	−1379	−2086	+462126	+102509
−404	−2466	+69330	+496836	−1235	−2271	+70947	+499521	−428	−522	+11154	+79596	−218	−6006	+11418	+186582	−1358	−2758	+438046	+133454
−395	−3045	+63879	+599667	−1205	−2739	+64767	+587631	−419	−645	+10329	+96261	−213	−7249	+10483	+219967	−1331	−3410	+407506	+161689
−384	−3599	+57345	+688708	−1169	−3185	+57505	+662519	−408	−763	+9339	+110836	−207	−8437	+9383	+248677	−1298	−4037	+370821	+186604
−371	−4123	+49805	+761663	−1127	−3605	+49245	+722249	−395	−875	+8195	+122969	−200	−9560	+8130	+272018	−1259	−4634	+328376	+207634
−356	−4612	+41350	+816488	−1079	−3995	+40085	+765119	−380	−980	+6910	+132344	−192	−10608	+6738	+289374	−1214	−5196	+280626	+224268
−339	−5061	+32085	+851433	−1025	−4351	+30137	+789697	−363	−1077	+5499	+138687	−183	−11571	+5223	+300219	−1163	−5718	+228096	+236058
−320	−5465	+22129	+865084	−965	−4669	+19527	+794857	−344	−1165	+3979	+141772	−173	−12439	+3603	+304129	−1106	−6195	+171381	+242628
−299	−5819	+11615	+856405	−899	−4945	+8395	+779815	−323	−1243	+2369	+141427	−162	−13202	+1898	+300794	−1043	−6622	+111146	+243683
−276	−6118	+690	+824780	−827	−5175	−3105	+744165	−300	−1310	+690	+137540	−150	−13850	+130	+290030	−974	−6994	+48126	+239018
−251	−6357	−10485	+770055	−749	−5355	−14805	+687915	−275	−1365	−1035	+130065	−137	−14373	−1677	+271791	−899	−7306	−16874	+228527
−224	−6531	−21735	+692580	−665	−5481	−26523	+611523	−248	−1407	−2781	+119028	−123	−14761	−3497	+246181	−818	−7553	−82979	+212212
−195	−6635	−32871	+593251	−575	−5549	−38063	+515933	−219	−1435	−4521	+104533	−108	−15004	−5302	+213466	−731	−7730	−149244	+190192
−164	−6664	−43690	+473552	−479	−5555	−49215	+402611	−188	−1448	−6226	+86768	−92	−15092	−7062	+174086	−638	−7832	−214654	+162712
−131	−6613	−53975	+335597	−377	−5495	−59755	+273581	−155	−1445	−7865	+66011	−75	−15015	−8745	+128667	−539	−7854	−278124	+130152
−96	−6477	−63495	+182172	−269	−5365	−69445	+131461	−120	−1425	−9405	+42636	−57	−14763	−10317	+78033	−434	−7791	−338499	+93036
−59	−6251	−72005	+16777	−155	−5161	−78033	−20501	−83	−1387	−10811	+17119	−38	−14326	−11742	+23218	−323	−7638	−394554	+52041
−20	−5930	−79246	−156332	−35	−4879	−85253	−178391	−44	−1330	−12046	−9956	−18	−13694	−12982	−34522	−206	−7390	−444994	+8006
+21	−5509	−84945	−332101	+91	−4515	−90825	−337593	−3	−1253	−13071	−37891	+3	−12857	−13997	−93697	−83	−7042	−488454	−38059
+64	−4983	−88815	−504636	+223	−4065	−94455	−492753	+40	−1155	−13845	−65868	+25	−11805	−14745	−152571	+46	−6589	−523499	−84964
+109	−4347	−90555	−667161	+361	−3525	−95835	−637743	+85	−1035	−14325	−92943	+48	−10528	−15182	−209150	+181	−6026	−548624	−131330
+156	−3596	−89850	−811976	+505	−2891	−94643	−765625	+132	−892	−14466	−118040	+72	−9016	−15262	−261170	+322	−5348	−562254	−175580
+205	−2725	−86371	−930415	+655	−2159	−90543	−868615	+181	−725	−14221	−139945	+97	−7259	−14937	−306085	+469	−4550	−562744	−215930
+256	−1729	−79775	−1012804	+811	−1325	−83185	−938047	+232	−533	−13541	−157300	+123	−5247	−14157	−341055	+622	−3627	−548379	−250380
+309	−603	−69705	−1048419	+973	−385	−72205	−964337	+285	−315	−12375	−168597	+150	−2970	−12870	−362934	+781	−2574	−517374	−276705
+364	+658	−55790	−1025444	+1141	+665	−57225	−936947	+340	−70	−10670	−172172	+178	−418	−11022	−368258	+946	−1386	−467874	−292446
+421	+2059	−37645	−930929	+1315	+1829	−37853	−844349	+397	+203	−8371	−166199	+207	+2419	−8557	−353233	+1117	−58	−397954	−294901
+480	+3605	−14871	−750748	+1495	+3111	−13683	−673989	+456	+505	−5421	−148684	+237	+5551	−5417	−313723	+1294	+1415	−305619	−281116
+541	+5301	+12945	−469557	+1681	+4515	+15705	−412251	+517	+837	−1761	−117459	+268	+8988	−1542	−245238	+1477	+3038	−188804	−247876
+604	+7152	+46230	−70752	+1873	+6045	+50745	−44421	+580	+1200	+2670	−70176	+300	+12740	+3130	−142922	+1666	+4816	−45374	−191696
+669	+9163	+85425	+463573	+2071	+7705	+91885	+445349	+645	+1595	+7935	−4301	+333	+16817	+8663	−1541	+1861	+6754	+126876	−108812
+736	+11339	+130985	+1152668	+2275	+9499	+139587	+1074077	+712	+2023	+14099	+82892	+367	+21229	+15123	+184529	+2062	+8857	+330221	+4828
+805	+13685	+183379	+2017169	+2485	+11431	+194327	+1859987	+781	+2485	+21229	+194327	+402	+25986	+22578	+421314	+2269	+11130	+567006	+153573
								+852	+2982	+29394	+333132	+438	+31098	+31098	+715254	+2482	+13578	+839646	+342078
																+2701	+16206	+1150626	+575313
	2,249,473,830		40,090,122,598,260		1,587,863,880		34,362,962,227,080		109,309,470		1,111,240,072,020		12,024,041,700		5,163,997,981,740		3,302,377,650		3,367,824,770,700
10,013,556		351,667,742,090		96,652,584		398,906,692,520		11,506,260		9,218,431,970		3,079,140		10,420,836,140		118,546,890		14,406,805,963,550	
1	$\frac{5}{6}$	$\frac{7}{12}$	$\frac{7}{20}$	3	$\frac{2}{3}$	$\frac{7}{12}$	$\frac{3}{10}$	1	$\frac{1}{6}$	$\frac{1}{12}$	$\frac{1}{20}$	$\frac{1}{2}$	$\frac{5}{3}$	$\frac{1}{12}$	$\frac{1}{10}$	3	$\frac{5}{6}$	$\frac{35}{12}$	$\frac{3}{40}$

Table XXIV. Calculation of Integrals from equally spaced Ordinates

Each panel over which the integral is required is divided by calculated ordinates into a number of sub-panels, usually an even number, and often a power of 2. The formula may involve only internal and terminal ordinates, or also adjacent external ordinates. Ordinates equidistant from the centre have equal coefficients. The central coefficient is the last given.

Number of sub-panels.	No external ordinates. Coefficients.	Divisor.	One external ordinate. Coefficients.	Divisor.
2	1, 4	6	−1, 34, 114	180
4	7, 32, 12	90	−4, 171, 612, 332	1890
6	41, 216, 27, 272	840	−9, 482, 1908, 774, 2090	8400

Equally we may use the central ordinate and its even differences with the following coefficients:

Number of sub-panels.	δ^0	δ^2	δ^4	δ^6	δ^8	δ^{10}
2	1	$0\cdot1\dot{6}$	$-\cdot00\dot{5}$	$\cdot000\dot{6}6\ 137\dot{5}$	$-\cdot00010\ 141$	$\cdot00001\ 76$
4	1	$0\cdot6\dot{6}$	$\cdot07\dot{7}$	$-\cdot002\dot{1}1\ 640\dot{2}$	$\cdot00022\ 928$	$-\cdot00003\ 31$
6	1	$1\cdot50$	$\cdot550$	$\cdot048\dot{8}0\ 952\dot{3}$	$-\cdot00107\ 143$	$\cdot00010\ 28$
8	1	$2\cdot6\dot{6}$	$1\cdot91\dot{1}$	$\cdot4867\dot{7}7\ 248\dot{6}$	$\cdot03488\ 536$	$-\cdot00063\ 28$
10	1	$4\cdot1\dot{6}$	$4\cdot86\dot{1}$	$2\cdot278\dot{4}3\ 915\dot{3}$	$\cdot44587\ 743$	$\cdot02683\ 41$

When the number of sub-panels is large, Gregory's formula, in terms of terminal differences, is valuable:

$$\int_0^r f(x)dx-(\tfrac{1}{2}f_0+f_1+f_2\ \ldots\ +f_{r-1}+\tfrac{1}{2}f_r)$$
$$=-a_1(\Delta f_{r-1}-\Delta f_0)-a_2(\Delta^2 f_{r-2}+\Delta^2 f_0)-a_3(\Delta^3 f_{r-3}-\Delta^3 f_0)-a_4(\Delta^4 f_{r-4}+\Delta^4 f_0)\ \ldots$$

where a_r is the coefficient of x^r in the expansion of $-1/\log(1-x)$.

Values of a_1 to a_{16}

1	$\cdot0833\dot{3}$	5	·01426 91799	9	·00678 585	13	·00421 5
2	$\cdot0416\dot{6}$	6	·01136 73942	10	·00592 406	14	·00382 7
3	$\cdot0263\dot{8}$	7	·00935 6537	11	·00523 67	15	·00350
4	·01875	8	·00789 2554	12	·00467 75	16	·00321

	0	1	2	3	4	5	6	7	8	9	1	2	3	4	5
10	·00000	00432	00860	01284	01703	—	—	—	—	—	43	85	128	170	213
	—	—	—	—	—	02119	02531	02938	03342	03743	41	81	122	162	203
11	·04139	04532	04922	05308	05690	—	—	—	—	—	39	78	116	155	194
	—	—	—	—	—	06070	06446	06819	07188	07555	37	74	111	148	186
12	·07918	08279	08636	08991	09342	—	—	—	—	—	36	71	107	142	178
	—	—	—	—	—	09691	10037	10380	10721	11059	34	68	103	137	171
13	·11394	11727	12057	12385	12710	—	—	—	—	—	33	66	99	132	164
	—	—	—	—	—	13033	13354	13672	13988	14301	32	63	95	127	158
14	·14613	14922	15229	15534	15836	—	—	—	—	—	31	61	92	122	153
	—	—	—	—	—	16137	16435	16732	17026	17319	30	59	89	118	148
15	·17609	17898	18184	18469	18752	—	—	—	—	—	29	57	86	114	143
	—	—	—	—	—	19033	19312	19590	19866	20140	28	55	83	111	138
16	·20412	20683	20952	21219	21484	—	—	—	—	—	27	54	80	107	134
	—	—	—	—	—	21748	22011	22272	22531	22789	26	52	78	104	130
17	·23045	23300	23553	23805	24055	—	—	—	—	—	25	50	76	101	126
	—	—	—	—	—	24304	24551	24797	25042	25285	25	49	74	98	123
18	·25527	25768	26007	26245	26482	—	—	—	—	—	24	48	72	96	119
	—	—	—	—	—	26717	26951	27184	27416	27646	23	46	70	93	116
19	·27875	28103	28330	28556	28780	—	—	—	—	—	23	45	69	90	113
	—	—	—	—	—	29003	29226	29447	29667	29885	22	44	66	88	110
20	·30103	30320	30535	30750	30963	31175	31387	31597	31806	32015	21	42	64	85	106
21	·32222	32428	32634	32838	33041	33244	33445	33646	33846	34044	20	40	61	81	101
22	·34242	34439	34635	34830	35025	35218	35411	35603	35793	35984	19	39	58	77	97
23	·36173	36361	36549	36736	36922	37107	37291	37475	37658	37840	19	37	56	74	93
24	·38021	38202	38382	38561	38739	38917	39094	39270	39445	39620	18	36	53	71	89
25	·39794	39967	40140	40312	40483	40654	40824	40993	41162	41330	17	34	51	68	85
26	·41497	41664	41830	41996	42160	42325	42488	42651	42813	42975	16	33	49	66	82
27	·43136	43297	43457	43616	43775	43933	44091	44248	44404	44560	16	32	47	63	79
28	·44716	44871	45025	45179	45332	45484	45637	45788	45939	46090	15	31	46	61	76
29	·46240	46389	46538	46687	46835	46982	47129	47276	47422	47567	15	29	44	59	74
30	·47712	47857	48001	48144	48287	48430	48572	48714	48855	48996	14	29	43	57	71
31	·49136	49276	49415	49554	49693	49831	49969	50106	50243	50379	14	28	41	55	69
32	·50515	50651	50786	50920	51055	51188	51322	51455	51587	51720	13	27	40	54	67
33	·51851	51983	52114	52244	52375	52504	52634	52763	52892	53020	13	26	39	52	65
34	·53148	53275	53403	53529	53656	53782	53908	54033	54158	54283	13	25	38	50	63
35	·54407	54531	54654	54777	54900	55023	55145	55267	55388	55509	12	24	37	49	61
36	·55630	55751	55871	55991	56110	56229	56348	56467	56585	56703	12	24	36	48	60
37	·56820	56937	57054	57171	57287	57403	57519	57634	57749	57864	12	23	35	46	58
38	·57978	58092	58206	58320	58433	58546	58659	58771	58883	58995	11	23	34	45	56
39	·59106	59218	59329	59439	59550	59660	59770	59879	59988	60097	11	22	33	44	55
40	·60206	60314	60423	60531	60638	60746	60853	60959	61066	61172	11	21	32	43	54
41	·61278	61384	61490	61595	61700	61805	61909	62014	62118	62221	10	21	31	42	52
42	·62325	62428	62531	62634	62737	62839	62941	63043	63144	63246	10	20	31	41	51
43	·63347	63448	63548	63649	63749	63849	63949	64048	64147	64246	10	20	30	40	50
44	·64345	64444	64542	64640	64738	64836	64933	65031	65128	65225	10	20	29	39	49
45	·65321	65418	65514	65610	65706	65801	65896	65992	66087	66181	10	19	29	38	48
46	·66276	66370	66464	66558	66652	66745	66839	66932	67025	67117	9	19	28	37	47
47	·67210	67302	67394	67486	67578	67669	67761	67852	67943	68034	9	18	27	37	46
48	·68124	68215	68305	68395	68485	68574	68664	68753	68842	68931	9	18	27	36	45
49	·69020	69108	69197	69285	69373	69461	69548	69636	69723	69810	9	18	26	35	44

	0	1	2	3	4	5	6	7	8	9	1	2	3	4	5
50	·69897	69984	70070	70157	70243	70329	70415	70501	70586	70672	9	17	26	34	43
51	·70757	70842	70927	71012	71096	71181	71265	71349	71433	71517	8	17	25	34	42
52	·71600	71684	71767	71850	71933	72016	72099	72181	72263	72346	8	17	25	33	41
53	·72428	72509	72591	72673	72754	72835	72916	72997	73078	73159	8	16	24	32	41
54	·73239	73320	73400	73480	73560	73640	73719	73799	73878	73957	8	16	24	32	40
55	·74036	74115	74194	74273	74351	74429	74507	74586	74663	74741	8	16	23	31	39
56	·74819	74896	74974	75051	75128	75205	75282	75358	75435	75511	8	15	23	31	38
57	·75587	75664	75740	75815	75891	75967	76042	76118	76193	76268	8	15	23	30	38
58	·76343	76418	76492	76567	76641	76716	76790	76864	76938	77012	7	15	22	30	37
59	·77085	77159	77232	77305	77379	77452	77525	77597	77670	77743	7	15	22	29	37
60	·77815	77887	77960	78032	78104	78176	78247	78319	78390	78462	7	14	22	29	36
61	·78533	78604	78675	78746	78817	78888	78958	79029	79099	79169	7	14	21	28	35
62	·79239	79309	79379	79449	79518	79588	79657	79727	79796	79865	7	14	21	28	35
63	·79934	80003	80072	80140	80209	80277	80346	80414	80482	80550	7	14	21	27	34
64	·80618	80686	80754	80821	80889	80956	81023	81090	81158	81224	7	13	20	27	34
65	·81291	81358	81425	81491	81558	81624	81690	81757	81823	81889	7	13	20	27	33
66	·81954	82020	82086	82151	82217	82282	82347	82413	82478	82543	7	13	20	26	33
67	·82607	82672	82737	82802	82866	82930	82995	83059	83123	83187	6	13	19	26	32
68	·83251	83315	83378	83442	83506	83569	83632	83696	83759	83822	6	13	19	25	32
69	·83885	83948	84011	84073	84136	84198	84261	84323	84386	84448	6	13	19	25	31
70	·84510	84572	84634	84696	84757	84819	84880	84942	85003	85065	6	12	19	25	31
71	·85126	85187	85248	85309	85370	85431	85491	85552	85612	85673	6	12	18	24	30
72	·85733	85794	85854	85914	85974	86034	86094	86153	86213	86273	6	12	18	24	30
73	·86332	86392	86451	86510	86570	86629	86688	86747	86806	86864	6	12	18	24	30
74	·86923	86982	87040	87099	87157	87216	87274	87332	87390	87448	6	12	17	23	29
75	·87506	87564	87622	87679	87737	87795	87852	87910	87967	88024	6	12	17	23	29
76	·88081	88138	88195	88252	88309	88366	88423	88480	88536	88593	6	11	17	23	28
77	·88649	88705	88762	88818	88874	88930	88986	89042	89098	89154	6	11	17	22	28
78	·89209	89265	89321	89376	89432	89487	89542	89597	89653	89708	6	11	17	22	28
79	·89763	89818	89873	89927	89982	90037	90091	90146	90200	90255	5	11	16	22	27
80	·90309	90363	90417	90472	90526	90580	90634	90687	90741	90795	5	11	16	22	27
81	·90849	90902	90956	91009	91062	91116	91169	91222	91275	91328	5	11	16	21	27
82	·91381	91434	91487	91540	91593	91645	91698	91751	91803	91855	5	11	16	21	26
83	·91908	91960	92012	92065	92117	92169	92221	92273	92324	92376	5	10	16	21	26
84	·92428	92480	92531	92583	92634	92686	92737	92788	92840	92891	5	10	15	21	26
85	·92942	92993	93044	93095	93146	93197	93247	93298	93349	93399	5	10	15	20	25
86	·93450	93500	93551	93601	93651	93702	93752	93802	93852	93902	5	10	15	20	25
87	·93952	94002	94052	94101	94151	94201	94250	94300	94349	94399	5	10	15	20	25
88	·94448	94498	94547	94596	94645	94694	94743	94792	94841	94890	5	10	15	20	25
89	·94939	94988	95036	95085	95134	95182	95231	95279	95328	95376	5	10	15	19	24
90	·95424	95472	95521	95569	95617	95665	95713	95761	95809	95856	5	10	14	19	24
91	·95904	95952	95999	96047	96095	96142	96190	96237	96284	96332	5	10	14	19	24
92	·96379	96426	96473	96520	96567	96614	96661	96708	96755	96802	5	9	14	19	24
93	·96848	96895	96942	96988	97035	97081	97128	97174	97220	97267	5	9	14	19	23
94	·97313	97359	97405	97451	97497	97543	97589	97635	97681	97727	5	9	14	18	23
95	·97772	97818	97864	97909	97955	98000	98046	98091	98137	98182	5	9	14	18	23
96	·98227	98272	98318	98363	98408	98453	98498	98543	98588	98632	4	9	14	18	22
97	·98677	98722	98767	98811	98856	98900	98945	98989	99034	99078	4	9	13	18	22
98	·99123	99167	99211	99255	99300	99344	99388	99432	99476	99520	4	9	13	18	22
99	·99564	99607	99651	99695	99739	99782	99826	99870	99913	99957	4	9	13	17	22

Table XXVI. Natural Logarithms

Negative Values, Decreasing

	0	1	2	3	4	5	6	7	8	9	1	2	3	4	5
·000		9·2103	8·5172	8·1117	7·8240	7·6009	7·4186	7·2644	7·1309	7·0131					
·001	6·90776	81245	72543	64539	57128	50229	43775	37713	31997	26590					
·002	6·21461	16582	11930	07485	03229	**99146**	**95224**	**91450**	**87814**	**84304**					
·003	5·80914	77635	74460	71383	68398	65499	62682	59942	57275	54678					
·004	5·52146	49677	47267	44914	42615	40368	38170	36019	33914	31852					
·005	5·29832	27851	25910	24005	22136	20301	18499	16729	14990	13280					
·006	5·11600	09947	08321	06721	05146	03595	02069	00565	**99083**	**97623**					
·007	4·96185	94766	93367	91988	90628	89285	87961	86653	85363	84089					
·008	4·82831	81589	80362	79150	77952	76769	75599	74443	73300	72170					
·009	4·71053	69948	68855	67774	66705	65646	64599	63563	62537	61522					
·010	4·60517	59522	58537	57561	56595	55638	54690	53751	52821	51899					
·011	4·50986	50081	49184	48295	47414	46541	45675	44817	43966	43122					
·012	4·42285	41455	40632	39816	39006	38203	37406	36615	35831	35053					
·013	4·34281	33514	32754	31999	31250	30507	29769	29036	28309	27587					
·014	4·26870	26158	25451	24750	24053	23361	22673	21991	21313	20639					
·015	4·19971	19306	18646	17990	17339	16692	16048	15409	14775	14144					
·016	4·13517	12894	12274	11659	11047	10439	09835	09235	08638	08044					
·017	4·07454	06868	06285	05705	05129	04555	03986	03419	02856	02295					
·018	4·01738	01184	00633	00085	**99540**	**98998**	**98459**	**97923**	**97390**	**96859**					
·019	3·96332	95807	95284	94765	94248	93734	93223	92714	92207	91704					
·020	3·91202	90704	90207	89713	89222	88733	88246	87762	87280	86801	49	98	147	196	244
·021	3·86323	85848	85375	84905	84436	83970	83506	83044	82585	82127	47	93	140	186	233
·022	3·81671	81218	80766	80317	79869	79424	78981	78539	78099	77662	45	89	134	178	223
·023	3·77226	76792	76360	75930	75502	75075	74651	74228	73807	73388	43	85	128	171	213
·024	3·72970	72554	72140	71728	71317	70908	70501	70095	69691	69289	41	82	123	164	204
·025	3·68888	68489	68091	67695	67301	66908	66516	66126	65738	65351	39	79	118	157	196
·026	3·64966	64582	64200	63819	63439	63061	62684	62309	61935	61563	38	76	113	151	189
·027	3·61192	60822	60454	60087	59721	59357	58994	58632	58272	57913	36	72	109	146	182
·028	3·57555	57199	56843	56489	56137	55785	55435	55086	54738	54391	35	70	105	141	176
·029	3·54046	53702	53359	53017	52676	52337	51998	51661	51325	50990	34	68	102	136	170
·030	3·50656	50323	49991	49661	49331	49003	48676	48349	48024	47700	33	66	99	131	164
·031	3·47377	47055	46734	46414	46095	45777	45460	45144	44829	44515	32	64	95	127	159
·032	3·44202	43890	43579	43269	42960	42652	42344	42038	41733	41428	31	62	92	123	154
·033	3·41125	40822	40521	40220	39920	39621	39323	39026	38729	38434	30	60	90	120	150
·034	3·38139	37846	37553	37261	36970	36680	36390	36102	35814	35527	29	58	87	116	145
·035	3·35241	34955	34671	34387	34104	33822	33541	33260	32981	32702	28	56	85	113	141
·036	3·32424	32146	31870	31594	31319	31044	30771	30498	30226	29954	27	55	82	110	137
·037	3·29684	29414	29145	28876	28608	28341	28075	27810	27545	27280	27	53	80	107	134
·038	3·27017	26754	26492	26231	25970	25710	25450	25192	24934	24676	26	52	78	104	130
·039	3·24419	24163	23908	23653	23399	23145	22893	22640	22389	22138	25	51	76	101	127
·040	3·21888	21638	21389	21140	20893	20645	20399	20153	19907	19663	25	49	74	99	124
·041	3·19418	19175	18932	18689	18447	18206	17966	17725	17486	17247	24	48	72	96	121
·042	3·17009	16771	16534	16297	16061	15825	15590	15356	15122	14888	24	47	71	94	118
·043	3·14656	14423	14191	13960	13730	13499	13270	13041	12812	12584	23	46	69	92	115
·044	3·12357	12130	11903	11677	11452	11227	11002	10778	10555	10332	22	45	68	90	112
·045	3·10109	09887	09666	09445	09224	09004	08785	08566	08347	08129	22	44	66	88	110
·046	3·07911	07694	07478	07261	07046	06830	06615	06401	06187	05974	22	43	65	86	108
·047	3·05761	05548	05336	05124	04913	04703	04492	04282	04073	03864	21	42	63	84	105
·048	3·03655	03447	03240	03032	02826	02619	02413	02208	02002	01798	21	41	62	83	103
·049	3·01593	01390	01186	00983	00780	00578	00376	00175	**99974**	**99773**	20	40	61	81	101

For intermediate values use other parts of the table, taking $\log_e x = \log_e 10^p x - \log_e 10^p$.

$\log_e 10 = 2{\cdot}30259$, $\log_e 100 = 4{\cdot}60517$, $\log_e 1000 = 6{\cdot}90776$

TABLE XXVI. NATURAL LOGARITHMS—*continued*

Negative Values, Decreasing

	0	1	2	3	4	5	6	7	8	9	1	2	3	4	5
·050	2·99573	99373	99174	98975	98776	98578	98380	98183	97986	97789	20	40	59	79	99
·051	2·97593	97397	97202	97006	96812	96617	96423	96230	96037	95844	19	39	58	78	97
·052	2·95651	95459	95267	95076	94885	94694	94504	94314	94124	93935	19	38	57	76	95
·053	2·93746	93558	93370	93182	92994	92807	92621	92434	92248	92062	19	37	56	75	94
·054	2·91877	91692	91507	91323	91139	90955	90772	90589	90407	90224	18	37	55	73	92
·055	2·90042	89861	89679	89498	89318	89137	88957	88778	88598	88419	18	36	54	72	90
·056	2·88240	88062	87884	87706	87529	87351	87175	86998	86822	86646	18	35	53	71	89
·057	2·86470	86295	86120	85945	85771	85597	85423	85250	85077	84904	17	35	52	70	87
·058	2·84731	84559	84387	84215	84044	83873	83702	83532	83361	83191	17	34	51	68	86
·059	2·83022	82852	82683	82515	82346	82178	82010	81842	81675	81508	17	34	50	67	84
·060	2·81341	81175	81008	80842	80677	80511	80346	80181	80017	79852	17	33	50	66	83
·061	2·79688	79524	79361	79198	79035	78872	78709	78547	78385	78224	16	33	49	65	81
·062	2·78062	77901	77740	77579	77419	77259	77099	76939	76780	76621	16	32	48	64	80
·063	2·76462	76303	76145	75987	75829	75672	75514	75357	75200	75044	16	32	47	63	79
·064	2·74887	74731	74575	74420	74264	74109	73954	73799	73645	73491	16	31	47	62	78
·065	2·73337	73183	73030	72876	72723	72571	72418	72266	72114	71962	15	31	46	61	76
·066	2·71810	71659	71507	71357	71206	71055	70905	70755	70605	70456	15	30	45	60	75
·067	2·70306	70157	70008	69860	69711	69563	69415	69267	69119	68972	15	30	44	59	74
·068	2·68825	68678	68531	68385	68238	68092	67946	67801	67655	67510	15	29	44	58	73
·069	2·67365	67220	67075	66931	66787	66643	66499	66355	66212	66069	14	29	43	58	72
·070	2·65926	65783	65641	65498	65356	65214	65073	64931	64790	64648	14	28	43	57	71
·071	2·64508	64367	64226	64086	63946	63806	63666	63526	63387	63248	14	28	42	56	70
·072	2·63109	62970	62832	62693	62555	62417	62279	62141	62004	61867	14	28	41	55	69
·073	2·61730	61593	61456	61319	61183	61047	60911	60775	60640	60504	14	27	41	54	68
·074	2·60369	60234	60099	59964	59830	59696	59561	59428	59294	59160	13	27	40	54	67
·075	2·59027	58893	58760	58628	58495	58362	58230	58098	57966	57834	13	27	40	53	66
·076	2·57702	57571	57439	57308	57177	57046	56916	56785	56655	56525	13	26	39	52	65
·077	2·56395	56265	56136	56006	55877	55748	55619	55490	55361	55233	13	26	39	52	65
·078	2·55105	54977	54849	54721	54593	54466	54338	54211	54084	53957	13	25	38	51	64
·079	2·53831	53704	53578	53452	53326	53200	53074	52949	52823	52698	13	25	38	50	63
·080	2·52573	52448	52323	52199	52074	51950	51826	51702	51578	51454	12	25	37	50	62
·081	2·51331	51207	51084	50961	50838	50715	50593	50470	50348	50226	12	25	37	49	61
·082	2·50104	49982	49860	49738	49617	49496	49375	49254	49133	49012	12	24	36	49	61
·083	2·48891	48771	48651	48531	48411	48291	48171	48052	47932	47813	12	24	36	48	60
·084	2·47694	47575	47456	47337	47219	47100	46982	46864	46746	46628	12	24	36	47	59
·085	2·46510	46393	46275	46158	46041	45924	45807	45690	45574	45457	12	23	35	47	58
·086	2·45341	45225	45109	44993	44877	44761	44646	44530	44415	44300	12	23	35	46	58
·087	2·44185	44070	43955	43840	43726	43612	43497	43383	43269	43156	11	23	34	46	57
·088	2·43042	42928	42815	42702	42588	42475	42362	42250	42137	42024	11	23	34	45	56
·089	2·41912	41800	41687	41575	41463	41352	41240	41128	41017	40906	11	22	34	45	56
·090	2·40795	40684	40573	40462	40351	40241	40130	40020	39910	39800	11	22	33	44	55
·091	2·39690	39580	39470	39360	39251	39142	39032	38923	38814	38705	11	22	33	44	55
·092	2·38597	38488	38380	38271	38163	38055	37947	37839	37731	37623	11	22	32	43	54
·093	2·37516	37408	37301	37194	37086	36979	36872	36766	36659	36552	11	21	32	43	54
·094	2·36446	36340	36234	36127	36021	35916	35810	35704	35599	35493	11	21	32	42	53
·095	2·35388	35283	35178	35073	34968	34863	34758	34654	34549	34445	10	21	31	42	52
·096	2·34341	34237	34133	34029	33925	33821	33718	33614	33511	33408	10	21	31	41	52
·097	2·33304	33201	33098	32996	32893	32790	32688	32585	32483	32381	10	21	31	41	51
·098	2·32279	32177	32075	31973	31871	31770	31668	31567	31466	31365	10	20	30	41	51
·099	2·31264	31163	31062	30961	30860	30760	30659	30559	30459	30359	10	20	30	40	50

$\log_e 10{,}000 = 9{\cdot}21034$, $\log_e 100{,}000 = 11{\cdot}51293$, $\log_e 1{,}000{,}000 = 13{\cdot}81551$

Table XXVI. Natural Logarithms—*continued*

Negative Values, Decreasing

	0	1	2	3	4	5	6	7	8	9	1	2	3	4	5
·10	2·30259	29263	28278	27303	26336	—	—	—	—	—	98	196	294	392	490
	—	—	—	—	—	25379	24432	23493	22562	21641	93	187	280	374	467
·11	2·20727	19823	18926	18037	17156	—	—	—	—	—	89	179	268	357	446
	—	—	—	—	—	16282	15417	14558	13707	12863	85	171	256	342	427
·12	2·12026	11196	10373	09557	08747	—	—	—	—	—	82	164	246	328	410
	—	—	—	—	—	07944	07147	06357	05573	04794	79	158	236	315	394
·13	2·04022	03256	02495	01741	00992	—	—	—	—	—	76	152	227	303	379
	—	—	—	—	—	00248	**99510**	**98777**	**98050**	**97328**	73	146	219	292	365
·14	1·96611	95900	95193	94491	93794	—	—	—	—	—	70	141	211	282	352
	—	—	—	—	—	93102	92415	91732	91054	90381	68	136	204	272	340
·15	1·89712	89048	88387	87732	87080	—	—	—	—	—	66	132	197	263	329
	—	—	—	—	—	86433	85790	85151	84516	83885	64	127	191	255	318
·16	1·83258	82635	82016	81401	80789	—	—	—	—	—	62	123	185	247	309
	—	—	—	—	—	80181	79577	78976	78379	77786	60	120	180	240	299
·17	1·77196	76609	76026	75446	74870	—	—	—	—	—	58	116	174	233	291
	—	—	—	—	—	74297	73727	73161	72597	72037	56	113	170	226	282
·18	1·71480	70926	70375	69827	69282	—	—	—	—	—	55	110	165	220	275
	—	—	—	—	—	68740	68201	67665	67131	66601	53	107	160	214	267
·19	1·66073	65548	65026	64507	63990	—	—	—	—	—	52	104	156	208	260
	—	—	—	—	—	63476	62964	62455	61949	61445	51	102	152	203	254
·20	1·60944	60445	59949	59455	58964	58475	57988	57504	57022	56542	49	98	147	196	245
·21	1·56065	55590	55117	54646	54178	53712	53248	52786	52326	51868	47	93	140	187	233
·22	1·51413	50959	50508	50058	49611	49165	48722	48281	47841	47403	45	89	134	178	223
·23	1·46968	46534	46102	45672	45243	44817	44392	43970	43548	43129	43	85	128	171	213
·24	1·42712	42296	41882	41469	41059	40650	40242	39837	39433	39030	41	82	123	164	205
·25	1·38629	38230	37833	37437	37042	36649	36258	35868	35480	35093	39	79	118	157	196
·26	1·34707	34323	33941	33560	33181	32803	32426	32051	31677	31304	38	76	113	151	189
·27	1·30933	30564	30195	29828	29463	29098	28735	28374	28013	27654	36	73	109	146	182
·28	1·27297	26940	26585	26231	25878	25527	25176	24827	24479	24133	35	70	105	141	176
·29	1·23787	23443	23100	22758	22418	22078	21740	21402	21066	20731	34	68	102	136	170
·30	1·20397	20065	19733	19402	19073	18744	18417	18091	17766	17441	33	66	99	131	164
·31	1·17118	16796	16475	16155	15836	15518	15201	14885	14570	14256	32	64	95	127	159
·32	1·13943	13631	13320	13010	12701	12393	12086	11780	11474	11170	31	62	92	123	154
·33	1·10866	10564	10262	09961	09661	09362	09064	08767	08471	08176	30	60	90	120	149
·34	1·07881	07587	07294	07002	06711	06421	06132	05843	05555	05268	29	58	87	116	145
·35	1·04982	04697	04412	04129	03846	03564	03282	03002	02722	02443	28	56	85	113	141
·36	1·02165	01888	01611	01335	01060	00786	00512	00239	**99967**	**99696**	27	55	82	110	137
·37	0·99425	99155	98886	98618	98350	98083	97817	97551	97286	97022	27	53	80	107	134
·38	0·96758	96496	96233	95972	95711	95451	95192	94933	94675	94418	26	52	78	104	130
·39	0·94161	93905	93649	93395	93140	92887	92634	92382	92130	91879	25	51	76	101	127
·40	0·91629	91379	91130	90882	90634	90387	90140	89894	89649	89404	25	49	74	99	124
·41	0·89160	88916	88673	88431	88189	87948	87707	87467	87227	86988	24	48	72	96	121
·42	0·86750	86512	86275	86038	85802	85567	85332	85097	84863	84630	24	47	71	94	118
·43	0·84397	84165	83933	83702	83471	83241	83011	82782	82554	82326	23	46	69	92	115
·44	0·82098	81871	81645	81419	81193	80968	80744	80520	80296	80073	22	45	68	90	112
·45	0·79851	79629	79407	79186	78966	78746	78526	78307	78089	77871	22	44	66	88	110
·46	0·77653	77436	77219	77003	76787	76572	76357	76143	75929	75715	22	43	65	86	108
·47	0·75502	75290	75078	74866	74655	74444	74234	74024	73814	73605	21	42	63	84	105
·48	0·73397	73189	72981	72774	72567	72361	72155	71949	71744	71539	21	41	62	83	103
·49	0·71335	71131	70928	70725	70522	70320	70118	69917	69716	69515	20	40	61	81	101

$\log_e 10 = 2{\cdot}30259$, $\log_e 100 = 4{\cdot}60517$, $\log_e 1000 = 6{\cdot}90776$

Table XXVI. Natural Logarithms—*continued*

Negative Values, Decreasing

	0	1	2	3	4	5	6	7	8	9	1	2	3	4	5
·50	0·69315	69115	68916	68717	68518	68320	68122	67924	67727	67531	20	40	59	79	99
·51	0·67334	67139	66943	66748	66553	66359	66165	65971	65778	65585	19	39	58	78	97
·52	0·65393	65201	65009	64817	64626	64436	64245	64055	63866	63677	19	38	57	76	95
·53	0·63488	63299	63111	62923	62736	62549	62362	62176	61990	61804	19	37	56	75	94
·54	0·61619	61434	61249	61065	60881	60697	60514	60331	60148	59966	18	37	55	73	92
·55	0·59784	59602	59421	59240	59059	58879	58699	58519	58340	58161	18	36	54	72	90
·56	0·57982	57803	57625	57448	57270	57093	56916	56740	56563	56387	18	35	53	71	89
·57	0·56212	56037	55862	55687	55513	55339	55165	54991	54818	54645	17	35	52	70	87
·58	0·54473	54300	54128	53957	53785	53614	53444	53273	53103	52933	17	34	51	68	86
·59	0·52763	52594	52425	52256	52088	51919	51751	51584	51416	51249	17	34	50	67	84
·60	0·51083	50916	50750	50584	50418	50253	50088	49923	49758	49594	17	33	50	66	83
·61	0·49430	49266	49102	48939	48776	48613	48451	48289	48127	47965	16	33	49	65	81
·62	0·47804	47642	47482	47321	47160	47000	46840	46681	46522	46362	16	32	48	64	80
·63	0·46204	46045	45887	45728	45571	45413	45256	45099	44942	44785	16	32	47	63	79
·64	0·44629	44473	44317	44161	44006	43850	43696	43541	43386	43232	16	31	47	62	78
·65	0·43078	42925	42771	42618	42465	42312	42159	42007	41855	41703	15	31	46	61	76
·66	0·41552	41400	41249	41098	40947	40797	40647	40497	40347	40197	15	30	45	60	75
·67	0·40048	39899	39750	39601	39453	39304	39156	39008	38861	38713	15	30	44	59	74
·68	0·38566	38419	38273	38126	37980	37834	37688	37542	37397	37251	15	29	44	58	73
·69	0·37106	36962	36817	36673	36528	36384	36241	36097	35954	35810	14	29	43	58	72
·70	0·35668	35525	35382	35240	35098	34956	34814	34672	34531	34390	14	28	43	57	71
·71	0·34249	34108	33968	33827	33687	33547	33408	33268	33129	32989	14	28	42	56	70
·72	0·32850	32712	32573	32435	32296	32158	32021	31883	31745	31608	14	28	41	55	69
·73	0·31471	31334	31197	31061	30925	30788	30653	30517	30381	30246	14	27	41	54	68
·74	0·30111	29975	29841	29706	29571	29437	29303	29169	29035	28902	13	27	40	54	67
·75	0·28768	28635	28502	28369	28236	28104	27971	27839	27707	27575	13	27	40	53	66
·76	0·27444	27312	27181	27050	26919	26788	26657	26527	26397	26266	13	26	39	52	65
·77	0·26136	26007	25877	25748	25618	25489	25360	25231	25103	24974	13	26	39	52	65
·78	0·24846	24718	24590	24462	24335	24207	24080	23953	23826	23699	13	25	38	51	64
·79	0·23572	23446	23319	23193	23067	22941	22816	22690	22565	22439	13	25	38	50	63
·80	0·22314	22189	22065	21940	21816	21691	21567	21443	21319	21196	12	25	37	50	62
·81	0·21072	20949	20825	20702	20579	20457	20334	20212	20089	19967	12	25	37	49	61
·82	0·19845	19723	19601	19480	19358	19237	19116	18995	18874	18754	12	24	36	49	61
·83	0·18633	18513	18392	18272	18152	18032	17913	17793	17674	17555	12	24	36	48	60
·84	0·17435	17316	17198	17079	16960	16842	16724	16605	16487	16370	12	24	36	47	59
·85	0·16252	16134	16017	15900	15782	15665	15548	15432	15315	15199	12	23	35	47	58
·86	0·15082	14966	14850	14734	14618	14503	14387	14272	14156	14041	12	23	35	46	58
·87	0·13926	13811	13697	13582	13467	13353	13239	13125	13011	12897	11	23	34	46	57
·88	0·12783	12670	12556	12443	12330	12217	12104	11991	11878	11766	11	23	34	45	56
·89	0·11653	11541	11429	11317	11205	11093	10981	10870	10759	10647	11	22	34	45	56
·90	0·10536	10425	10314	10203	10093	09982	09872	09761	09651	09541	11	22	33	44	55
·91	0·09431	09321	09212	09102	08992	08883	08774	08665	08556	08447	11	22	33	44	55
·92	0·08338	08230	08121	08013	07904	07796	07688	07580	07472	07365	11	22	32	43	54
·93	0·07257	07150	07042	06935	06828	06721	06614	06507	06401	06294	11	21	32	43	54
·94	0·06188	06081	05975	05869	05763	05657	05551	05446	05340	05235	11	21	32	42	53
·95	0·05129	05024	04919	04814	04709	04604	04500	04395	04291	04186	10	21	31	42	52
·96	0·04082	03978	03874	03770	03666	03563	03459	03356	03252	03149	10	21	31	41	52
·97	0·03046	02943	02840	02737	02634	02532	02429	02327	02225	02122	10	21	31	41	51
·98	0·02020	01918	01816	01715	01613	01511	01410	01309	01207	01106	10	20	30	41	51
·99	0·01005	00904	00803	00702	00602	00501	00401	00300	00200	00100	10	20	30	40	50

$\log_e 10{,}000 = 9{\cdot}21034$, $\log_e 100{,}000 = 11{\cdot}51293$, $\log_e 1{,}000{,}000 = 13{\cdot}81551$

	0	1	2	3	4	5	6	7	8	9	1	2	3	4	5
1·0	0·00000	00995	01980	02956	03922	—	—	—	—	—	98	196	294	392	490
	—	—	—	—	—	04879	05827	06766	07696	08618	93	187	280	374	467
1·1	0·09531	10436	11333	12222	13103	—	—	—	—	—	89	179	268	357	446
	—	—	—	—	—	13976	14842	15700	16551	17395	85	171	256	342	427
1·2	0·18232	19062	19885	20701	21511	—	—	—	—	—	82	164	246	328	410
	—	—	—	—	—	22314	23111	23902	24686	25464	79	158	236	315	394
1·3	0·26236	27003	27763	28518	29267	—	—	—	—	—	76	152	227	303	379
	—	—	—	—	—	30010	30748	31481	32208	32930	73	146	219	292	365
1·4	0·33647	34359	35066	35767	36464	—	—	—	—	—	70	141	211	282	352
	—	—	—	—	—	37156	37844	38526	39204	39878	68	136	204	272	340
1·5	0·40547	41211	41871	42527	43178	—	—	—	—	—	66	132	197	263	329
	—	—	—	—	—	43825	44469	45108	45742	46373	64	127	191	255	318
1·6	0·47000	47623	48243	48858	49470	—	—	—	—	—	62	124	185	247	309
	—	—	—	—	—	50078	50682	51282	51879	52473	60	120	180	240	299
1·7	0·53063	53649	54232	54812	55389	—	—	—	—	—	58	116	174	233	291
	—	—	—	—	—	55962	56531	57098	57661	58222	56	113	169	226	282
1·8	0·58779	59333	59884	60432	60977	—	—	—	—	—	55	110	165	220	275
	—	—	—	—	—	61519	62058	62594	63127	63658	53	107	160	214	267
1·9	0·64185	64710	65233	65752	66269	—	—	—	—	—	52	104	156	208	260
	—	—	—	—	—	66783	67294	67803	68310	68813	51	102	152	203	254
2·0	0·69315	69813	70310	70804	71295	71784	72271	72755	73237	73716	49	98	147	196	244
2·1	0·74194	74669	75142	75612	76081	76547	77011	77473	77932	78390	47	93	140	186	233
2·2	0·78846	79299	79751	80200	80648	81093	81536	81978	82418	82855	45	89	134	178	223
2·3	0·83291	83725	84157	84587	85015	85442	85866	86289	86710	87129	43	85	128	171	213
2·4	0·87547	87963	88377	88789	89200	89609	90016	90422	90826	91228	41	82	123	164	204
2·5	0·91629	92028	92426	92822	93216	93609	94001	94391	94779	95166	39	79	118	157	196
2·6	0·95551	95935	96317	96698	97078	97456	97833	98208	98582	98954	38	76	113	151	189
2·7	0·99325	99695	**00063**	**00430**	**00796**	**01160**	**01523**	**01885**	**02245**	**02604**	36	73	109	146	182
2·8	1·02962	03318	03674	04028	04380	04732	05082	05431	05779	06126	35	70	105	141	176
2·9	1·06471	06815	07158	07500	07841	08181	08519	08856	09192	09527	34	68	102	136	170
3·0	1·09861	10194	10526	10856	11186	11514	11841	12168	12493	12817	33	66	99	131	164
3·1	1·13140	13462	13783	14103	14422	14740	15057	15373	15688	16002	32	64	95	127	159
3·2	1·16315	16627	16938	17248	17557	17865	18173	18479	18784	19089	31	62	92	123	154
3·3	1·19392	19695	19996	20297	20597	20896	21194	21491	21788	22083	30	60	90	120	150
3·4	1·22378	22671	22964	23256	23547	23837	24127	24415	24703	24990	29	58	87	116	145
3·5	1·25276	25562	25846	26130	26413	26695	26976	27257	27536	27815	28	56	85	113	141
3·6	1·28093	28371	28647	28923	29198	29473	29746	30019	30291	30563	27	55	82	110	137
3·7	1·30833	31103	31372	31641	31909	32176	32442	32708	32972	33237	27	53	80	107	134
3·8	1·33500	33763	34025	34286	34547	34807	35067	35325	35584	35841	26	52	78	104	130
3·9	1·36098	36354	36609	36864	37118	37372	37624	37877	38128	38379	25	51	76	101	127
4·0	1·38629	38879	39128	39377	39624	39872	40118	40364	40610	40854	25	49	74	99	124
4·1	1·41099	41342	41585	41828	42070	42311	42552	42792	43031	43270	24	48	72	96	121
4·2	1·43508	43746	43984	44220	44456	44692	44927	45161	45395	45629	24	47	71	94	118
4·3	1·45862	46094	46326	46557	46787	47018	47247	47476	47705	47933	23	46	69	92	115
4·4	1·48160	48387	48614	48840	49065	49290	49515	49739	49962	50185	22	45	68	90	112
4·5	1·50408	50630	50851	51072	51293	51513	51732	51951	52170	52388	22	44	66	88	110
4·6	1·52606	52823	53039	53256	53471	53687	53902	54116	54330	54543	22	43	65	86	108
4·7	1·54756	54969	55181	55393	55604	55814	56025	56235	56444	56653	21	42	63	84	105
4·8	1·56862	57070	57277	57485	57691	57898	58104	58309	58515	58719	21	41	62	83	103
4·9	1·58924	59127	59331	59534	59737	59939	60141	60342	60543	60744	20	40	61	81	101

$\log_e 10 = 2{\cdot}30259$, $\log_e 100 = 4{\cdot}60517$, $\log_e 1000 = 6{\cdot}90776$

TABLE XXVI. NATURAL LOGARITHMS—*continued*

	0	1	2	3	4	5	6	7	8	9	1	2	3	4	5
5·0	1·60944	61144	61343	61542	61741	61939	62137	62334	62531	62728	20	40	59	79	99
5·1	1·62924	63120	63315	63511	63705	63900	64094	64287	64481	64673	19	39	58	78	97
5·2	1·64866	65058	65250	65441	65632	65823	66013	66203	66393	66582	19	38	57	76	95
5·3	1·66771	66959	67147	67335	67523	67710	67896	68083	68269	68455	19	37	56	75	94
5·4	1·68640	68825	69010	69194	69378	69562	69745	69928	70111	70293	18	37	55	73	92
5·5	1·70475	70656	70838	71019	71199	71380	71560	71740	71919	72098	18	36	54	72	90
5·6	1·72277	72455	72633	72811	72988	73166	73342	73519	73695	73871	18	35	53	71	89
5·7	1·74047	74222	74397	74572	74746	74920	75094	75267	75440	75613	17	35	52	70	87
5·8	1·75786	75958	76130	76302	76473	76644	76815	76985	77156	77326	17	34	51	68	86
5·9	1·77495	77665	77834	78002	78171	78339	78507	78675	78842	79009	17	34	50	67	84
6·0	1·79176	79342	79509	79675	79840	80006	80171	80336	80500	80665	17	33	50	66	83
6·1	1·80829	80993	81156	81319	81482	81645	81808	81970	82132	82294	16	33	49	65	81
6·2	1·82455	82616	82777	82938	83098	83258	83418	83578	83737	83896	16	32	48	64	80
6·3	1·84055	84214	84372	84530	84688	84845	85003	85160	85317	85473	16	32	47	63	79
6·4	1·85630	85786	85942	86097	86253	86408	86563	86718	86872	87026	16	31	47	62	78
6·5	1·87180	87334	87487	87641	87794	87947	88099	88251	88403	88555	15	31	46	61	76
6·6	1·88707	88858	89010	89160	89311	89462	89612	89762	89912	90061	15	30	45	60	75
6·7	1·90211	90360	90509	90658	90806	90954	91102	91250	91398	91545	15	30	44	59	74
6·8	1·91692	91839	91986	92132	92279	92425	92571	92716	92862	93007	15	29	44	58	73
6·9	1·93152	93297	93442	93586	93730	93874	94018	94162	94305	94448	14	29	43	58	72
7·0	1·94591	94734	94876	95019	95161	95303	95445	95586	95727	95869	14	28	43	57	71
7·1	1·96009	96150	96291	96431	96571	96711	96851	96991	97130	97269	14	28	42	56	70
7·2	1·97408	97547	97685	97824	97962	98100	98238	98376	98513	98650	14	28	41	55	69
7·3	1·98787	98924	99061	99198	99334	99470	99606	99742	99877	**00013**	14	27	41	54	68
7·4	2·00148	00283	00418	00553	00687	00821	00956	01089	01223	01357	13	27	40	54	67
7·5	2·01490	01624	01757	01890	02022	02155	02287	02419	02551	02683	13	27	40	53	66
7·6	2·02815	02946	03078	03209	03340	03471	03601	03732	03862	03992	13	26	39	52	65
7·7	2·04122	04252	04381	04511	04640	04769	04898	05027	05156	05284	13	26	39	52	65
7·8	2·05412	05540	05668	05796	05924	06051	06179	06306	06433	06560	13	25	38	51	64
7·9	2·06686	06813	06939	07065	07191	07317	07443	07568	07694	07819	13	25	38	50	63
8·0	2·07944	08069	08194	08318	08443	08567	08691	08815	08939	09063	12	25	37	50	62
8·1	2·09186	09310	09433	09556	09679	09802	09924	10047	10169	10291	12	25	37	49	61
8·2	2·10413	10535	10657	10779	10900	11021	11142	11263	11384	11505	12	24	36	49	61
8·3	2·11626	11746	11866	11986	12106	12226	12346	12465	12585	12704	12	24	36	48	60
8·4	2·12823	12942	13061	13180	13298	13417	13535	13653	13771	13889	12	24	36	47	59
8·5	2·14007	14124	14242	14359	14476	14593	14710	14827	14943	15060	12	23	35	47	58
8·6	2·15176	15292	15409	15524	15640	15756	15871	15987	16102	16217	12	23	35	46	58
8·7	2·16332	16447	16562	16677	16791	16905	17020	17134	17248	17361	11	23	34	46	57
8·8	2·17475	17589	17702	17816	17929	18042	18155	18267	18380	18493	11	23	34	45	56
8·9	2·18605	18717	18830	18942	19054	19165	19277	19389	19500	19611	11	22	34	45	56
9·0	2·19722	19834	19944	20055	20166	20276	20387	20497	20607	20717	11	22	33	44	55
9·1	2·20827	20937	21047	21157	21266	21375	21485	21594	21703	21812	11	22	33	44	55
9·2	2·21920	22029	22138	22246	22354	22462	22570	22678	22786	22894	11	22	32	43	54
9·3	2·23001	23109	23216	23324	23431	23538	23645	23751	23858	23965	11	21	32	43	54
9·4	2·24071	24177	24284	24390	24496	24601	24707	24813	24918	25024	11	21	32	42	53
9·5	2·25129	25234	25339	25444	25549	25654	25759	25863	25968	26072	10	21	31	42	52
9·6	2·26176	26280	26384	26488	26592	26696	26799	26903	27006	27109	10	21	31	41	52
9·7	2·27213	27316	27419	27521	27624	27727	27829	27932	28034	28136	10	21	31	41	51
9·8	2·28238	28340	28442	28544	28646	28747	28849	28950	29051	29152	10	20	30	41	51
9·9	2·29253	29354	29455	29556	29657	29757	29858	29958	30058	30158	10	20	30	40	50

$\log_e 10{,}000 = 9{\cdot}21034$, $\log_e 100{,}000 = 11{\cdot}51293$, $\log_e 1{,}000{,}000 = 13{\cdot}81551$

TABLE XXVI. NATURAL LOGARITHMS—*continued*

	0	1	2	3	4	5	6	7	8	9	1	2	3	4	5
10	2·30259	31254	32239	33214	34181	—	—	—	—	—	98	196	294	392	490
	—	—	—	—	—	35138	36085	37024	37955	38876	93	187	280	374	467
11	2·39790	40695	41591	42480	43361	—	—	—	—	—	89	179	268	357	446
	—	—	—	—	—	44235	45101	45959	46810	47654	85	171	256	342	427
12	2·48491	49321	50144	50960	51770	—	—	—	—	—	82	164	246	328	410
	—	—	—	—	—	52573	53370	54160	54945	55723	79	158	236	315	394
13	2·56495	57261	58022	58776	59525	—	—	—	—	—	76	152	227	303	379
	—	—	—	—	—	60269	61007	61740	62467	63189	73	146	219	292	365
14	2·63906	64617	65324	66026	66723	—	—	—	—	—	70	141	211	282	352
	—	—	—	—	—	67415	68102	68785	69463	70136	68	136	204	272	340
15	2·70805	71469	72130	72785	73437	—	—	—	—	—	66	132	197	263	329
	—	—	—	—	—	74084	74727	75366	76001	76632	64	127	191	255	318
16	2·77259	77882	78501	79117	79728	—	—	—	—	—	62	124	185	247	309
	—	—	—	—	—	80336	80940	81541	82138	82731	60	120	180	240	299
17	2·83321	83908	84491	85071	85647	—	—	—	—	—	58	116	174	233	291
	—	—	—	—	—	86220	86790	87356	87920	88480	56	113	169	226	282
18	2·89037	89591	90142	90690	91235	—	—	—	—	—	55	110	165	220	275
	—	—	—	—	—	91777	92316	92852	93386	93916	53	107	160	214	267
19	2·94444	94969	95491	96011	96527	—	—	—	—	—	52	104	156	208	260
	—	—	—	—	—	97041	97553	98062	98568	99072	51	102	152	203	254
20	2·99573	**00072**	**00568**	**01062**	**01553**	**02042**	**02529**	**03013**	**03495**	**03975**	49	98	147	196	244
21	3·04452	04927	05400	05871	06339	06805	07269	07731	08191	08649	47	93	140	186	233
22	3·09104	09558	10009	10459	10906	11352	11795	12237	12676	13114	45	89	134	178	223
23	3·13549	13983	14415	14845	15274	15700	16125	16548	16969	17388	43	85	128	171	213
24	3·17805	18221	18635	19048	19458	19867	20275	20680	21084	21487	41	82	123	164	204
25	3·21888	22287	22684	23080	23475	23868	24259	24649	25037	25424	39	79	118	157	196
26	3·25810	26194	26576	26957	27336	27714	28091	28466	28840	29213	38	76	113	151	189
27	3·29584	29953	30322	30689	31054	31419	31782	32143	32504	32863	36	73	109	146	182
28	3·33220	33577	33932	34286	34639	34990	35341	35690	36038	36384	35	70	105	141	176
29	3·36730	37074	37417	37759	38099	38439	38777	39115	39451	39786	34	68	102	136	170
30	3·40120	40453	40784	41115	41444	41773	42100	42426	42751	43076	33	66	99	131	164
31	3·43399	43721	44042	44362	44681	44999	45316	45632	45947	46261	32	64	95	127	159
32	3·46574	46886	47197	47507	47816	48124	48431	48738	49043	49347	31	62	92	123	154
33	3·49651	49953	50255	50556	50856	51155	51453	51750	52046	52342	30	60	90	120	150
34	3·52636	52930	53223	53515	53806	54096	54385	54674	54962	55249	29	58	87	116	145
35	3·55535	55820	56105	56388	56671	56953	57235	57515	57795	58074	28	56	85	113	141
36	3·58352	58629	58906	59182	59457	59731	60005	60278	60550	60821	27	55	82	110	137
37	3·61092	61362	61631	61899	62167	62434	62700	62966	63231	63495	27	53	80	107	134
38	3·63759	64021	64284	64545	64806	65066	65325	65584	65842	66099	26	52	78	104	130
39	3·66356	66612	66868	67122	67377	67630	67883	68135	68387	68638	25	51	76	101	127
40	3·68888	69138	69387	69635	69883	70130	70377	70623	70868	71113	25	49	74	99	124
41	3·71357	71601	71844	72086	72328	72569	72810	73050	73290	73529	24	48	72	96	121
42	3·73767	74005	74242	74479	74715	74950	75185	75420	75654	75887	24	47	71	94	118
43	3·76120	76352	76584	76815	77046	77276	77506	77735	77963	78191	23	46	69	92	115
44	3·78419	78646	78872	79098	79324	79549	79773	79997	80221	80444	22	45	68	90	112
45	3·80666	80888	81110	81331	81551	81771	81991	82210	82428	82647	22	44	66	88	110
46	3·82864	83081	83298	83514	83730	83945	84160	84374	84588	84802	22	43	65	86	108
47	3·85015	85227	85439	85651	85862	86073	86283	86493	86703	86912	21	42	63	84	105
48	3·87120	87328	87536	87743	87950	88156	88362	88568	88773	88978	21	41	62	83	103
49	3·89182	89386	89589	89792	89995	90197	90399	90600	90801	91002	20	40	61	81	101

$\log_e 10 = 2{\cdot}30259$, $\log_e 100 = 4{\cdot}60517$, $\log_e 1000 = 6{\cdot}90776$

	0	1	2	3	4	5	6	7	8	9	1	2	3	4	5
50	3·91202	91402	91601	91801	91999	92197	92395	92593	92790	92986	20	40	59	79	99
51	3·93183	93378	93574	93769	93964	94158	94352	94546	94739	94932	19	39	58	78	97
52	3·95124	95316	95508	95700	95891	96081	96272	96462	96651	96840	19	38	57	76	95
53	3·97029	97218	97406	97594	97781	97968	98155	98341	98527	98713	19	37	56	75	94
54	3·98898	99083	99268	99452	99636	99820	**00003**	**00186**	**00369**	**00551**	18	37	55	73	92
55	4·00733	00915	01096	01277	01458	01638	01818	01998	02177	02356	18	36	54	72	90
56	4·02535	02714	02892	03069	03247	03424	03601	03777	03954	04130	18	35	53	71	89
57	4·04305	04480	04655	04830	05004	05178	05352	05526	05699	05872	17	35	52	70	87
58	4·06044	06217	06389	06560	06732	06903	07073	07244	07414	07584	17	34	51	68	86
59	4·07754	07923	08092	08261	08429	08598	08766	08933	09101	09268	17	34	50	67	84
60	4·09434	09601	09767	09933	10099	10264	10429	10594	10759	10923	17	33	50	66	83
61	4·11087	11251	11415	11578	11741	11904	12066	12228	12390	12552	16	33	49	65	81
62	4·12713	12875	13036	13196	13357	13517	13677	13836	13995	14155	16	32	48	64	80
63	4·14313	14472	14630	14789	14946	15104	15261	15418	15575	15732	16	32	47	63	79
64	4·15888	16044	16200	16356	16511	16667	16821	16976	17131	17285	16	31	47	62	78
65	4·17439	17592	17746	17899	18052	18205	18358	18510	18662	18814	15	31	46	61	76
66	4·18965	19117	19268	19419	19570	19720	19870	20020	20170	20320	15	30	45	60	75
67	4·20469	20618	20767	20916	21065	21213	21361	21509	21656	21804	15	30	44	59	74
68	4·21951	22098	22244	22391	22537	22683	22829	22975	23120	23266	15	29	44	58	73
69	4·23411	23555	23700	23844	23989	24133	24276	24420	24563	24707	14	29	43	58	72
70	4·24850	24992	25135	25277	25419	25561	25703	25845	25986	26127	14	28	43	57	71
71	4·26268	26409	26549	26690	26830	26970	27110	27249	27388	27528	14	28	42	56	70
72	4·27667	27805	27944	28082	28221	28359	28496	28634	28772	28909	14	28	41	55	69
73	4·29046	29183	29320	29456	29592	29729	29865	30000	30136	30271	14	27	41	54	68
74	4·30407	30542	30676	30811	30946	31080	31214	31348	31482	31615	13	27	40	54	67
75	4·31749	31882	32015	32148	32281	32413	32546	32678	32810	32942	13	27	40	53	66
76	4·33073	33205	33336	33467	33598	33729	33860	33990	34120	34251	13	26	39	52	65
77	4·34381	34510	34640	34769	34899	35028	35157	35286	35414	35543	13	26	39	52	65
78	4·35671	35799	35927	36055	36182	36310	36437	36564	36691	36818	13	25	38	51	64
79	4·36945	37071	37198	37324	37450	37576	37701	37827	37952	38078	13	25	38	50	63
80	4·38203	38328	38452	38577	38701	38826	38950	39074	39198	39321	12	25	37	50	62
81	4·39445	39568	39692	39815	39938	40060	40183	40305	40428	40550	12	25	37	49	61
82	4·40672	40794	40916	41037	41159	41280	41401	41522	41643	41763	12	24	36	49	61
83	4·41884	42004	42125	42245	42365	42485	42604	42724	42843	42963	12	24	36	48	60
84	4·43082	43201	43319	43438	43557	43675	43793	43912	44030	44147	12	24	36	47	59
85	4·44265	44383	44500	44617	44735	44852	44969	45085	45202	45318	12	23	35	47	58
86	4·45435	45551	45667	45783	45899	46014	46130	46245	46361	46476	12	23	35	46	58
87	4·46591	46706	46820	46935	47050	47164	47278	47392	47506	47620	11	23	34	46	57
88	4·47734	47847	47961	48074	48187	48300	48413	48526	48639	48751	11	23	34	45	56
89	4·48864	48976	49088	49200	49312	49424	49536	49647	49758	49870	11	22	34	45	56
90	4·49981	50092	50203	50314	50424	50535	50645	50756	50866	50976	11	22	33	44	55
91	4·51086	51196	51305	51415	51525	51634	51743	51852	51961	52070	11	22	33	44	55
92	4·52179	52287	52396	52504	52613	52721	52829	52937	53045	53152	11	22	32	43	54
93	4·53260	53367	53475	53582	53689	53796	53903	54010	54116	54223	11	21	32	43	54
94	4·54329	54436	54542	54648	54754	54860	54966	55071	55177	55282	11	21	32	42	53
95	4·55388	55493	55598	55703	55808	55913	56017	56122	56226	56331	10	21	31	42	52
96	4·56435	56539	56643	56747	56851	56954	57058	57161	57265	57368	10	21	31	41	52
97	4·57471	57574	57677	57780	57883	57985	58088	58190	58292	58395	10	21	31	41	51
98	4·58497	58599	58701	58802	58904	59006	59107	59209	59310	59411	10	20	30	41	51
99	4·59512	59613	59714	59815	59915	60016	60116	60217	60317	60417	10	20	30	40	50

$\log_e 10{,}000 = 9{\cdot}21034$, $\log_e 100{,}000 = 11{\cdot}51293$, $\log_e 1{,}000{,}000 = 13{\cdot}81551$

Table XXVII. Squares

	0	1	2	3	4	5	6	7	8	9
0	0	1	4	9	16	25	36	49	64	81
1	100	121	144	169	196	225	256	289	324	361
2	400	441	484	529	576	625	676	729	784	841
3	900	961	1024	1089	1156	1225	1296	1369	1444	1521
4	1600	1681	1764	1849	1936	2025	2116	2209	2304	2401
5	2500	2601	2704	2809	2916	3025	3136	3249	3364	3481
6	3600	3721	3844	3969	4096	4225	4356	4489	4624	4761
7	4900	5041	5184	5329	5476	5625	5776	5929	6084	6241
8	6400	6561	6724	6889	7056	7225	7396	7569	7744	7921
9	8100	8281	8464	8649	8836	9025	9216	9409	9604	9801
10	10000	10201	10404	10609	10816	11025	11236	11449	11664	11881
11	12100	12321	12544	12769	12996	13225	13456	13689	13924	14161
12	14400	14641	14884	15129	15376	15625	15876	16129	16384	16641
13	16900	17161	17424	17689	17956	18225	18496	18769	19044	19321
14	19600	19881	20164	20449	20736	21025	21316	21609	21904	22201
15	22500	22801	23104	23409	23716	24025	24336	24649	24964	25281
16	25600	25921	26244	26569	26896	27225	27556	27889	28224	28561
17	28900	29241	29584	29929	30276	30625	30976	31329	31684	32041
18	32400	32761	33124	33489	33856	34225	34596	34969	35344	35721
19	36100	36481	36864	37249	37636	38025	38416	38809	39204	39601
20	40000	40401	40804	41209	41616	42025	42436	42849	43264	43681
21	44100	44521	44944	45369	45796	46225	46656	47089	47524	47961
22	48400	48841	49284	49729	50176	50625	51076	51529	51984	52441
23	52900	53361	53824	54289	54756	55225	55696	56169	56644	57121
24	57600	58081	58564	59049	59536	60025	60516	61009	61504	62001
25	62500	63001	63504	64009	64516	65025	65536	66049	66564	67081
26	67600	68121	68644	69169	69696	70225	70756	71289	71824	72361
27	72900	73441	73984	74529	75076	75625	76176	76729	77284	77841
28	78400	78961	79524	80089	80656	81225	81796	82369	82944	83521
29	84100	84681	85264	85849	86436	87025	87616	88209	88804	89401
30	90000	90601	91204	91809	92416	93025	93636	94249	94864	95481
31	96100	96721	97344	97969	98596	99225	99856	100489	101124	101761
32	102400	103041	103684	104329	104976	105625	106276	106929	107584	108241
33	108900	109561	110224	110889	111556	112225	112896	113569	114244	114921
34	115600	116281	116964	117649	118336	119025	119716	120409	121104	121801
35	122500	123201	123904	124609	125316	126025	126736	127449	128164	128881
36	129600	130321	131044	131769	132496	133225	133956	134689	135424	136161
37	136900	137641	138384	139129	139876	140625	141376	142129	142884	143641
38	144400	145161	145924	146689	147456	148225	148996	149769	150544	151321
39	152100	152881	153664	154449	155236	156025	156816	157609	158404	159201
40	160000	160801	161604	162409	163216	164025	164836	165649	166464	167281
41	168100	168921	169744	170569	171396	172225	173056	173889	174724	175561
42	176400	177241	178084	178929	179776	180625	181476	182329	183184	184041
43	184900	185761	186624	187489	188356	189225	190096	190969	191844	192721
44	193600	194481	195364	196249	197136	198025	198916	199809	200704	201601
45	202500	203401	204304	205209	206116	207025	207936	208849	209764	210681
46	211600	212521	213444	214369	215296	216225	217156	218089	219024	219961
47	220900	221841	222784	223729	224676	225625	226576	227529	228484	229441
48	230400	231361	232324	233289	234256	235225	236196	237169	238144	239121
49	240100	241081	242064	243049	244036	245025	246016	247009	248004	249001

Exact squares of 4 figure numbers can be quickly calculated from the identity $(a \pm b)^2 = a^2 \pm 2ab + b^2$.

Table XXVII. Squares—*continued*

	0	1	2	3	4	5	6	7	8	9
50	250000	251001	252004	253009	254016	255025	256036	257049	258064	259081
51	260100	261121	262144	263169	264196	265225	266256	267289	268324	269361
52	270400	271441	272484	273529	274576	275625	276676	277729	278784	279841
53	280900	281961	283024	284089	285156	286225	287296	288369	289444	290521
54	291600	292681	293764	294849	295936	297025	298116	299209	300304	301401
55	302500	303601	304704	305809	306916	308025	309136	310249	311364	312481
56	313600	314721	315844	316969	318096	319225	320356	321489	322624	323761
57	324900	326041	327184	328329	329476	330625	331776	332929	334084	335241
58	336400	337561	338724	339889	341056	342225	343396	344569	345744	346921
59	348100	349281	350464	351649	352836	354025	355216	356409	357604	358801
60	360000	361201	362404	363609	364816	366025	367236	368449	369664	370881
61	372100	373321	374544	375769	376996	378225	379456	380689	381924	383161
62	384400	385641	386884	388129	389376	390625	391876	393129	394384	395641
63	396900	398161	399424	400689	401956	403225	404496	405769	407044	408321
64	409600	410881	412164	413449	414736	416025	417316	418609	419904	421201
65	422500	423801	425104	426409	427716	429025	430336	431649	432964	434281
66	435600	436921	438244	439569	440896	442225	443556	444889	446224	447561
67	448900	450241	451584	452929	454276	455625	456976	458329	459684	461041
68	462400	463761	465124	466489	467856	469225	470596	471969	473344	474721
69	476100	477481	478864	480249	481636	483025	484416	485809	487204	488601
70	490000	491401	492804	494209	495616	497025	498436	499849	501264	502681
71	504100	505521	506944	508369	509796	511225	512656	514089	515524	516961
72	518400	519841	521284	522729	524176	525625	527076	528529	529984	531441
73	532900	534361	535824	537289	538756	540225	541696	543169	544644	546121
74	547600	549081	550564	552049	553536	555025	556516	558009	559504	561001
75	562500	564001	565504	567009	568516	570025	571536	573049	574564	576081
76	577600	579121	580644	582169	583696	585225	586756	588289	589824	591361
77	592900	594441	595984	597529	599076	600625	602176	603729	605284	606841
78	608400	609961	611524	613089	614656	616225	617796	619369	620944	622521
79	624100	625681	627264	628849	630436	632025	633616	635209	636804	638401
80	640000	641601	643204	644809	646416	648025	649636	651249	652864	654481
81	656100	657721	659344	660969	662596	664225	665856	667489	669124	670761
82	672400	674041	675684	677329	678976	680625	682276	683929	685584	687241
83	688900	690561	692224	693889	695556	697225	698896	700569	702244	703921
84	705600	707281	708964	710649	712336	714025	715716	717409	719104	720801
85	722500	724201	725904	727609	729316	731025	732736	734449	736164	737881
86	739600	741321	743044	744769	746496	748225	749956	751689	753424	755161
87	756900	758641	760384	762129	763876	765625	767376	769129	770884	772641
88	774400	776161	777924	779689	781456	783225	784996	786769	788544	790321
89	792100	793881	795664	797449	799236	801025	802816	804609	806404	808201
90	810000	811801	813604	815409	817216	819025	820836	822649	824464	826281
91	828100	829921	831744	833569	835396	837225	839056	840889	842724	844561
92	846400	848241	850084	851929	853776	855625	857476	859329	861184	863041
93	864900	866761	868624	870489	872356	874225	876096	877969	879844	881721
94	883600	885481	887364	889249	891136	893025	894916	896809	898704	900601
95	902500	904401	906304	908209	910116	912025	913936	915849	917764	919681
96	921600	923521	925444	927369	929296	931225	933156	935089	937024	938961
97	940900	942841	944784	946729	948676	950625	952576	954529	956484	958441
98	960400	962361	964324	966289	968256	970225	972196	974169	976144	978121
99	980100	982081	984064	986049	988036	990025	992016	994009	996004	998001

Thus $693{\cdot}3^2 = 480249 + 415{\cdot}8 + 0{\cdot}09 = 480664{\cdot}89$.

Table XXVIII. Square Roots

	0	1	2	3	4	5	6	7	8	9	1	2	3	4	5
10	10000	10050	10100	10149	10198	10247	10296	10344	10392	10440	5	10	15	20	24
	31623	31780	31937	32094	32249	32404	32558	32711	32863	33015	15	31	46	62	77
11	10488	10536	10583	10630	10677	10724	10770	10817	10863	10909	5	9	14	19	23
	33166	33317	33466	33615	33764	33912	34059	34205	34351	34496	15	30	44	59	74
12	10954	11000	11045	11091	11136	11180	11225	11269	11314	11358	4	9	13	18	22
	34641	34785	34928	35071	35214	35355	35496	35637	35777	35917	14	28	43	57	71
13	11402	11446	11489	11533	11576	11619	11662	11705	11747	11790	4	9	13	17	22
	36056	36194	36332	36469	36606	36742	36878	37014	37148	37283	14	27	41	55	68
14	11832	11874	11916	11958	12000	12042	12083	12124	12166	12207	4	8	13	17	21
	37417	37550	37683	37815	37947	38079	38210	38341	38471	38601	13	26	39	53	66
15	12247	12288	12329	12369	12410	12450	12490	12530	12570	12610	4	8	12	16	20
	38730	38859	38987	39115	39243	39370	39497	39623	39749	39875	13	25	38	51	64
16	12649	12689	12728	12767	12806	12845	12884	12923	12961	13000	4	8	12	16	20
	40000	40125	40249	40373	40497	40620	40743	40866	40988	41110	12	25	37	49	62
17	13038	13077	13115	13153	13191	13229	13266	13304	13342	13379	4	8	11	15	19
	41231	41352	41473	41593	41713	41833	41952	42071	42190	42308	12	24	36	48	60
18	13416	13454	13491	13528	13565	13601	13638	13675	13711	13748	4	7	11	15	18
	42426	42544	42661	42778	42895	43012	43128	43243	43359	43474	12	23	35	47	58
19	13784	13820	13856	13892	13928	13964	14000	14036	14071	14107	4	7	11	14	18
	43589	43704	43818	43932	44045	44159	44272	44385	44497	44609	11	23	34	45	57
20	14142	14177	14213	14248	14283	14318	14353	14387	14422	14457	4	7	10	14	18
	44721	44833	44944	45056	45166	45277	45387	45497	45607	45717	11	22	33	44	55
21	14491	14526	14560	14595	14629	14663	14697	14731	14765	14799	3	7	10	14	17
	45826	45935	46043	46152	46260	46368	46476	46583	46690	46797	11	22	32	43	54
22	14832	14866	14900	14933	14967	15000	15033	15067	15100	15133	3	7	10	13	17
	46904	47011	47117	47223	47329	47434	47539	47645	47749	47854	11	21	32	42	53
23	15166	15199	15232	15264	15297	15330	15362	15395	15427	15460	3	7	10	13	16
	47958	48062	48166	48270	48374	48477	48580	48683	48785	48888	10	21	31	41	52
24	15492	15524	15556	15588	15620	15652	15684	15716	15748	15780	3	6	10	13	16
	48990	49092	49193	49295	49396	49497	49598	49699	49800	49900	10	20	30	40	51
25	15811	15843	15875	15906	15937	15969	16000	16031	16062	16093	3	6	9	13	16
	50000	50100	50200	50299	50398	50498	50596	50695	50794	50892	10	20	30	40	50
26	16125	16155	16186	16217	16248	16279	16310	16340	16371	16401	3	6	9	12	15
	50990	51088	51186	51284	51381	51478	51575	51672	51769	51865	10	19	29	39	49
27	16432	16462	16492	16523	16553	16583	16613	16643	16673	16703	3	6	9	12	15
	51962	52058	52154	52249	52345	52440	52536	52631	52726	52820	10	19	29	38	48
28	16733	16763	16793	16823	16852	16882	16912	16941	16971	17000	3	6	9	12	15
	52915	53009	53104	53198	53292	53385	53479	53572	53666	53759	9	19	28	38	47
29	17029	17059	17088	17117	17146	17176	17205	17234	17263	17292	3	6	9	12	15
	53852	53944	54037	54129	54222	54314	54406	54498	54589	54681	9	18	28	37	46

Table XXVIII. Square Roots—*continued*

	0	1	2	3	4	5	6	7	8	9	1	2	3	4	5
30	17321	17349	17378	17407	17436	17464	17493	17521	17550	17578	3	6	9	11	14
	54772	54863	54955	55045	55136	55227	55317	55408	55498	55588	9	18	27	36	45
31	17607	17635	17664	17692	17720	17748	17776	17804	17833	17861	3	6	8	11	14
	55678	55767	55857	55946	56036	56125	56214	56303	56391	56480	9	18	27	36	45
32	17889	17916	17944	17972	18000	18028	18055	18083	18111	18138	3	6	8	11	14
	56569	56657	56745	56833	56921	57009	57096	57184	57271	57359	9	18	26	35	44
33	18166	18193	18221	18248	18276	18303	18330	18358	18385	18412	3	5	8	11	14
	57446	57533	57619	57706	57793	57879	57966	58052	58138	58224	9	17	26	35	43
34	18439	18466	18493	18520	18547	18574	18601	18628	18655	18682	3	5	8	11	14
	58310	58395	58481	58566	58652	58737	58822	58907	58992	59076	9	17	26	34	43
35	18708	18735	18762	18788	18815	18841	18868	18894	18921	18947	3	5	8	11	13
	59161	59245	59330	59414	59498	59582	59666	59749	59833	59917	8	17	25	34	42
36	18974	19000	19026	19053	19079	19105	19131	19157	19183	19209	3	5	8	10	13
	60000	60083	60166	60249	60332	60415	60498	60581	60663	60745	8	17	25	33	41
37	19235	19261	19287	19313	19339	19365	19391	19416	19442	19468	3	5	8	10	13
	60828	60910	60992	61074	61156	61237	61319	61400	61482	61563	8	16	25	33	41
38	19494	19519	19545	19570	19596	19621	19647	19672	19698	19723	3	5	8	10	13
	61644	61725	61806	61887	61968	62048	62129	62209	62290	62370	8	16	24	32	40
39	19748	19774	19799	19824	19849	19875	19900	19925	19950	19975	3	5	8	10	13
	62450	62530	62610	62690	62769	62849	62929	63008	63087	63166	8	16	24	32	40
40	20000	20025	20050	20075	20100	20125	20149	20174	20199	20224	2	5	7	10	12
	63246	63325	63403	63482	63561	63640	63718	63797	63875	63953	8	16	24	31	39
41	20248	20273	20298	20322	20347	20372	20396	20421	20445	20469	2	5	7	10	12
	64031	64109	64187	64265	64343	64420	64498	64576	64653	64730	8	16	23	31	39
42	20494	20518	20543	20567	20591	20616	20640	20664	20688	20712	2	5	7	10	12
	64807	64885	64962	65038	65115	65192	65269	65345	65422	65498	8	15	23	31	38
43	20736	20761	20785	20809	20833	20857	20881	20905	20928	20952	2	5	7	10	12
	65574	65651	65727	65803	65879	65955	66030	66106	66182	66257	8	15	23	30	38
44	20976	21000	21024	21048	21071	21095	21119	21142	21166	21190	2	5	7	10	12
	66332	66408	66483	66558	66633	66708	66783	66858	66933	67007	8	15	22	30	38
45	21213	21237	21260	21284	21307	21331	21354	21378	21401	21424	2	5	7	9	12
	67082	67157	67231	67305	67380	67454	67528	67602	67676	67750	7	15	22	30	37
46	21448	21471	21494	21517	21541	21564	21587	21610	21633	21656	2	5	7	9	12
	67823	67897	67971	68044	68118	68191	68264	68337	68411	68484	7	15	22	29	37
47	21679	21703	21726	21749	21772	21794	21817	21840	21863	21886	2	5	7	9	12
	68557	68629	68702	68775	68848	68920	68993	69065	69138	69210	7	15	22	29	36
48	21909	21932	21954	21977	22000	22023	22045	22068	22091	22113	2	5	7	9	11
	69282	69354	69426	69498	69570	69642	69714	69785	69857	69929	7	14	22	29	36
49	22136	22159	22181	22204	22226	22249	22271	22293	22316	22338	2	4	7	9	11
	70000	70071	70143	70214	70285	70356	70427	70498	70569	70640	7	14	21	28	36

Table XXVIII. Square Roots—*continued*

	0	1	2	3	4	5	6	7	8	9	1	2	3	4	5
50	22361	22383	22405	22428	22450	22472	22494	22517	22539	22561	2	4	7	9	11
	70711	70781	70852	70922	70993	71063	71134	71204	71274	71344	7	14	21	28	35
51	22583	22605	22627	22650	22672	22694	22716	22738	22760	22782	2	4	7	9	11
	71414	71484	71554	71624	71694	71764	71833	71903	71972	72042	7	14	21	28	35
52	22804	22825	22847	22869	22891	22913	22935	22956	22978	23000	2	4	7	9	11
	72111	72180	72250	72319	72388	72457	72526	72595	72664	72732	7	14	21	28	34
53	23022	23043	23065	23087	23108	23130	23152	23173	23195	23216	2	4	6	9	11
	72801	72870	72938	73007	73075	73144	73212	73280	73348	73417	7	14	21	27	34
54	23238	23259	23281	23302	23324	23345	23367	23388	23409	23431	2	4	6	9	11
	73485	73553	73621	73689	73756	73824	73892	73959	74027	74095	7	14	20	27	34
55	23452	23473	23495	23516	23537	23558	23580	23601	23622	23643	2	4	6	8	11
	74162	74229	74297	74364	74431	74498	74565	74632	74699	74766	7	13	20	27	34
56	23664	23685	23707	23728	23749	23770	23791	23812	23833	23854	2	4	6	8	11
	74833	74900	74967	75033	75100	75166	75233	75299	75366	75432	7	13	20	27	33
57	23875	23896	23917	23937	23958	23979	24000	24021	24042	24062	2	4	6	8	10
	75498	75565	75631	75697	75763	75829	75895	75961	76026	76092	7	13	20	26	33
58	24083	24104	24125	24145	24166	24187	24207	24228	24249	24269	2	4	6	8	10
	76158	76223	76289	76354	76420	76485	76551	76616	76681	76746	7	13	20	26	33
59	24290	24310	24331	24352	24372	24393	24413	24434	24454	24474	2	4	6	8	10
	76811	76877	76942	77006	77071	77136	77201	77266	77330	77395	6	13	19	26	32
60	24495	24515	24536	24556	24576	24597	24617	24637	24658	24678	2	4	6	8	10
	77460	77524	77589	77653	77717	77782	77846	77910	77974	78038	6	13	19	26	32
61	24698	24718	24739	24759	24779	24799	24819	24839	24860	24880	2	4	6	8	10
	78102	78166	78230	78294	78358	78422	78486	78549	78613	78677	6	13	19	26	32
62	24900	24920	24940	24960	24980	25000	25020	25040	25060	25080	2	4	6	8	10
	78740	78804	78867	78930	78994	79057	79120	79183	79246	79310	6	13	19	25	32
63	25100	25120	25140	25159	25179	25199	25219	25239	25259	25278	2	4	6	8	10
	79373	79436	79498	79561	79624	79687	79750	79812	79875	79937	6	13	19	25	31
64	25298	25318	25338	25357	25377	25397	25417	25436	25456	25475	2	4	6	8	10
	80000	80062	80125	80187	80250	80312	80374	80436	80498	80561	6	12	19	25	31
65	25495	25515	25534	25554	25573	25593	25612	25632	25652	25671	2	4	6	8	10
	80623	80685	80747	80808	80870	80932	80994	81056	81117	81179	6	12	19	25	31
66	25690	25710	25729	25749	25768	25788	25807	25826	25846	25865	2	4	6	8	10
	81240	81302	81363	81425	81486	81548	81609	81670	81731	81792	6	12	18	25	31
67	25884	25904	25923	25942	25962	25981	26000	26019	26038	26058	2	4	6	8	10
	81854	81915	81976	82037	82098	82158	82219	82280	82341	82401	6	12	18	24	30
68	26077	26096	26115	26134	26153	26173	26192	26211	26230	26249	2	4	6	8	10
	82462	82523	82583	82644	82704	82765	82825	82885	82946	83006	6	12	18	24	30
69	26268	26287	26306	26325	26344	26363	26382	26401	26420	26439	2	4	6	8	10
	83066	83126	83187	83247	83307	83367	83427	83487	83546	83606	6	12	18	24	30
70	26458	26476	26495	26514	26533	26552	26571	26589	26608	26627	2	4	6	8	9
	83666	83726	83785	83845	83905	83964	84024	84083	84143	84202	6	12	18	24	30
71	26646	26665	26683	26702	26721	26739	26758	26777	26796	26814	2	4	6	7	9
	84261	84321	84380	84439	84499	84558	84617	84676	84735	84794	6	12	18	24	30
72	26833	26851	26870	26889	26907	26926	26944	26963	26981	27000	2	4	6	7	9
	84853	84912	84971	85029	85088	85147	85206	85264	85323	85381	6	12	18	23	29
73	27019	27037	27055	27074	27092	27111	27129	27148	27166	27185	2	4	6	7	9
	85440	85499	85557	85615	85674	85732	85790	85849	85907	85965	6	12	17	23	29
74	27203	27221	27240	27258	27276	27295	27313	27331	27350	27368	2	4	5	7	9
	86023	86081	86139	86197	86255	86313	86371	86429	86487	86545	6	12	17	23	29

TABLE XXVIII. SQUARE ROOTS—*continued*

	0	1	2	3	4	5	6	7	8	9	1	2	3	4	5
75	27386	27404	27423	27441	27459	27477	27495	27514	27532	27550	2	4	5	7	9
	86603	86660	86718	86776	86833	86891	86948	87006	87063	87121	6	12	17	23	29
76	27568	27586	27604	27622	27641	27659	27677	27695	27713	27731	2	4	5	7	9
	87178	87235	87293	87350	87407	87464	87521	87579	87636	87693	6	11	17	23	29
77	27749	27767	27785	27803	27821	27839	27857	27875	27893	27911	2	4	5	7	9
	87750	87807	87864	87920	87977	88034	88091	88148	88204	88261	6	11	17	23	28
78	27928	27946	27964	27982	28000	28018	28036	28054	28071	28089	2	4	5	7	9
	88318	88374	88431	88487	88544	88600	88657	88713	88769	88826	6	11	17	23	28
79	28107	28125	28142	28160	28178	28196	28213	28231	28249	28267	2	4	5	7	9
	88882	88938	88994	89051	89107	89163	89219	89275	89331	89387	6	11	17	22	28
80	28284	28302	28320	28337	28355	28373	28390	28408	28425	28443	2	4	5	7	9
	89443	89499	89554	89610	89666	89722	89778	89833	89889	89944	6	11	17	22	28
81	28460	28478	28496	28513	28531	28548	28566	28583	28601	28618	2	4	5	7	9
	90000	90056	90111	90167	90222	90277	90333	90388	90443	90499	6	11	17	22	28
82	28636	28653	28671	28688	28705	28723	28740	28758	28775	28792	2	3	5	7	9
	90554	90609	90664	90719	90774	90830	90885	90940	90995	91049	6	11	16	22	28
83	28810	28827	28844	28862	28879	28896	28914	28931	28948	28965	2	3	5	7	9
	91104	91159	91214	91269	91324	91378	91433	91488	91542	91597	5	11	16	22	27
84	28983	29000	29017	29034	29052	29069	29086	29103	29120	29138	2	3	5	7	9
	91652	91706	91761	91815	91869	91924	91978	92033	92087	92141	5	11	16	22	27
85	29155	29172	29189	29206	29223	29240	29257	29275	29292	29309	2	3	5	7	9
	92195	92250	92304	92358	92412	92466	92520	92574	92628	92682	5	11	16	22	27
86	29326	29343	29360	29377	29394	29411	29428	29445	29462	29479	2	3	5	7	8
	92736	92790	92844	92898	92952	93005	93059	93113	93167	93220	5	11	16	22	27
87	29496	29513	29530	29547	29563	29580	29597	29614	29631	29648	2	3	5	7	8
	93274	93327	93381	93434	93488	93541	93595	93648	93702	93755	5	11	16	21	27
88	29665	29682	29698	29715	29732	29749	29766	29783	29799	29816	2	3	5	7	8
	93808	93862	93915	93968	94021	94074	94128	94181	94234	94287	5	11	16	21	27
89	29833	29850	29866	29883	29900	29917	29933	29950	29967	29983	2	3	5	7	8
	94340	94393	94446	94499	94552	94604	94657	94710	94763	94816	5	11	16	21	26
90	30000	30017	30033	30050	30067	30083	30100	30116	30133	30150	2	3	5	7	8
	94868	94921	94974	95026	95079	95131	95184	95237	95289	95341	5	11	16	21	26
91	30166	30183	30199	30216	30232	30249	30265	30282	30299	30315	2	3	5	7	8
	95394	95446	95499	95551	95603	95656	95708	95760	95812	95864	5	10	16	21	26
92	30332	30348	30364	30381	30397	30414	30430	30447	30463	30480	2	3	5	7	8
	95917	95969	96021	96073	96125	96177	96229	96281	96333	96385	5	10	16	21	26
93	30496	30512	30529	30545	30561	30578	30594	30610	30627	30643	2	3	5	7	8
	96437	96488	96540	96592	96644	96695	96747	96799	96850	96902	5	10	16	21	26
94	30659	30676	30692	30708	30725	30741	30757	30773	30790	30806	2	3	5	7	8
	96954	97005	97057	97108	97160	97211	97263	97314	97365	97417	5	10	15	21	26
95	30822	30838	30854	30871	30887	30903	30919	30935	30952	30968	2	3	5	6	8
	97468	97519	97570	97622	97673	97724	97775	97826	97877	97929	5	10	15	20	26
96	30984	31000	31016	31032	31048	31064	31081	31097	31113	31129	2	3	5	6	8
	97980	98031	98082	98133	98184	98234	98285	98336	98387	98438	5	10	15	20	25
97	31145	31161	31177	31193	31209	31225	31241	31257	31273	31289	2	3	5	6	8
	98489	98539	98590	98641	98691	98742	98793	98843	98894	98944	5	10	15	20	25
98	31305	31321	31337	31353	31369	31385	31401	31417	31432	31448	2	3	5	6	8
	98995	99045	99096	99146	99197	99247	99298	99348	99398	99448	5	10	15	20	25
99	31464	31480	31496	31512	31528	31544	31559	31575	31591	31607	2	3	5	6	8
	99499	99549	99599	99649	99700	99750	99800	99850	99900	99950	5	10	15	20	25

TABLE XXIX. RECIPROCALS

	0	1	2	3	4	5	6	7	8	9	Diffs.*
1·0	1·000000	990099	980392	970874	961538	952381	943396	934579	925926	917431	
	−9901	−9707	−9518	−9336	−9157	−8985	−8817	−8653	−8495	−8340	
1·1	·909091	900901	892857	884956	877193	869565	862069	854701	847458	840336	
	−8190	−8044	−7901	−7763	−7628	−7496	−7368	−7243	−7122	−7003	
1·2	·833333	826446	819672	813008	806452	800000	793651	787402	781250	775194	
	−6887	−6774	−6664	−6556	−6452	−6349	−6249	−6152	−6056	−5963	
1·3	·769231	763359	757576	751880	746269	740741	735294	729927	724638	719424	
	−5872	−5783	−5696	−5611	−5528	−5447	−5367	−5289	−5214	−5138	
1·4	·714286	709220	704225	699301	694444	689655	684932	680272	675676	671141	
	−5066	−4995	−4924	−4857	−4789	−4723	−4660	−4596	−4535	−4474	
1·5	·666667	662252	657895	653595	649351	645161	641026	636943	632911	628931	
	−4415	−4357	−4300	−4244	−4190	−4135	−4083	−4032	−3980	−3931	
1·6	·625000	621118	617284	613497	609756	606061	602410	598802	595238	591716	
	−3882	−3834	−3787	−3741	−3695	−3651	−3608	−3564	−3522	−3481	
1·7	·588235	584795	581395	578035	574713	571429	568182	564972	561798	558659	
	−3440	−3400	−3360	−3322	−3284	−3247	−3210	−3174	−3139	−3103	
1·8	·555556	552486	549451	546448	543478	540541	537634	534759	531915	529101	
	−3070	−3035	−3003	−2970	−2937	−2907	−2875	−2844	−2814	−2785	
1·9	·526316	523560	520833	518135	515464	512821	510204	507614	505051	502513	
	−2756	−2727	−2698	−2671	−2643	−2617	−2590	−2563	−2538	−2513	
2·0	·500000	497512	495050	492611	490196	487805	485437	483092	480769	478469	
	−2488	−2462	−2439	−2415	−2391	−2368	−2345	−2323	−2300	−2279	
2·1	·476190	473934	471698	469484	467290	465116	462963	460829	458716	456621	
	−2256	−2236	−2214	−2194	−2174	−2153	−2134	−2113	−2095	−2076	
2·2	·454545	452489	450450	448430	446429	444444	442478	440529	438596	436681	
	−2056	−2039	−2020	−2001	−1985	−1966	−1949	−1933	−1915	−1898	
2·3	·434783	432900	431034	429185	427350	425532	423729	421941	420168	418410	
	−1883	−1866	−1849	−1835	−1818	−1803	−1788	−1773	−1758	−1743	
2·4	·416667	414938	413223	411523	409836	408163	406504	404858	403226	401606	
	−1729	−1715	−1700	−1687	−1673	−1659	−1646	−1632	−1620	−1606	
2·5	·400000	398406	396825	395257	393701	392157	390625	389105	387597	386100	−1544
2·6	·384615	383142	381679	380228	378788	377358	375940	374532	373134	371747	−1430
2·7	·370370	369004	367647	366300	364964	363636	362319	361011	359712	358423	−1327
2·8	·357143	355872	354610	353357	352113	350877	349650	348432	347222	346021	−1236
2·9	·344828	343643	342466	341297	340136	338983	337838	336700	335570	334448	−1153
3·0	·333333	332226	331126	330033	328947	327869	326797	325733	324675	323625	−1079
3·1	·322581	321543	320513	319489	318471	317460	316456	315457	314465	313480	−1011
3·2	·312500	311526	310559	309598	308642	307692	306748	305810	304878	303951	−950
3·3	·303030	302115	301205	300300	299401	298507	297619	296736	295858	294985	−894
3·4	·294118	293255	292398	291545	290698	289855	289017	288184	287356	286533	−843
3·5	·285714	284900	284091	283286	282486	281690	280899	280112	279330	278552	−796
3·6	·277778	277008	276243	275482	274725	273973	273224	272480	271739	271003	−753
3·7	·270270	269542	268817	268097	267380	266667	265957	265252	264550	263852	−713
3·8	·263158	262467	261780	261097	260417	259740	259067	258398	257732	257069	−677
3·9	·256410	255754	255102	254453	253807	253165	252525	251889	251256	250627	−643
4·0	·250000	249377	248756	248139	247525	246914	246305	245700	245098	244499	−611
4·1	·243902	243309	242718	242131	241546	240964	240385	239808	239234	238663	−582
4·2	·238095	237530	236967	236407	235849	235294	234742	234192	233645	233100	−555
4·3	·232558	232019	231481	230947	230415	229885	229358	228833	228311	227790	−530
4·4	·227273	226757	226244	225734	225225	224719	224215	223714	223214	222717	−506
4·5	·222222	221729	221239	220751	220264	219780	219298	218818	218341	217865	−484
4·6	·217391	216920	216450	215983	215517	215054	214592	214133	213675	213220	−463
4·7	·212766	212314	211864	211416	210970	210526	210084	209644	209205	208768	−444
4·8	·208333	207900	207469	207039	206612	206186	205761	205339	204918	204499	−426
4·9	·204082	203666	203252	202840	202429	202020	201613	201207	200803	200401	−409

* Tabular differences up to 2·5, and mean differences thereafter. Differences are *negative*.

TABLE XXIX. RECIPROCALS—*continued*

	0	1	2	3	4	5	6	7	8	9	Diffs.*
5·0	·200000	199601	199203	198807	198413	198020	197628	197239	196850	196464	−393
5·1	·196078	195695	195312	194932	194553	194175	193798	193424	193050	192678	−378
5·2	·192308	191939	191571	191205	190840	190476	190114	189753	189394	189036	−364
5·3	·188679	188324	187970	187617	187266	186916	186567	186220	185874	185529	−350
5·4	·185185	184843	184502	184162	183824	183486	183150	182815	182482	182149	−337
5·5	·181818	181488	181159	180832	180505	180180	179856	179533	179211	178891	−325
5·6	·178571	178253	177936	177620	177305	176991	176678	176367	176056	175747	−314
5·7	·175439	175131	174825	174520	174216	173913	173611	173310	173010	172712	−303
5·8	·172414	172117	171821	171527	171233	170940	170648	170358	170068	169779	−293
5·9	·169492	169205	168919	168634	168350	168067	167785	167504	167224	166945	−283
6·0	·166667	166389	166113	165837	165563	165289	165017	164745	164474	164204	−274
6·1	·163934	163666	163399	163132	162866	162602	162338	162075	161812	161551	−265
6·2	·161290	161031	160772	160514	160256	160000	159744	159490	159236	158983	−256
6·3	·158730	158479	158228	157978	157729	157480	157233	156986	156740	156495	−248
6·4	·156250	156006	155763	155521	155280	155039	154799	154560	154321	154083	−241
6·5	·153846	153610	153374	153139	152905	152672	152439	152207	151976	151745	−233
6·6	·151515	151286	151057	150830	150602	150376	150150	149925	149701	149477	−226
6·7	·149254	149031	148810	148588	148368	148148	147929	147710	147493	147275	−220
6·8	·147059	146843	146628	146413	146199	145985	145773	145560	145349	145138	−213
6·9	·144928	144718	144509	144300	144092	143885	143678	143472	143266	143062	−207
7·0	·142857	142653	142450	142248	142045	141844	141643	141443	141243	141044	−201
7·1	·140845	140647	140449	140252	140056	139860	139665	139470	139276	139082	−196
7·2	·138889	138696	138504	138313	138122	137931	137741	137552	137363	137174	−191
7·3	·136986	136799	136612	136426	136240	136054	135870	135685	135501	135318	−185
7·4	·135135	134953	134771	134590	134409	134228	134048	133869	133690	133511	−180
7·5	·133333	133156	132979	132802	132626	132450	132275	132100	131926	131752	−176
7·6	·131579	131406	131234	131062	130890	130719	130548	130378	130208	130039	−171
7·7	·129870	129702	129534	129366	129199	129032	128866	128700	128535	128370	−167
7·8	·128205	128041	127877	127714	127551	127389	127226	127065	126904	126743	−162
7·9	·126582	126422	126263	126103	125945	125786	125628	125471	125313	125156	−158
8·0	·125000	124844	124688	124533	124378	124224	124069	123916	123762	123609	−155
8·1	·123457	123305	123153	123001	122850	122699	122549	122399	122249	122100	−151
8·2	·121951	121803	121655	121507	121359	121212	121065	120919	120773	120627	−147
8·3	·120482	120337	120192	120048	119904	119760	119617	119474	119332	119190	−144
8·4	·119048	118906	118765	118624	118483	118343	118203	118064	117925	117786	−140
8·5	·117647	117509	117371	117233	117096	116959	116822	116686	116550	116414	−137
8·6	·116279	116144	116009	115875	115741	115607	115473	115340	115207	115075	−134
8·7	·114943	114811	114679	114548	114416	114286	114155	114025	113895	113766	−131
8·8	·113636	113507	113379	113250	113122	112994	112867	112740	112613	112486	−128
8·9	·112360	112233	112108	111982	111857	111732	111607	111483	111359	111235	−125
9·0	·111111	110988	110865	110742	110619	110497	110375	110254	110132	110011	−122
9·1	·109890	109769	109649	109529	109409	109290	109170	109051	108932	108814	−120
9·2	·108696	108578	108460	108342	108225	108108	107991	107875	107759	107643	−117
9·3	·107527	107411	107296	107181	107066	106952	106838	106724	106610	106496	−115
9·4	·106383	106270	106157	106045	105932	105820	105708	105597	105485	105374	−112
9·5	·105263	105152	105042	104932	104822	104712	104603	104493	104384	104275	−110
9·6	·104167	104058	103950	103842	103734	103627	103520	103413	103306	103199	−108
9·7	·103093	102987	102881	102775	102669	102564	102459	102354	102249	102145	−105
9·8	·102041	101937	101833	101729	101626	101523	101420	101317	101215	101112	−103
9·9	·101010	100908	100806	100705	100604	100503	100402	100301	100200	100100	−101

Thus 1/4·214 = 0·237530—0·000222 = 0·237308, since 0·4×555 = 222 (correct value 0·237304).

Table XXX. Factorials

No.	Factorial.	Logarithm.	No.	Factorial.	Logarithm.	No.	Factorial.	Logarithm.
1	1	0·000 0000	51	1·55112	66·190 6450	101	9·42595	159·974 3250
2	2	0·301 0300	52	8·06582	67·906 6484	102	9·61447	161·982 9252
3	6	0·778 1513	53	4·27488	69·630 9243	103	9·90290	163·995 7624
4	24	1·380 2112	54	2·30844	71·363 3180	104	1·02990	166·012 7958
5	120	2·079 1812	55	1·26964	73·103 6807	105	1·08140	168·033 9851
6	720	2·857 3325	56	7·10999	74·851 8687	106	1·14628	170·059 2909
7	5040	3·702 4305	57	4·05269	76·607 7436	107	1·22652	172·088 6747
8	40320	4·605 5205	58	2·35056	78·371 1716	108	1·32464	174·122 0985
9	362880	5·559 7630	59	1·38683	80·142 0236	109	1·44386	176·159 5250
10	3·62880	6·559 7630	60	8·32099	81·920 1748	110	1·58825	178·200 9176
11	3·99168	7·601 1557	61	5·07580	83·705 5047	111	1·76295	180·246 2406
12	4·79002	8·680 3370	62	3·14700	85·497 8964	112	1·97451	182·295 4586
13	6·22702	9·794 2803	63	1·98261	87·297 2369	113	2·23119	184·348 5371
14	8·71783	10·940 4084	64	1·26887	89·103 4169	114	2·54356	186·405 4419
15	1·30767	12·116 4996	65	8·24765	90·916 3303	115	2·92509	188·466 1398
16	2·09228	13·320 6196	66	5·44345	92·735 8742	116	3·39311	190·530 5978
17	3·55687	14·551 0685	67	3·64711	94·561 9490	117	3·96994	192·598 7836
18	6·40237	15·806 3410	68	2·48004	96·394 4579	118	4·68453	194·670 6656
19	1·21645	17·085 0946	69	1·71122	98·233 3070	119	5·57459	196·746 2126
20	2·43290	18·386 1246	70	1·19786	100·078 4050	120	6·68950	198·825 3938
21	5·10909	19·708 3439	71	8·50479	101·929 6634	121	8·09430	200·908 1792
22	1·12400	21·050 7666	72	6·12345	103·786 9959	122	9·87504	202·994 5390
23	2·58520	22·412 4944	73	4·47012	105·650 3187	123	1·21463	205·084 4442
24	6·20448	23·792 7057	74	3·30789	107·519 5505	124	1·50614	207·177 8658
25	1·55112	25·190 6457	75	2·48091	109·394 6117	125	1·88268	209·274 7759
26	4·03291	26·605 6190	76	1·88549	111·275 4253	126	2·37217	211·375 1464
27	1·08889	28·036 9828	77	1·45183	113·161 9160	127	3·01266	213·478 9501
28	3·04888	29·484 1408	78	1·13243	115·054 0106	128	3·85620	215·586 1601
29	8·84176	30·946 5388	79	8·94618	116·951 6377	129	4·97450	217·696 7498
30	2·65253	32·423 6601	80	7·15695	118·854 7277	130	6·46686	219·810 6932
31	8·22284	33·915 0218	81	5·79713	120·763 2127	131	8·47158	221·927 9645
32	2·63131	35·420 1717	82	4·75364	122·677 0266	132	1·11825	224·048 5384
33	8·68332	36·938 6857	83	3·94552	124·596 1047	133	1·48727	226·172 3900
34	2·95233	38·470 1646	84	3·31424	126·520 3840	134	1·99294	228·299 4948
35	1·03331	40·014 2326	85	2·81710	128·449 8029	135	2·69047	230·429 8286
36	3·71993	41·570 5351	86	2·42271	130·384 3013	136	3·65904	232·563 3675
37	1·37638	43·138 7369	87	2·10776	132·323 8206	137	5·01289	234·700 0881
38	5·23023	44·718 5205	88	1·85483	134·268 3033	138	6·91779	236·839 9672
39	2·03979	46·309 5851	89	1·65080	136·217 6933	139	9·61572	238·982 9820
40	8·15915	47·911 6451	90	1·48572	138·171 9358	140	1·34620	241·129 1100
41	3·34525	49·524 4289	91	1·35200	140·130 9772	141	1·89814	243·278 3291
42	1·40501	51·147 6782	92	1·24384	142·094 7650	142	2·69536	245·430 6174
43	6·04153	52·781 1467	93	1·15677	144·063 2480	143	3·85437	247·585 9535
44	2·65827	54·424 5993	94	1·08737	146·036 3758	144	5·55029	249·744 3160
45	1·19622	56·077 8119	95	1·03300	148·014 0994	145	8·04793	251·905 6840
46	5·50262	57·740 5697	96	9·91678	149·996 3707	146	1·17500	254·070 0368
47	2·58623	59·412 6676	97	9·61928	151·983 1424	147	1·72725	256·237 3542
48	1·24139	61·093 9088	98	9·42689	153·974 3685	148	2·55632	258·407 6159
49	6·08282	62·784 1049	99	9·33262	155·970 0037	149	3·80892	260·580 8022
50	3·04141	64·483 0749	100	9·33262	157·970 0037	150	5·71338	262·756 8934

The power of 10 by which to multiply the factorial is given by the whole number of the logarithm.

Table XXX. Factorials—*continued*

No.	Factorial.	Logarithm.	No.	Factorial.	Logarithm.	No.	Factorial.	Logarithm.
151	8·62721	264·935 8704	201	1·58520	377·200 0847	251	8·11447	494·909 2601
152	1·31134	267·117 7139	202	3·20211	379·505 4361	252	2·04485	497·310 6607
153	2·00634	269·302 4054	203	6·50028	381·812 9321	253	5·17346	499·713 7812
154	3·08977	271·489 9261	204	1·32606	384·122 5623	254	1·31406	502·118 6149
155	4·78914	273·680 2578	205	2·71842	386·434 3161	255	3·35085	504·525 1551
156	7·47106	275·873 3824	206	5·59994	388·748 1834	256	8·57818	506·933 3950
157	1·17296	278·069 2820	207	1·15919	391·064 1537	257	2·20459	509·343 3282
158	1·85327	280·267 9391	208	2·41111	393·382 2170	258	5·68785	511·754 9479
159	2·94670	282·469 3363	209	5·03922	395·702 3633	259	1·47315	514·168 2476
160	4·71472	284·673 4562	210	1·05824	398·024 5826	260	3·83020	516·583 2210
161	7·59071	286·880 2821	211	2·23288	400·348 8651	261	9·99681	518·999 8615
162	1·22969	289·089 7971	212	4·73370	402·675 2009	262	2·61916	521·418 1628
163	2·00440	291·301 9847	213	1·00828	405·003 5805	263	6·88840	523·838 1185
164	3·28722	293·516 8286	214	2·15772	407·333 9943	264	1·81854	526·259 7225
165	5·42391	295·734 3125	215	4·63909	409·666 4328	265	4·81913	528·682 9683
166	9·00369	297·954 4206	216	1·00204	412·000 8865	266	1·28189	531·107 8500
167	1·50362	300·177 1371	217	2·17443	414·337 3463	267	3·42264	533·534 3612
168	2·52608	302·402 4464	218	4·74027	416·675 8027	268	9·17268	535·962 4960
169	4·26907	304·630 3331	219	1·03812	419·016 2469	269	2·46745	538·392 2483
170	7·25742	306·860 7820	220	2·28386	421·358 6695	270	6·66211	540·823 6121
171	1·24102	309·093 7781	221	5·04733	423·703 0618	271	1·80543	543·256 5814
172	2·13455	311·329 3066	222	1·12051	426·049 4148	272	4·91078	545·691 1503
173	3·69277	313·567 3527	223	2·49873	428·397 7197	273	1·34064	548·127 3129
174	6·42543	315·807 9019	224	5·59716	430·747 9677	274	3·67336	550·565 0635
175	1·12445	318·050 9400	225	1·25936	433·100 1502	275	1·01017	553·004 3962
176	1·97903	320·296 4526	226	2·84616	435·454 2586	276	2·78808	555·445 3052
177	3·50289	322·544 4259	227	6·46077	437·810 2845	277	7·72298	557·887 7850
178	6·23514	324·794 8459	228	1·47306	440·168 2193	278	2·14699	560·331 8298
179	1·11609	327·047 6989	229	3·37330	442·528 0548	279	5·99010	562·777 4340
180	2·00896	329·302 9714	230	7·75859	444·889 7827	280	1·67723	565·224 5920
181	3·63622	331·560 6500	231	1·79223	447·253 3946	281	4·71301	567·673 2984
182	6·61792	333·820 7214	232	4·15798	449·618 8826	282	1·32907	570·123 5475
183	1·21108	336·083 1725	233	9·68810	451·986 2385	283	3·76126	572·575 3339
184	2·22839	338·347 9903	234	2·26702	454·355 4544	284	1·06820	575·028 6523
185	4·12251	340·615 1620	235	5·32749	456·726 5223	285	3·04437	577·483 4971
186	7·66787	342·884 6750	236	1·25729	459·099 4343	286	8·70689	579·939 8631
187	1·43389	345·156 5166	237	2·97977	461·474 1826	287	2·49888	582·397 7450
188	2·69572	347·430 6744	238	7·09185	463·850 7596	288	7·19677	584·857 1375
189	5·09491	349·707 1362	239	1·69495	466·229 1575	289	2·07987	587·318 0354
190	9·68032	351·985 8898	240	4·06789	468·609 3687	290	6·03161	589·780 4334
191	1·84894	354·266 9232	241	9·80360	470·991 3857	291	1·75520	592·244 3264
192	3·54997	356·550 2244	242	2·37247	473·375 2011	292	5·12518	594·709 7092
193	6·85144	358·835 7817	243	5·76511	475·760 8074	293	1·50168	597·176 5768
194	1·32918	361·123 5835	244	1·40669	478·148 1972	294	4·41493	599·644 9242
195	2·59190	363·413 6181	245	3·44638	480·537 3633	295	1·30241	602·114 7462
196	5·08012	365·705 8742	246	8·47810	482·928 2984	296	3·85512	604·586 0379
197	1·00078	368·000 3404	247	2·09409	485·320 9954	297	1·14497	607·058 7943
198	1·98155	370·297 0056	248	5·19334	487·715 4470	298	3·41201	609·533 0106
199	3·94329	372·595 8586	249	1·29314	490·111 6464	299	1·02019	612·008 6818
200	7·88658	374·896 8886	250	3·23286	492·509 5864	300	3·06058	614·485 8030

For higher values, and non-integral $x>6$ (or $x \geqslant 1{\cdot}5$ for 4-figure accuracy), use :

$$\log x! = (x+\tfrac{1}{2})\log x + \tfrac{1}{2}\log 2\pi - \left(x - \frac{1}{12x} + \frac{1}{360x^3}\right)\log e.$$

Table XXXI. Natural Sines

	0′	6′	12′	18′	24′	30′	36′	42′	48′	54′	1′	2′	3′
0	·00000	00175	00349	00524	00698	00873	01047	01222	01396	01571	29	58	87
1	·01745	01920	02094	02269	02443	02618	02792	02967	03141	03316	29	58	87
2	·03490	03664	03839	04013	04188	04362	04536	04711	04885	05059	29	58	87
3	·05234	05408	05582	05756	05931	06105	06279	06453	06627	06802	29	58	87
4	·06976	07150	07324	07498	07672	07846	08020	08194	08368	08542	29	58	87
5	·08716	08889	09063	09237	09411	09585	09758	09932	10106	10279	29	58	87
6	·10453	10626	10800	10973	11147	11320	11494	11667	11840	12014	29	58	87
7	·12187	12360	12533	12706	12880	13053	13226	13399	13572	13744	29	58	86
8	·13917	14090	14263	14436	14608	14781	14954	15126	15299	15471	29	58	86
9	·15643	15816	15988	16160	16333	16505	16677	16849	17021	17193	29	57	86
10	·17365	17537	17708	17880	18052	18224	18395	18567	18738	18910	29	57	86
11	·19081	19252	19423	19595	19766	19937	20108	20279	20450	20620	28	57	86
12	·20791	20962	21132	21303	21474	21644	21814	21985	22155	22325	28	57	85
13	·22495	22665	22835	23005	23175	23345	23514	23684	23853	24023	28	57	85
14	·24192	24362	24531	24700	24869	25038	25207	25376	25545	25713	28	56	85
15	·25882	26050	26219	26387	26556	26724	26892	27060	27228	27396	28	56	84
16	·27564	27731	27899	28067	28234	28402	28569	28736	28903	29070	28	56	84
17	·29237	29404	29571	29737	29904	30071	30237	30403	30570	30736	28	56	83
18	·30902	31068	31233	31399	31565	31730	31896	32061	32227	32392	28	55	83
19	·32557	32722	32887	33051	33216	33381	33545	33710	33874	34038	27	55	82
20	·34202	34366	34530	34694	34857	35021	35184	35347	35511	35674	27	55	82
21	·35837	36000	36162	36325	36488	36650	36812	36975	37137	37299	27	54	81
22	·37461	37622	37784	37946	38107	38268	38430	38591	38752	38912	27	54	81
23	·39073	39234	39394	39555	39715	39875	40035	40195	40355	40514	27	53	80
24	·40674	40833	40992	41151	41310	41469	41628	41787	41945	42104	26	53	79
25	·42262	42420	42578	42736	42894	43051	43209	43366	43523	43680	26	53	79
26	·43837	43994	44151	44307	44464	44620	44776	44932	45088	45243	26	52	78
27	·45399	45554	45710	45865	46020	46175	46330	46484	46639	46793	26	52	77
28	·46947	47101	47255	47409	47562	47716	47869	48022	48175	48328	26	51	77
29	·48481	48634	48786	48938	49090	49242	49394	49546	49697	49849	25	51	76
30	·50000	50151	50302	50453	50603	50754	50904	51054	51204	51354	25	50	75
31	·51504	51653	51803	51952	52101	52250	52399	52547	52696	52844	25	50	74
32	·52992	53140	53288	53435	53583	53730	53877	54024	54171	54317	25	49	74
33	·54464	54610	54756	54902	55048	55194	55339	55484	55630	55775	24	49	73
34	·55919	56064	56208	56353	56497	56641	56784	56928	57071	57215	24	48	72
35	·57358	57501	57643	57786	57928	58070	58212	58354	58496	58637	24	47	71
36	·58779	58920	59061	59201	59342	59482	59622	59763	59902	60042	23	47	70
37	·60182	60321	60460	60599	60738	60876	61015	61153	61291	61429	23	46	69
38	·61566	61704	61841	61978	62115	62251	62388	62524	62660	62796	23	46	68
39	·62932	63068	63203	63338	63473	63608	63742	63877	64011	64145	22	45	67
40	·64279	64412	64546	64679	64812	64945	65077	65210	65342	65474	22	44	66
41	·65606	65738	65869	66000	66131	66262	66393	66523	66653	66783	22	44	65
42	·66913	67043	67172	67301	67430	67559	67688	67816	67944	68072	21	43	64
43	·68200	68327	68455	68582	68709	68835	68962	69088	69214	69340	21	42	63
44	·69466	69591	69717	69842	69966	70091	70215	70339	70463	70587	21	42	62

	0′	6′	12′	18′	24′	30′	36′	42′	48′	54′	1′	2′	3′
45	·70711	70834	70957	71080	71203	71325	71447	71569	71691	71813	20	41	61
46	·71934	72055	72176	72297	72417	72537	72657	72777	72897	73016	20	40	60
47	·73135	73254	73373	73491	73610	73728	73846	73963	74080	74198	20	39	59
48	·74314	74431	74548	74664	74780	74896	75011	75126	75241	75356	19	39	58
49	·75471	75585	75700	75813	75927	76041	76154	76267	76380	76492	19	38	57
50	·76604	76717	76828	76940	77051	77162	77273	77384	77494	77605	19	37	56
51	·77715	77824	77934	78043	78152	78261	78369	78478	78586	78694	18	36	54
52	·78801	78908	79016	79122	79229	79335	79441	79547	79653	79758	18	35	53
53	·79864	79968	80073	80178	80282	80386	80489	80593	80696	80799	17	35	52
54	·80902	81004	81106	81208	81310	81412	81513	81614	81714	81815	17	34	51
55	·81915	82015	82115	82214	82314	82413	82511	82610	82708	82806	16	33	50
56	·82904	83001	83098	83195	83292	83389	83485	83581	83676	83772	16	32	48
57	·83867	83962	84057	84151	84245	84339	84433	84526	84619	84712	16	31	47
58	·84805	84897	84989	85081	85173	85264	85355	85446	85536	85627	15	30	46
59	·85717	85806	85896	85985	86074	86163	86251	86340	86427	86515	15	30	44
60	·86603	86690	86777	86863	86949	87036	87121	87207	87292	87377	14	29	43
61	·87462	87546	87631	87715	87798	87882	87965	88048	88130	88213	14	28	42
62	·88295	88377	88458	88539	88620	88701	88782	88862	88942	89021	13	27	40
63	·89101	89180	89259	89337	89415	89493	89571	89649	89726	89803	13	26	39
64	·89879	89956	90032	90108	90183	90259	90334	90408	90483	90557	13	25	38
65	·90631	90704	90778	90851	90924	90996	91068	91140	91212	91283	12	24	36
66	·91355	91425	91496	91566	91636	91706	91775	91845	91914	91982	12	23	35
67	·92050	92119	92186	92254	92321	92388	92455	92521	92587	92653	11	22	34
68	·92718	92784	92849	92913	92978	93042	93106	93169	93232	93295	11	21	32
69	·93358	93420	93483	93544	93606	93667	93728	93789	93849	93909	10	20	31
70	·93969	94029	94088	94147	94206	94264	94322	94380	94438	94495	10	19	29
71	·94552	94609	94665	94721	94777	94832	94888	94943	94997	95052	9	19	28
72	·95106	95159	95213	95266	95319	95372	95424	95476	95528	95579	9	18	26
73	·95630	95681	95732	95782	95832	95882	95931	95981	96029	96078	8	17	25
74	·96126	96174	96222	96269	96316	96363	96410	96456	96502	96547	8	16	23
75	·96593	96638	96682	96727	96771	96815	96858	96902	96945	96987	7	15	22
76	·97030	97072	97113	97155	97196	97237	97278	97318	97358	97398	7	14	20
77	·97437	97476	97515	97553	97592	97630	97667	97705	97742	97778	6	13	19
78	·97815	97851	97887	97922	97958	97992	98027	98061	98096	98129	6	12	17
79	·98163	98196	98229	98261	98294	98325	98357	98389	98420	98450	5	11	16
80	·98481	98511	98541	98570	98600	98629	98657	98686	98714	98741	5	10	14
81	·98769	98796	98823	98849	98876	98902	98927	98953	98978	99002	4	9	13
82	·99027	99051	99075	99098	99122	99144	99167	99189	99211	99233	4	8	11
83	·99255	99276	99297	99317	99337	99357	99377	99396	99415	99434	3	7	10
84	·99452	99470	99488	99506	99523	99540	99556	99572	99588	99604	3	6	8
85	·99619	99635	99649	99664	99678	99692	99705	99719	99731	99744	2	5	7
86	·99756	99768	99780	99792	99803	99813	99824	99834	99844	99854	2	4	5
87	·99863	99872	99881	99889	99897	99905	99912	99919	99926	99933	1	3	4
88	·99939	99945	99951	99956	99961	99966	99970	99974	99978	99982	1	2	2
89	·99985	99988	99990	99993	99995	99996	99998	99999	99999	100000	0	1	1

Table XXXII. Natural Tangents

	0′	6′	12′	18′	24′	30′	36′	42′	48′	54′	1′	2′	3′
0	0·00000	·00175	·00349	·00524	·00698	·00873	·01047	·01222	·01396	·01571	29	58	87
1	0·01746	·01920	·02095	·02269	·02444	·02619	·02793	·02968	·03143	·03317	29	58	87
2	0·03492	·03667	·03842	·04016	·04191	·04366	·04541	·04716	·04891	·05066	29	58	87
3	0·05241	·05416	·05591	·05766	·05941	·06116	·06291	·06467	·06642	·06817	29	58	88
4	0·06993	·07168	·07344	·07519	·07695	·07870	·08046	·08221	·08397	·08573	29	59	88
5	0·08749	·08925	·09101	·09277	·09453	·09629	·09805	·09981	·10158	·10334	29	59	88
6	0·10510	·10687	·10863	·11040	·11217	·11394	·11570	·11747	·11924	·12101	29	59	88
7	0·12278	·12456	·12633	·12810	·12988	·13165	·13343	·13521	·13698	·13876	30	59	89
8	0·14054	·14232	·14410	·14588	·14767	·14945	·15124	·15302	·15481	·15660	30	59	89
9	0·15838	·16017	·16196	·16376	·16555	·16734	·16914	·17093	·17273	·17453	30	60	90
10	0·17633	·17813	·17993	·18173	·18353	·18534	·18714	·18895	·19076	·19257	30	60	90
11	0·19438	·19619	·19801	·19982	·20164	·20345	·20527	·20709	·20891	·21073	30	61	91
12	0·21256	·21438	·21621	·21804	·21986	·22169	·22353	·22536	·22719	·22903	30	61	92
13	0·23087	·23271	·23455	·23639	·23823	·24008	·24193	·24377	·24562	·24747	31	61	92
14	0·24933	·25118	·25304	·25490	·25676	·25862	·26048	·26235	·26421	·26608	31	62	93
15	0·26795	·26982	·27169	·27357	·27545	·27732	·27921	·28109	·28297	·28486	31	63	94
16	0·28675	·28864	·29053	·29242	·29432	·29621	·29811	·30001	·30192	·30382	32	63	95
17	0·30573	·30764	·30955	·31147	·31338	·31530	·31722	·31914	·32106	·32299	32	64	96
18	0·32492	·32685	·32878	·33072	·33266	·33460	·33654	·33848	·34043	·34238	32	65	97
19	0·34433	·34628	·34824	·35020	·35216	·35412	·35608	·35805	·36002	·36199	33	65	98
20	0·36397	·36595	·36793	·36991	·37190	·37388	·37588	·37787	·37986	·38186	33	66	99
21	0·38386	·38587	·38787	·38988	·39190	·39391	·39593	·39795	·39997	·40200	34	67	101
22	0·40403	·40606	·40809	·41013	·41217	·41421	·41626	·41831	·42036	·42242	34	68	102
23	0·42447	·42654	·42860	·43067	·43274	·43481	·43689	·43897	·44105	·44314	35	69	104
24	0·44523	·44732	·44942	·45152	·45362	·45573	·45784	·45995	·46206	·46418	35	70	105
25	0·46631	·46843	·47056	·47270	·47483	·47698	·47912	·48127	·48342	·48557	36	71	107
26	0·48773	·48989	·49206	·49423	·49640	·49858	·50076	·50295	·50514	·50733	36	73	109
27	0·50953	·51173	·51393	·51614	·51835	·52057	·52279	·52501	·52724	·52947	37	74	111
28	0·53171	·53395	·53620	·53844	·54070	·54296	·54522	·54748	·54975	·55203	38	75	113
29	0·55431	·55659	·55888	·56117	·56347	·56577	·56808	·57039	·57271	·57503	38	77	115
30	0·57735	·57968	·58201	·58435	·58670	·58905	·59140	·59376	·59612	·59849	39	78	117
31	0·60086	·60324	·60562	·60801	·61040	·61280	·61520	·61761	·62003	·62245	40	80	120
32	0·62487	·62730	·62973	·63217	·63462	·63707	·63953	·64199	·64446	·64693	41	82	123
33	0·64941	·65189	·65438	·65688	·65938	·66189	·66440	·66692	·66944	·67197	42	84	125
34	0·67451	·67705	·67960	·68215	·68471	·68728	·68985	·69243	·69502	·69761	43	86	128
35	0·70021	·70281	·70542	·70804	·71066	·71329	·71593	·71857	·72122	·72388	44	88	131
36	0·72654	·72921	·73189	·73457	·73726	·73996	·74267	·74538	·74810	·75082	45	90	135
37	0·75355	·75629	·75904	·76180	·76456	·76733	·77010	·77289	·77568	·77848	46	92	139
38	0·78129	·78410	·78692	·78975	·79259	·79544	·79829	·80115	·80402	·80690	47	95	142
39	0·80978	·81268	·81558	·81849	·82141	·82434	·82727	·83022	·83317	·83613	49	98	146
40	0·83910	·84208	·84507	·84806	·85107	·85408	·85710	·86014	·86318	·86623	50	100	151
41	0·86929	·87236	·87543	·87852	·88162	·88473	·88784	·89097	·89410	·89725	52	104	155
42	0·90040	·90357	·90674	·90993	·91313	·91633	·91955	·92277	·92601	·92926	53	107	160
43	0·93252	·93578	·93906	·94235	·94565	·94896	·95229	·95562	·95897	·96232	55	110	166
44	0·96569	·96907	·97246	·97586	·97927	·98270	·98613	·98958	·99304	·99652	57	114	171

Table XXXII. Natural Tangents—*continued*

	0′	6′	12′	18′	24′	30′	36′	42′	48′	54′	1′	2′	3′
45	1·00000	·00350	·00701	·01053	·01406	·01761	·02117	·02474	·02832	·03192	59	118	177
46	1·03553	·03915	·04279	·04644	·05010	·05378	·05747	·06117	·06489	·06862	61	123	184
47	1·07237	·07613	·07990	·08369	·08749	·09131	·09514	·09899	·10285	·10672	64	127	191
48	1·11061	·11452	·11844	·12238	·12633	·13029	·13428	·13828	·14229	·14632	66	132	198
49	1·15037	·15443	·15851	·16261	·16672	·17085	·17500	·17916	·18334	·18754	69	138	206
50	1·19175	·19599	·20024	·20451	·20879	·21310	·21742	·22176	·22612	·23050	72	144	215
51	1·23490	·23931	·24375	·24820	·25268	·25717	·26169	·26622	·27077	·27535	75	150	225
52	1·27994	·28456	·28919	·29385	·29853	·30323	·30795	·31269	·31745	·32224	78	157	235
53	1·32704	·33187	·33673	·34160	·34650	·35142	·35637	·36134	·36633	·37134	82	164	246
54	1·37638	·38145	·38653	·39165	·39679	·40195	·40714	·41235	·41759	·42286	86	172	258
55	1·42815	·43347	·43881	·44418	·44958	·45501	·46046	·46595	·47146	·47699	90	181	271
56	1·48256	·48816	·49378	·49944	·50512	·51084	·51658	·52235	·52816	·53400	95	191	286
57	1·53986	·54576	·55170	·55766	·56366	·56969	·57575	·58184	·58797	·59414	100	201	302
58	1·60033	·60657	61283	·61914	·62548	·63185	·63826	·64471	·65120	·65772	106	213	319
59	1·66428	·67088	·67752	·68419	·69091	·69766	·70446	·71129	·71817	·72509	113	225	338
60	1·73205	·73905	·74610	·75319	·76032	·76749	·77471	·78198	·78929	·79665	—	—	—

(90°—θ) tan θ

	0′	6′	12′	18′	24′	30′	36′	42′	48′	54′	1′	2′	3′	tan θ
60	51·962	51·998	52·034	52·070	52·105	52·141	52·177	52·212	52·247	52·282	6	12	18	1·73
61	52·317	52·352	52·387	52·422	52·456	52·490	52·525	52·559	52·593	52·627	6	11	17	1·80
62	52·660	52·694	52·727	52·761	52·794	52·827	52·860	52·893	52·925	52·958	6	11	17	1·88
63	52·990	53·023	53·055	53·087	53·119	53·151	53·182	53·214	53·245	53·277	5	11	16	1·96
64	53·308	53·339	53·370	53·401	53·431	53·462	53·492	53·523	53·553	53·583	5	10	15	2·05
65	53·613	53·642	53·672	53·702	53·731	53·760	53·790	53·819	53·847	53·876	5	10	15	2·14
66	53·905	53·933	53·962	53·990	54·018	54·046	54·074	54·102	54·130	54·157	5	9	14	2·25
67	54·185	54·212	54·239	54·266	54·293	54·320	54·346	54·373	54·399	54·426	4	9	13	2·36
68	54·452	54·478	54·504	54·530	54·555	54·581	54·606	54·632	54·657	54·682	4	9	13	2·48
69	54·707	54·732	54·756	54·781	54·805	54·830	54·854	54·878	54·902	54·926	4	8	12	2·61
70	54·950	54·973	54·997	55·020	55·043	55·066	55·089	55·112	55·135	55·158	4	8	12	2·75
71	55·180	55·202	55·225	55·247	55·269	55·291	55·312	55·334	55·356	55·377	4	7	11	2·90
72	55·398	55·419	55·441	55·461	55·482	55·503	55·523	55·544	55·564	55·584	3	7	10	3·08
73	55·604	55·624	55·644	55·664	55·684	55·703	55·722	55·742	55·761	55·780	3	7	10	3·27
74	55·799	55·817	55·836	55·855	55·873	55·891	55·909	55·927	55·945	55·963	3	6	9	3·49
75	55·981	55·998	56·016	56·033	56·050	56·067	56·084	56·101	56·118	56·134	3	6	8	3·73
76	56·151	56·167	56·184	56·200	56·216	56·232	56·247	56·263	56·278	56·294	3	5	8	4·01
77	56·309	56·324	56·339	56·354	56·369	56·384	56·398	56·413	56·427	56·441	2	5	7	4·33
78	56·456	56·470	56·483	56·497	56·511	56·524	56·538	56·551	56·564	56·577	2	4	7	4·70
79	56·590	56·603	56·616	56·628	56·641	56·653	56·665	56·677	56·689	56·701	2	4	6	5·14
80	56·713	56·724	56·736	56·747	56·759	56·770	56·781	56·792	56·803	56·813	2	4	6	5·67
81	56·824	56·834	56·845	56·855	56·865	56·875	56·885	56·894	56·904	56·914	2	3	5	6·31
82	56·923	56·932	56·941	56·950	56·959	56·968	56·977	56·985	56·994	57·002	1	3	4	7·12
83	57·010	57·019	57·027	57·034	57·042	57·050	57·057	57·065	57·072	57·079	1	3	4	8·14
84	57·086	57·093	57·100	57·107	57·113	57·120	57·126	57·132	57·138	57·144	1	2	3	9·51
85	57·150	57·156	57·162	57·167	57·173	57·178	57·183	57·188	57·193	57·198	1	2	3	11·4
86	57·203	57·207	57·212	57·216	57·220	57·224	57·229	57·232	57·236	57·240	1	1	2	14·3
87	57·243	57·247	57·250	57·253	57·256	57·259	57·262	57·265	57·268	57·270	0	1	2	19·1
88	57·273	57·275	57·277	57·279	57·281	57·283	57·284	57·286	57·287	57·289	0	1	1	28·6
89	57·290	57·291	57·292	57·293	57·294	57·294	57·295	57·295	57·296	57·296	0	0	0	**57·3**

Table XXXIII. Random Numbers (I)

03 47 43 73 86	36 96 47 36 61	46 98 63 71 62	33 26 16 80 45	60 11 14 10 95
97 74 24 67 62	42 81 14 57 20	42 53 32 37 32	27 07 36 07 51	24 51 79 89 73
16 76 62 27 66	56 50 26 71 07	32 90 79 78 53	13 55 38 58 59	88 97 54 14 10
12 56 85 99 26	96 96 68 27 31	05 03 72 93 15	57 12 10 14 21	88 26 49 81 76
55 59 56 35 64	38 54 82 46 22	31 62 43 09 90	06 18 44 32 53	23 83 01 30 30
16 22 77 94 39	49 54 43 54 82	17 37 93 23 78	87 35 20 96 43	84 26 34 91 64
84 42 17 53 31	57 24 55 06 88	77 04 74 47 67	21 76 33 50 25	83 92 12 06 76
63 01 63 78 59	16 95 55 67 19	98 10 50 71 75	12 86 73 58 07	44 39 52 38 79
33 21 12 34 29	78 64 56 07 82	52 42 07 44 38	15 51 00 13 42	99 66 02 79 54
57 60 86 32 44	09 47 27 96 54	49 17 46 09 62	90 52 84 77 27	08 02 73 43 28
18 18 07 92 46	44 17 16 58 09	79 83 86 19 62	06 76 50 03 10	55 23 64 05 05
26 62 38 97 75	84 16 07 44 99	83 11 46 32 24	20 14 85 88 45	10 93 72 88 71
23 42 40 64 74	82 97 77 77 81	07 45 32 14 08	32 98 94 07 72	93 85 79 10 75
52 36 28 19 95	50 92 26 11 97	00 56 76 31 38	80 22 02 53 53	86 60 42 04 53
37 85 94 35 12	83 39 50 08 30	42 34 07 96 88	54 42 06 87 98	35 85 29 48 39
70 29 17 12 13	40 33 20 38 26	13 89 51 03 74	17 76 37 13 04	07 74 21 19 30
56 62 18 37 35	96 83 50 87 75	97 12 25 93 47	70 33 24 03 54	97 77 46 44 80
99 49 57 22 77	88 42 95 45 72	16 64 36 16 00	04 43 18 66 79	94 77 24 21 90
16 08 15 04 72	33 27 14 34 09	45 59 34 68 49	12 72 07 34 45	99 27 72 95 14
31 16 93 32 43	50 27 89 87 19	20 15 37 00 49	52 85 66 60 44	38 68 88 11 80
68 34 30 13 70	55 74 30 77 40	44 22 78 84 26	04 33 46 09 52	68 07 97 06 57
74 57 25 65 76	59 29 97 68 60	71 91 38 67 54	13 58 18 24 76	15 54 55 95 52
27 42 37 86 53	48 55 90 65 72	96 57 69 36 10	96 46 92 42 45	97 60 49 04 91
00 39 68 29 61	66 37 32 20 30	77 84 57 03 29	10 45 65 04 26	11 04 96 67 24
29 94 98 94 24	68 49 69 10 82	53 75 91 93 30	34 25 20 57 27	40 48 73 51 92
16 90 82 66 59	83 62 64 11 12	67 19 00 71 74	60 47 21 29 68	02 02 37 03 31
11 27 94 75 06	06 09 19 74 66	02 94 37 34 02	76 70 90 30 86	38 45 94 30 38
35 24 10 16 20	33 32 51 26 38	79 78 45 04 91	16 92 53 56 16	02 75 50 95 98
38 23 16 86 38	42 38 97 01 50	87 75 66 81 41	40 01 74 91 62	48 51 84 08 32
31 96 25 91 47	96 44 33 49 13	34 86 82 53 91	00 52 43 48 85	27 55 26 89 62
56 67 40 67 14	64 05 71 95 86	11 05 65 09 68	76 83 20 37 90	57 16 00 11 66
14 90 84 45 11	75 73 88 05 90	52 27 41 14 86	22 98 12 22 08	07 52 74 95 80
68 05 51 18 00	33 96 02 75 19	07 60 62 93 55	59 33 82 43 90	49 37 38 44 59
20 46 78 73 90	97 51 40 14 02	04 02 33 31 08	39 54 16 49 36	47 95 93 13 30
64 19 58 97 79	15 06 15 93 20	01 90 10 75 06	40 78 78 89 62	02 67 74 17 33
05 26 93 70 60	22 35 85 15 13	92 03 51 59 77	59 56 78 06 83	52 91 05 70 74
07 97 10 88 23	09 98 42 99 64	61 71 62 99 15	06 51 29 16 93	58 05 77 09 51
68 71 86 85 85	54 87 66 47 54	73 32 08 11 12	44 95 92 63 16	29 56 24 29 48
26 99 61 65 53	58 37 78 80 70	42 10 50 67 42	32 17 55 85 74	94 44 67 16 94
14 65 52 68 75	87 59 36 22 41	26 78 63 06 55	13 08 27 01 50	15 29 39 39 43
17 53 77 58 71	71 41 61 50 72	12 41 94 96 26	44 95 27 36 99	02 96 74 30 83
90 26 59 21 19	23 52 23 33 12	96 93 02 18 39	07 02 18 36 07	25 99 32 70 23
41 23 52 55 99	31 04 49 69 96	10 47 48 45 88	13 41 43 89 20	97 17 14 49 17
60 20 50 81 69	31 99 73 68 68	35 81 33 03 76	24 30 12 48 60	18 99 10 72 34
91 25 38 05 90	94 58 28 41 36	45 37 59 03 09	90 35 57 29 12	82 62 54 65 60
34 50 57 74 37	98 80 33 00 91	09 77 93 19 82	74 94 80 04 04	45 07 31 66 49
85 22 04 39 43	73 81 53 94 79	33 62 46 86 28	08 31 54 46 31	53 94 13 38 47
09 79 13 77 48	73 82 97 22 21	05 03 27 24 83	72 89 44 05 60	35 80 39 94 88
88 75 80 18 14	22 95 75 42 49	39 32 82 22 49	02 48 07 70 37	16 04 61 67 87
90 96 23 70 00	39 00 03 06 90	55 85 78 38 36	94 37 30 69 32	90 89 00 76 33

Table XXXIII. Random Numbers (II)

53 74 23 99 67	61 32 28 69 84	94 62 67 86 24	98 33 41 19 95	47 53 53 38 09
63 38 06 86 54	99 00 65 26 94	02 82 90 23 07	79 62 67 80 60	75 91 12 81 19
35 30 58 21 46	06 72 17 10 94	25 21 31 75 96	49 28 24 00 49	55 65 79 78 07
63 43 36 82 69	65 51 18 37 88	61 38 44 12 45	32 92 85 88 65	54 34 81 85 35
98 25 37 55 26	01 91 82 81 46	74 71 12 94 97	24 02 71 37 07	03 92 18 66 75
02 63 21 17 69	71 50 80 89 56	38 15 70 11 48	43 40 45 86 98	00 83 26 91 03
64 55 22 21 82	48 22 28 06 00	61 54 13 43 91	82 78 12 23 29	06 66 24 12 27
85 07 26 13 89	01 10 07 82 04	59 63 69 36 03	69 11 15 83 80	13 29 54 19 28
58 54 16 24 15	51 54 44 82 00	62 61 65 04 69	38 18 65 18 97	85 72 13 49 21
34 85 27 84 87	61 48 64 56 26	90 18 48 13 26	37 70 15 42 57	65 65 80 39 07
03 92 18 27 46	57 99 16 96 56	30 33 72 85 22	84 64 38 56 98	99 01 30 98 64
62 95 30 27 59	37 75 41 66 48	86 97 80 61 45	23 53 04 01 63	45 76 08 64 27
08 45 93 15 22	60 21 75 46 91	98 77 27 85 42	28 88 61 08 84	69 62 03 42 73
07 08 55 18 40	45 44 75 13 90	24 94 96 61 02	57 55 66 83 15	73 42 37 11 61
01 85 89 95 66	51 10 19 34 88	15 84 97 19 75	12 76 39 43 78	64 63 91 08 25
72 84 71 14 35	19 11 58 49 26	50 11 17 17 76	86 31 57 20 18	95 60 78 46 75
88 78 28 16 84	13 52 53 94 53	75 45 69 30 96	73 89 65 70 31	99 17 43 48 76
45 17 75 65 57	28 40 19 72 12	25 12 74 75 67	60 40 60 81 19	24 62 01 61 16
96 76 28 12 54	22 01 11 94 25	71 96 16 16 88	68 64 36 74 45	19 59 50 88 92
43 31 67 72 30	24 02 94 08 63	38 32 36 66 02	69 36 38 25 39	48 03 45 15 22
50 44 66 44 21	66 06 58 05 62	68 15 54 35 02	42 35 48 96 32	14 52 41 52 48
22 66 22 15 86	26 63 75 41 99	58 42 36 72 24	58 37 52 18 51	03 37 18 39 11
96 24 40 14 51	23 22 30 88 57	95 67 47 29 83	94 69 40 06 07	18 16 36 78 86
31 73 91 61 19	60 20 72 93 48	98 57 07 23 69	65 95 39 69 58	56 80 30 19 44
78 60 73 99 84	43 89 94 36 45	56 69 47 07 41	90 22 91 07 12	78 35 34 08 72
84 37 90 61 56	70 10 23 98 05	85 11 34 76 60	76 48 45 34 60	01 64 18 39 96
36 67 10 08 23	98 93 35 08 86	99 29 76 29 81	33 34 91 58 93	63 14 52 32 52
07 28 59 07 48	89 64 58 89 75	83 85 62 27 89	30 14 78 56 27	86 63 59 80 02
10 15 83 87 60	79 24 31 66 56	21 48 24 06 93	91 98 94 05 49	01 47 59 38 00
55 19 68 97 65	03 73 52 16 56	00 53 55 90 27	33 42 29 38 87	22 13 88 83 34
53 81 29 13 39	35 01 20 71 34	62 33 74 82 14	53 73 19 09 03	56 54 29 56 93
51 86 32 68 92	33 98 74 66 99	40 14 71 94 58	45 94 19 38 81	14 44 99 81 07
35 91 70 29 13	80 03 54 07 27	96 94 78 32 66	50 95 52 74 33	13 80 55 62 54
37 71 67 95 13	20 02 44 95 94	64 85 04 05 72	01 32 90 76 14	53 89 74 60 41
93 66 13 83 27	92 79 64 64 72	28 54 96 53 84	48 14 52 98 94	56 07 93 89 30
02 96 08 45 65	13 05 00 41 84	93 07 54 72 59	21 45 57 09 77	19 48 56 27 44
49 83 43 48 35	82 88 33 69 96	72 36 04 19 76	47 45 15 18 60	82 11 08 95 97
84 60 71 62 46	40 80 81 30 37	34 39 23 05 38	25 15 35 71 30	88 12 57 21 77
18 17 30 88 71	44 91 14 88 47	89 23 30 63 15	56 34 20 47 89	99 82 93 24 98
79 69 10 61 78	71 32 76 95 62	87 00 22 58 40	92 54 01 75 25	43 11 71 99 31
75 93 36 57 83	56 20 14 82 11	74 21 97 90 65	96 42 68 63 86	74 54 13 26 94
38 30 92 29 03	06 28 81 39 38	62 25 06 84 63	61 29 08 93 67	04 32 92 08 09
51 29 50 10 34	31 57 75 95 80	51 97 02 74 77	76 15 48 49 44	18 55 63 77 09
21 31 38 86 24	37 79 81 53 74	73 24 16 10 33	52 83 90 94 76	70 47 14 54 36
29 01 23 87 88	58 02 39 37 67	42 10 14 20 92	16 55 23 42 45	54 96 09 11 06
95 33 95 22 00	18 74 72 00 18	38 79 58 69 32	81 76 80 26 92	82 80 84 25 39
90 84 60 79 80	24 36 59 87 38	82 07 53 89 35	96 35 23 79 18	05 98 90 07 35
46 40 62 98 82	54 97 20 56 95	15 74 80 08 32	16 46 70 50 80	67 72 16 42 79
20 31 89 03 43	38 46 82 68 72	32 14 82 99 70	80 60 47 18 97	63 49 30 21 30
71 59 73 05 50	08 22 23 71 77	91 01 93 20 49	82 96 59 26 94	66 39 67 98 60

Table XXXIII. Random Numbers (III)

22 17 68 65 84	68 95 23 92 35	87 02 22 57 51	61 09 43 95 06	58 24 82 03 47
19 36 27 59 46	13 79 93 37 55	39 77 32 77 09	85 52 05 30 62	47 83 51 62 74
16 77 23 02 77	09 61 87 25 21	28 06 24 25 93	16 71 13 59 78	23 05 47 47 25
78 43 76 71 61	20 44 90 32 64	97 67 63 99 61	46 38 03 93 22	69 81 21 99 21
03 28 28 26 08	73 37 32 04 05	69 30 16 09 05	88 69 58 28 99	35 07 44 75 47
93 22 53 64 39	07 10 63 76 35	87 03 04 79 88	08 13 13 85 51	55 34 57 72 69
78 76 58 54 74	92 38 70 96 92	52 06 79 79 45	82 63 18 27 44	69 66 92 19 09
23 68 35 26 00	99 53 93 61 28	52 70 05 48 34	56 65 05 61 86	90 92 10 70 80
15 39 25 70 99	93 86 52 77 65	15 33 59 05 28	22 87 26 07 47	86 96 98 29 06
58 71 96 30 24	18 46 23 34 27	85 13 99 24 44	49 18 09 79 49	74 16 32 23 02
57 35 27 33 72	24 53 63 94 09	41 10 76 47 91	44 04 95 49 66	39 60 04 59 81
48 50 86 54 48	22 06 34 72 52	82 21 15 65 20	33 29 94 71 11	15 91 29 12 03
61 96 48 95 03	07 16 39 33 66	98 56 10 56 79	77 21 30 27 12	90 49 22 23 62
36 93 89 41 26	29 70 83 63 51	99 74 20 52 36	87 09 41 15 09	98 60 16 03 03
18 87 00 42 31	57 90 12 02 07	23 47 37 17 31	54 08 01 88 63	39 41 88 92 10
88 56 53 27 59	33 35 72 67 47	77 34 55 45 70	08 18 27 38 90	16 95 86 70 75
09 72 95 84 29	49 41 31 06 70	42 38 06 45 18	64 84 73 31 65	52 53 37 97 15
12 96 88 17 31	65 19 69 02 83	60 75 86 90 68	24 64 19 35 51	56 61 87 39 12
85 94 57 24 16	92 09 84 38 76	22 00 27 69 85	29 81 94 78 70	21 94 47 90 12
38 64 43 59 98	98 77 87 68 07	91 51 67 62 44	40 98 05 93 78	23 32 65 41 18
53 44 09 42 72	00 41 86 79 79	68 47 22 00 20	35 55 31 51 51	00 83 63 22 55
40 76 66 26 84	57 99 99 90 37	36 63 32 08 58	37 40 13 68 97	87 64 81 07 83
02 17 79 18 05	12 59 52 57 02	22 07 90 47 03	28 14 11 30 79	20 69 22 40 98
95 17 82 06 53	31 51 10 96 46	92 06 88 07 77	56 11 50 81 69	40 23 72 51 39
35 76 22 42 92	96 11 83 44 80	34 68 35 48 77	33 42 40 90 60	73 96 53 97 86
26 29 13 56 41	85 47 04 66 08	34 72 57 59 13	82 43 80 46 15	38 26 61 70 04
77 80 20 75 82	72 82 32 99 90	63 95 73 76 63	89 73 44 99 05	48 67 26 43 18
46 40 66 44 52	91 36 74 43 53	30 82 13 54 00	78 45 63 98 35	55 03 36 67 68
37 56 08 18 09	77 53 84 46 47	31 91 18 95 58	24 16 74 11 53	44 10 13 85 57
61 65 61 68 66	37 27 47 39 19	84 83 70 07 48	53 21 40 06 71	95 06 79 88 54
93 43 69 64 07	34 18 04 52 35	56 27 09 24 86	61 85 53 83 45	19 90 70 99 00
21 96 60 12 99	11 20 99 45 18	48 13 93 55 34	18 37 79 49 90	65 97 38 20 46
95 20 47 97 97	27 37 83 28 71	00 06 41 41 74	45 89 09 39 84	51 67 11 52 49
97 86 21 78 73	10 65 81 92 59	58 76 17 14 97	04 76 62 16 17	17 95 70 45 80
69 92 06 34 13	59 71 74 17 32	27 55 10 24 19	23 71 82 13 74	63 52 52 01 41
04 31 17 21 56	33 73 99 19 87	26 72 39 27 67	53 77 57 68 93	60 61 97 22 61
61 06 98 03 91	87 14 77 43 96	43 00 65 98 50	45 60 33 01 07	98 99 46 50 47
85 93 85 86 88	72 87 08 62 40	16 06 10 89 20	23 21 34 74 97	76 38 03 29 63
21 74 32 47 45	73 96 07 94 52	09 65 90 77 47	25 76 16 19 33	53 05 70 53 30
15 69 53 82 80	79 96 23 53 10	65 39 07 16 29	45 33 02 43 70	02 87 40 41 45
02 89 08 04 49	20 21 14 68 86	87 63 93 95 17	11 29 01 95 80	35 14 97 35 33
87 18 15 89 79	85 43 01 72 73	08 61 74 51 69	89 74 39 82 15	94 51 33 41 67
98 83 71 94 22	59 97 50 99 52	08 52 85 08 40	87 80 61 65 31	91 51 80 32 44
10 08 58 21 66	72 68 49 29 31	89 85 84 46 06	59 73 19 85 23	65 09 29 75 63
47 90 56 10 08	88 02 84 27 83	42 29 72 23 19	66 56 45 65 79	20 71 53 20 25
22 85 61 68 90	49 64 92 85 44	16 40 12 89 88	50 14 49 81 06	01 82 77 45 12
67 80 43 79 33	12 83 11 41 16	25 58 19 68 70	77 02 54 00 52	53 43 37 15 26
27 62 50 96 72	79 44 61 40 15	14 53 40 65 39	27 31 58 50 28	11 39 03 34 25
33 78 80 87 15	38 30 06 38 21	14 47 47 07 26	54 96 87 53 32	40 36 40 96 76
13 13 92 66 99	47 24 49 57 74	32 25 43 62 17	10 97 11 69 84	99 63 22 32 98

Table XXXIII. Random Numbers (IV)

10 27 53 96 23	71 50 54 36 23	54 31 04 82 98	04 14 12 15 09	26 78 25 47 47
28 41 50 61 88	64 85 27 20 18	83 36 36 05 56	39 71 65 09 62	94 76 62 11 89
34 21 42 57 02	59 19 18 97 48	80 30 03 30 98	05 24 67 70 07	84 97 50 87 46
61 81 77 23 23	82 82 11 54 08	53 28 70 58 96	44 07 39 55 43	42 34 43 39 28
61 15 18 13 54	16 86 20 26 88	90 74 80 55 09	14 53 90 51 17	52 01 63 01 59
91 76 21 64 64	44 91 13 32 97	75 31 62 66 54	84 80 32 75 77	56 08 25 70 29
00 97 79 08 06	37 30 28 59 85	53 56 68 53 40	01 74 39 59 73	30 19 99 85 48
36 46 18 34 94	75 20 80 27 77	78 91 69 16 00	08 43 18 73 68	67 69 61 34 25
88 98 99 60 50	65 95 79 42 94	93 62 40 89 96	43 56 47 71 66	46 76 29 67 02
04 37 59 87 21	05 02 03 24 17	47 97 81 56 51	92 34 86 01 82	55 51 33 12 91
63 62 06 34 41	94 21 78 55 09	72 76 45 16 94	29 95 81 83 83	79 88 01 97 30
78 47 23 53 90	34 41 92 45 71	09 23 70 70 07	12 38 92 79 43	14 85 11 47 23
87 68 62 15 43	53 14 36 59 25	54 47 33 70 15	59 24 48 40 35	50 03 42 99 36
47 60 92 10 77	88 59 53 11 52	66 25 69 07 04	48 68 64 71 06	61 65 70 22 12
56 88 87 59 41	65 28 04 67 53	95 79 88 37 31	50 41 06 94 76	81 83 17 16 33
02 57 45 86 67	73 43 07 34 48	44 26 87 93 29	77 09 61 67 84	06 69 44 77 75
31 54 14 13 17	48 62 11 90 60	68 12 93 64 28	46 24 79 16 76	14 60 25 51 01
28 50 16 43 36	28 97 85 58 99	67 22 52 76 23	24 70 36 54 54	59 28 61 71 96
63 29 62 66 50	02 63 45 52 38	67 63 47 54 75	83 24 78 43 20	92 63 13 47 48
45 65 58 26 51	76 96 59 38 72	86 57 45 71 46	44 67 76 14 55	44 88 01 62 12
39 65 36 63 70	77 45 85 50 51	74 13 39 35 22	30 53 36 02 95	49 34 88 73 61
73 71 98 16 04	29 18 94 51 23	76 51 94 84 86	79 93 96 38 63	08 58 25 58 94
72 20 56 20 11	72 65 71 08 86	79 57 95 13 91	97 48 72 66 48	09 71 17 24 89
75 17 26 99 76	89 37 20 70 01	77 31 61 95 46	26 97 05 73 51	53 33 18 72 87
37 48 60 82 29	81 30 15 39 14	48 38 75 93 29	06 87 37 78 48	45 56 00 84 47
68 08 02 80 72	83 71 46 30 49	89 17 95 88 29	02 39 56 03 46	97 74 06 56 17
14 23 98 61 67	70 52 85 01 50	01 84 02 78 43	10 62 98 19 41	18 83 99 47 99
49 08 96 21 44	25 27 99 41 28	07 41 08 34 66	19 42 74 39 91	41 96 53 78 72
78 37 06 08 43	63 61 62 42 29	39 68 95 10 96	09 24 23 00 62	56 12 80 73 16
37 21 34 17 68	68 96 83 23 56	32 84 60 15 31	44 73 67 34 77	91 15 79 74 58
14 29 09 34 04	87 83 07 55 07	76 58 30 83 64	87 29 25 58 84	86 50 60 00 25
58 43 28 06 36	49 52 83 51 14	47 56 91 29 34	05 87 31 06 95	12 45 57 09 09
10 43 67 29 70	80 62 80 03 42	10 80 21 38 84	90 56 35 03 09	43 12 74 49 14
44 38 88 39 54	86 97 37 44 22	00 95 01 31 76	17 16 29 56 63	38 78 94 49 81
90 69 59 19 51	85 39 52 85 13	07 28 37 07 61	11 16 36 27 03	78 86 72 04 95
41 47 10 25 62	97 05 31 03 61	20 26 36 31 62	68 69 86 95 44	84 95 48 46 45
91 94 14 63 19	75 89 11 47 11	31 56 34 19 09	79 57 92 36 59	14 93 87 81 40
80 06 54 18 66	09 18 94 06 19	98 40 07 17 81	22 45 44 84 11	24 62 20 42 31
67 72 77 63 48	84 08 31 55 58	24 33 45 77 58	80 45 67 93 82	75 70 16 08 24
59 40 24 13 27	79 26 88 86 30	01 31 60 10 39	53 58 47 70 93	85 81 56 39 38
05 90 35 89 95	01 61 16 96 94	50 78 13 69 36	37 68 53 37 31	71 26 35 03 71
44 43 80 69 98	46 68 05 14 82	90 78 50 05 62	77 79 13 57 44	59 60 10 39 66
61 81 31 96 82	00 57 25 60 59	46 72 60 18 77	55 66 12 62 11	08 99 55 64 57
42 88 07 10 05	24 98 65 63 21	47 21 61 88 32	27 80 30 21 60	10 92 35 36 12
77 94 30 05 39	28 10 99 00 27	12 73 73 99 12	49 99 57 94 82	96 88 57 17 91
78 83 19 76 16	94 11 68 84 26	23 54 20 86 85	23 86 66 99 07	36 37 34 92 09
87 76 59 61 81	43 63 64 61 61	65 76 36 95 90	18 48 27 45 68	27 23 65 30 72
91 43 05 96 47	55 78 99 95 24	37 55 85 78 78	01 48 41 19 10	35 19 54 07 73
84 97 77 72 73	09 62 06 65 72	87 12 49 03 60	41 15 20 76 27	50 47 02 29 16
87 41 60 76 83	44 88 96 07 80	83 05 83 38 96	73 70 66 81 90	30 56 10 48 59

Table XXXIII. Random Numbers (V)

28 89 65 87 08	13 50 63 04 23	25 47 57 91 13	52 62 24 19 94	91 67 48 57 10
30 29 43 65 42	78 66 28 55 80	47 46 41 90 08	55 98 78 10 70	49 92 05 12 07
95 74 62 60 53	51 57 32 22 27	12 72 72 27 77	44 67 32 23 13	67 95 07 76 30
01 85 54 96 72	66 86 65 64 60	56 59 75 36 75	46 44 33 63 71	54 50 06 44 75
10 91 46 96 86	19 83 52 47 53	65 00 51 93 51	30 80 05 19 29	56 23 27 19 03
05 33 18 08 51	51 78 57 26 17	34 87 96 23 95	89 99 93 39 79	11 28 94 15 52
04 43 13 37 00	79 68 96 26 60	70 39 83 66 56	62 03 55 86 57	77 55 33 62 02
05 85 40 25 24	73 52 93 70 50	48 21 47 74 63	17 27 27 51 26	35 96 29 00 45
84 90 90 65 77	63 99 25 69 02	09 04 03 35 78	19 79 95 07 21	02 84 48 51 97
28 55 53 09 48	86 28 30 02 35	71 30 32 06 47	93 74 21 86 33	49 90 21 69 74
89 83 40 69 80	97 96 47 59 97	56 33 24 87 36	17 18 16 90 46	75 27 28 52 13
73 20 96 05 68	93 41 69 96 07	97 50 81 79 59	42 37 13 81 83	92 42 85 04 31
10 89 07 76 21	40 24 74 36 42	40 33 04 46 24	35 63 02 31 61	34 59 43 36 96
91 50 27 78 37	06 06 16 25 98	17 78 80 36 85	26 41 77 63 37	71 63 94 94 33
03 45 44 66 88	97 81 26 03 89	39 46 67 21 17	98 10 39 33 15	61 63 00 25 92
89 41 58 91 63	65 99 59 97 84	90 14 79 61 55	56 16 88 87 60	32 15 99 67 43
13 43 00 97 26	16 91 21 32 41	60 22 66 72 17	31 85 33 69 07	68 49 20 43 29
71 71 00 51 72	62 03 89 26 32	35 27 99 18 25	78 12 03 09 70	50 93 19 35 56
19 28 15 00 41	92 27 73 40 38	37 11 05 75 16	98 81 99 37 29	92 20 32 39 67
56 38 30 92 30	45 51 94 69 04	00 84 14 36 37	95 66 39 01 09	21 68 40 95 79
39 27 52 89 11	00 81 06 28 48	12 08 05 75 26	03 35 63 05 77	13 81 20 67 58
73 13 28 58 01	05 06 42 24 07	60 60 29 99 93	72 93 78 04 36	25 76 01 54 03
81 60 84 51 57	12 68 46 55 89	60 09 71 87 89	70 81 10 95 91	83 79 68 20 66
05 62 98 07 85	07 79 26 69 61	67 85 72 37 41	85 79 76 48 23	61 58 87 08 05
62 97 16 29 18	52 16 16 23 56	62 95 80 97 63	32 25 34 03 36	48 84 60 37 65
31 13 63 21 08	16 01 92 58 21	48 79 74 73 72	08 64 80 91 38	07 28 66 61 59
97 38 35 34 19	89 84 05 34 47	88 09 31 54 88	97 96 86 01 69	46 13 95 65 96
32 11 78 33 82	51 99 98 44 39	12 75 10 60 36	80 66 39 94 97	42 36 31 16 59
81 99 13 37 05	08 12 60 39 23	61 73 84 89 18	26 02 04 37 95	96 18 69 06 30
45 74 00 03 05	69 99 47 26 52	48 06 30 00 18	03 30 28 55 59	66 10 71 44 05
11 84 13 69 01	88 91 28 79 50	71 42 14 96 55	98 59 96 01 36	88 77 90 45 59
14 66 12 87 22	59 45 27 08 51	85 64 23 85 41	64 72 08 59 44	67 98 36 65 56
40 25 67 87 82	84 27 17 30 37	48 69 49 02 58	98 02 50 58 11	95 39 06 35 63
44 48 97 49 43	65 45 53 41 07	14 83 46 74 11	76 66 63 60 08	90 54 33 65 84
41 94 54 06 57	48 28 01 83 84	09 11 21 91 73	97 28 44 74 06	22 30 95 69 72
07 12 15 58 84	93 18 31 83 45	54 52 62 29 91	53 58 54 66 05	47 19 63 92 75
64 27 90 43 52	18 26 32 96 83	50 58 45 27 57	14 96 39 64 85	73 87 96 76 23
80 71 86 41 03	45 62 63 40 88	35 69 34 10 94	32 22 52 04 74	69 63 21 83 41
27 06 08 09 92	26 22 59 28 27	38 58 22 14 79	24 32 12 38 42	33 56 90 92 57
54 68 97 20 54	33 26 74 03 30	74 22 19 13 48	30 28 01 92 49	58 61 52 27 03
02 92 65 68 99	05 53 15 26 70	04 69 22 64 07	04 73 25 74 82	78 35 22 21 88
83 52 57 78 62	98 61 70 48 22	68 50 64 55 75	42 70 32 09 60	58 70 61 43 97
82 82 76 31 33	85 13 41 38 10	16 47 61 43 77	83 27 19 70 41	34 78 77 60 25
38 61 34 09 49	04 41 66 09 76	20 50 73 40 95	24 77 95 73 20	47 42 80 61 03
01 01 11 88 38	03 10 16 82 24	39 58 20 12 39	82 77 02 18 88	33 11 49 15 16
21 66 14 38 28	54 08 18 07 04	92 17 63 36 75	33 14 11 11 78	97 30 53 62 38
32 29 30 69 59	68 50 33 31 47	15 64 88 75 27	04 51 41 61 96	86 62 93 66 71
04 59 21 65 47	39 90 89 86 77	46 86 86 88 86	50 09 13 24 91	54 80 67 78 66
38 64 50 07 36	56 50 45 94 25	48 28 48 30 51	60 73 73 03 87	68 47 37 10 84
48 33 50 83 53	59 77 64 59 90	58 92 62 50 18	93 09 45 89 06	13 26 98 86 29

TABLE XXXIII. RANDOM NUMBERS (VI)

25 19 64 82 84	62 74 29 92 24	61 03 91 22 48	64 94 63 15 07	66 85 12 00 27
23 02 41 46 04	44 31 52 43 07	44 06 03 09 34	19 83 94 62 94	48 28 01 51 92
55 85 66 96 28	28 30 62 58 83	65 68 62 42 45	13 08 60 46 28	95 68 45 52 43
68 45 19 69 59	35 14 82 56 80	22 06 52 26 39	59 78 98 76 14	36 09 03 01 86
69 31 46 29 85	18 88 26 95 54	01 02 14 03 05	48 00 26 43 85	33 93 81 45 95
37 31 61 28 98	94 61 47 03 10	67 80 84 41 26	88 84 59 69 14	77 32 82 81 89
66 42 19 24 94	13 13 38 69 96	76 69 76 24 13	43 83 10 13 24	18 32 84 85 04
33 65 78 12 35	91 59 11 38 44	23 31 48 75 74	05 30 08 46 32	90 04 93 56 16
76 32 06 19 35	22 95 30 19 29	57 74 43 20 90	20 25 36 70 69	38 32 11 01 01
43 33 42 02 59	20 39 84 95 61	58 22 04 02 99	99 78 78 83 82	43 67 16 38 95
28 31 93 43 94	87 73 19 38 47	54 36 90 98 10	83 43 32 26 26	22 00 90 59 22
97 19 21 63 34	69 33 17 03 02	11 15 50 46 08	42 69 60 17 42	14 68 61 14 48
82 80 37 14 20	56 39 59 89 63	33 90 38 44 50	78 22 87 10 88	06 58 87 39 67
03 68 03 13 60	64 13 09 37 11	86 02 57 41 99	31 66 60 65 64	03 03 02 58 97
65 16 58 11 01	98 78 80 63 23	07 37 66 20 56	20 96 06 79 80	33 39 40 49 42
24 65 58 57 04	18 62 85 28 24	26 45 17 82 76	39 65 01 73 91	50 37 49 38 73
02 72 64 07 75	85 66 48 38 73	75 10 96 59 31	48 78 58 08 88	72 08 54 57 17
79 16 78 63 99	43 61 00 66 42	76 26 71 14 33	33 86 76 71 66	37 85 05 56 07
04 75 14 93 39	68 52 16 83 34	64 09 44 62 58	48 32 72 26 95	32 67 35 49 71
40 64 64 57 60	97 00 12 91 33	22 14 73 01 11	83 97 68 95 65	67 77 80 98 87
06 27 07 34 26	01 52 48 69 57	19 17 53 55 96	02 41 03 89 33	86 85 73 02 32
62 40 03 87 10	96 88 22 46 94	35 56 60 94 20	60 73 04 84 98	96 45 18 47 07
00 98 48 18 97	91 51 63 27 95	74 25 84 03 07	88 29 04 79 84	03 71 13 78 26
50 64 19 18 91	98 55 83 46 09	49 66 41 12 45	41 49 36 83 43	53 75 35 13 39
38 54 52 25 78	01 98 00 89 85	86 12 22 89 25	10 10 71 19 45	88 84 77 00 07
46 86 80 97 78	65 12 64 64 70	58 41 05 49 08	68 68 88 54 00	81 61 61 80 41
90 72 92 93 10	09 12 81 93 63	69 30 02 04 26	92 36 48 69 45	91 99 08 07 65
66 21 41 77 60	99 35 72 61 22	52 40 74 67 29	97 50 71 39 79	57 82 14 88 06
87 05 46 52 76	89 96 34 22 37	27 11 57 04 19	57 93 08 35 69	07 51 19 92 66
46 90 61 03 06	89 85 33 22 80	34 89 12 29 37	44 71 38 40 37	15 49 55 51 08
11 88 53 06 09	81 83 33 98 29	91 27 59 43 09	70 72 51 49 73	35 97 25 83 41
11 05 92 06 97	68 82 34 08 83	25 40 58 40 64	56 42 78 54 06	60 96 96 12 82
33 94 24 20 28	62 42 07 12 63	34 39 02 92 31	80 61 68 44 19	09 92 14 73 49
24 89 74 75 61	61 02 73 36 85	67 28 50 49 85	37 79 95 02 66	73 19 76 28 13
15 19 74 67 23	61 38 93 73 68	76 23 15 58 20	35 36 82 82 59	01 33 48 17 66
05 64 12 70 83	80 58 35 06 88	73 48 27 39 43	43 40 13 35 45	55 10 54 38 50
57 49 36 44 06	74 93 55 39 26	27 70 98 76 68	78 36 26 24 06	43 24 56 40 80
77 82 96 96 97	60 42 17 18 48	16 34 92 19 52	98 84 48 42 92	83 19 06 77 78
24 10 70 06 51	59 62 37 95 42	53 67 14 95 29	84 65 43 07 30	77 54 00 15 42
50 00 07 78 23	49 54 36 85 14	18 50 54 18 82	23 79 80 71 37	60 62 95 40 30
44 37 76 21 96	37 03 08 98 64	90 85 59 43 64	17 79 96 52 35	21 05 22 59 30
90 57 55 17 47	53 26 79 20 38	69 90 58 64 03	33 48 32 91 54	68 44 90 24 25
50 74 64 67 42	95 28 12 73 23	32 54 98 64 94	82 17 18 17 14	55 10 61 64 29
44 04 70 22 02	84 31 64 64 08	52 55 04 24 29	91 95 43 81 14	66 13 18 47 44
32 74 61 64 73	21 46 51 44 77	72 48 92 00 05	83 59 89 65 06	53 76 70 58 78
75 73 51 70 49	12 53 67 51 54	38 10 11 67 73	22 32 61 43 75	31 61 22 21 11
76 18 36 16 34	16 28 25 82 98	64 26 70 54 87	49 48 55 11 39	94 25 20 80 85
00 17 37 71 81	64 21 91 15 82	81 04 14 52 11	39 07 30 60 77	39 18 27 85 68
54 95 57 55 04	12 77 40 70 14	79 86 61 57 50	52 49 41 73 46	05 63 34 92 33
69 99 95 54 63	44 37 33 53 17	38 06 58 37 93	47 10 62 31 28	63 59 40 40 32

Table XXXIIIi. Random Permutations of 10 Numbers

	0 5	1 6	2 7	3 8	4 9	5 0	6 1	7 2	8 3	9 4	0 5	1 6	2 7	3 8	4 9
0	03528	28071	97041	45167	35421	71345	47286	83567	94170	46789	56471	65132	18294	73618	89714
	49761	39465	52683	82093	09867	06982	10359	49102	65328	02531	93208	74809	50376	04952	02365
15	80219	29485	85093	45687	32579	52867	70851	03621	51490	53180	76283	90287	34605	67295	82906
	43657	70361	24176	30129	48016	03194	29463	94587	32768	76924	90451	56314	12879	10834	45137
30	36524	07561	67820	20739	69045	54978	90317	64529	57319	74861	94816	10798	06431	53984	07196
	97801	98234	13459	58146	13782	36102	48562	78301	64802	39502	05723	32546	29857	16072	32584
45	56913	06289	64527	30265	43209	84152	69183	01278	41973	03471	85014	01628	42016	90164	90418
	84027	51734	80193	49718	81675	69307	47250	39546	25068	56829	27639	73954	58379	53827	73625
60	70523	83297	81239	05913	75206	18350	81623	59327	08317	39581	45736	12987	40568	20915	40819
	68194	60541	06745	68742	38491	67924	07549	64081	56492	60247	19802	54360	23917	48367	52367
75	59208	56078	74268	54981	12573	51246	43985	04628	54670	93108	86341	10678	70468	53920	91478
	13674	31492	09351	27630	68490	38097	27160	97531	32918	67542	95207	42593	51239	74618	32650
90	40329	15237	65897	18693	43709	10783	41237	72168	87194	80472	38125	23945	73981	31749	52831
	71568	98604	13402	20547	58612	95426	89605	35904	23506	69351	46970	10687	06254	26580	49760
105	78136	62531	92143	28935	98410	78059	76901	24618	85102	30642	10378	61739	41980	59761	65943
	09452	84097	05876	46107	75236	61432	35482	57093	93647	15987	24956	08245	25763	42803	80172
120	58236	29871	53107	50762	47103	81973	35716	83924	21593	98432	95260	83496	27810	82193	24851
	10974	64350	89264	41839	69852	65042	42089	67501	40678	57601	84371	17250	95634	67450	90367
135	83417	26179	19562	71358	78542	72896	74312	57346	16248	68102	76581	73294	02698	72850	35710
	29065	45038	38704	90246	13069	43105	85960	82019	97305	53794	24309	86015	43517	94316	98624
150	48210	25789	52193	36489	82167	01325	52839	25301	24150	71306	06417	23057	80129	82439	34620
	79365	31460	04687	17502	49053	74896	70614	78649	63987	94258	58329	94168	65374	16075	87915
165	67439	45961	21035	64530	84067	32516	57819	23461	15894	20935	37198	67983	94786	96408	58037
	10285	80723	64789	81972	51923	07498	04623	80957	06723	78641	65240	02514	02315	57123	92164
180	25481	69345	02391	32647	27845	92017	82401	95342	02538	15230	60125	97813	57493	27386	13450
	70963	17820	68574	85091	16093	64538	93756	01786	67419	84679	84937	04625	20681	90514	97286
195	59816	13897	31250	32940	28754	48721	31748	43960	93028	93802	26039	47605	84961	39274	13064
	47023	45260	47896	68751	61039	63905	96250	78521	54176	47651	78145	81239	75203	86105	58927
210	03957	87529	60895	27615	21875	74590	90758	29064	57408	28947	69210	89710	53076	10578	36809
	84612	61403	23714	40839	69403	31286	34621	87513	63219	16053	37485	54236	92184	34269	14275
225	69012	15203	94652	47395	93605	43895	60274	40519	06384	27149	58302	76523	89631	18594	41059
	74385	47689	18073	01628	72148	20167	18359	28736	95721	30586	41679	91408	07524	07236	62738
240	57693	32816	87162	61348	15237	17620	17096	50139	56192	31970	32781	79358	23695	13864	24538
	48012	50749	45930	72509	86094	98543	34258	74286	73048	28564	50469	42061	04817	07529	60917
255	69407	45982	71896	32891	67984	93618	54867	28730	12674	84715	74269	28940	50628	92810	49168
	52138	70316	20453	64057	01325	25470	30912	51946	03895	23069	03851	65317	39417	34576	53270
270	20531	59762	71692	25684	39425	90132	14379	16720	16732	07423	71893	48307	64192	76915	79431
	98746	81340	84305	17309	86107	68547	56802	45983	84590	89165	02546	15269	85307	84023	05628
285	03124	41920	20834	61239	45723	12539	76530	38720	23697	19875	31256	83914	79064	74296	71326
	58967	73865	69175	80547	06981	04768	84291	19564	58104	06324	70984	20657	15832	53801	04859
300	36574	78296	04258	92407	82051	14928	78423	92530	85791	96513	49580	59671	29413	01534	19532
	90182	13045	37169	38651	73469	63570	01659	84671	34260	42807	72163	40823	05786	72986	67804
315	42689	70964	36297	68294	93784	31920	03621	27610	95274	46982	03529	35964	30176	46087	46208
	71503	81235	58140	30715	20651	86475	57948	35948	63081	31075	18764	21087	98254	19532	37159
330	35709	90582	01382	81564	07952	97462	90584	52048	50176	40791	28059	49650	91023	84723	34820
	41286	64371	57496	02397	18436	35801	21736	63197	43928	65328	17634	78312	65478	96051	96751
345	50182	61045	68974	54673	43782	48517	68975	29783	48591	08793	41986	32607	30912	07253	89536
	43967	79328	13205	91208	65091	06329	13420	54016	76032	54216	57023	95481	45867	61984	74201
360	40756	86974	43075	12690	06234	42905	60351	56148	94237	31056	01274	91724	59136	51082	89132
	38921	35021	89621	83457	78915	31786	78429	73092	18065	82794	65398	36508	07842	34679	47056

Table XXXIII1. Random Permutations of 10 Numbers—*continued*

	5	6	7	8	9	0	1	2	3	4	5	6	7	8	9
	0	1	2	3	4	5	6	7	8	9	0	1	2	3	4
375	10759	51908	64982	87132	43718	03218	72186	62014	64389	84105	40735	40973	18329	31294	16402
	26348	42763	31075	94560	90256	94567	43059	87395	10572	37629	69218	26185	07645	76805	75398
390	63850	13069	57238	47958	41952	10798	43751	13508	32108	40913	27538	72104	57214	47068	06135
	94721	82457	16094	16302	08673	46352	86209	74269	47596	68725	04691	69358	06398	92135	78942
405	20459	59746	04659	46179	08913	13694	04398	47021	28937	50273	35410	82947	23841	10748	91385
	31786	18023	82137	23850	75624	02758	15267	38956	50461	94618	79862	01356	07596	52936	70246
420	51048	81502	98734	34015	79162	37298	94351	90485	42065	85290	31547	19864	28175	27319	96780
	32967	43679	21650	69278	83540	45160	06287	36721	91387	13674	06928	27530	64039	54806	12543
435	74302	34521	58640	87429	29536	68543	05179	75891	80574	19462	81537	45270	31270	08451	34295
	89561	96087	91732	36501	14870	79120	68243	36024	29316	50873	06294	16983	96854	72369	76801
450	86104	61943	12950	13045	81497	41092	83052	31854	48065	29485	86174	76485	79356	13809	86702
	72593	20857	68734	89267	02365	37685	61479	60792	32719	36710	25039	02913	14028	25647	35149
465	06853	48592	10835	42765	46792	70156	34876	02359	54736	08453	02967	54063	45139	37854	76014
	12749	61307	64297	38091	81530	23498	50912	68147	01289	21967	45183	91872	80726	61920	25893
480	87912	70324	69037	30815	51673	24897	14590	08297	95217	36581	14360	01829	50324	68271	49172
	03546	61985	54281	24967	82094	61503	73862	31645	48360	09427	29587	36475	69187	39540	03586
495	60294	25316	24936	48193	74318	78629	31609	92045	50729	34792	42795	56780	87056	87204	39208
	83175	80974	70815	57602	52609	41530	85427	68731	81643	60518	13608	21493	13924	56391	54761
510	67943	94068	91452	97510	17362	57093	83109	27364	96237	21395	87159	27138	32105	35067	51298
	52108	57132	60837	42386	49805	62418	27645	95180	51480	64078	02463	69540	48796	89241	03746
525	18506	61842	73260	86702	46293	67039	97625	17304	18456	50291	60784	21074	13520	47518	36795
	97342	09357	81954	19453	80751	21458	01843	98562	07239	67834	31592	83695	46978	30296	10428
540	08613	93286	47801	32769	58296	39021	10894	61238	72103	71943	56384	42136	73610	79318	94513
	95742	50471	56239	85014	70143	45678	57326	54079	96854	02568	01792	08795	52498	20465	27680
555	61270	80264	52689	05479	38512	54602	93052	51406	51342	78190	51937	40615	76238	87935	69048
	94835	95173	70314	86231	49076	81937	17468	27938	07689	36425	68420	83972	45190	60142	73215
570	54901	81326	58063	94053	36978	93048	90168	76102	60275	14935	04385	82453	08941	24385	74513
	86372	57490	24917	62781	50124	26715	74523	34958	93814	27860	91627	90671	25367	79610	98260
585	43027	59671	09746	20837	12436	87543	16482	75294	47038	42183	32509	06952	27589	35012	60287
	69815	80324	52813	49561	58790	16029	90753	30186	29516	56970	87614	38174	01436	98674	34591
600	12487	73960	09485	26340	75402	03279	17932	53160	73924	96823	52134	18532	49027	57218	14903
	56039	12458	62731	51897	18369	46851	56804	78294	15068	04517	08769	47960	58163	43906	85276
615	40928	47028	46120	63172	61283	15409	49752	85172	03126	13894	35980	34078	27509	95243	90746
	51376	31965	75398	84950	45907	83672	36801	46093	89475	67250	17426	16259	14863	78016	52138
630	08564	57064	53692	46273	87014	70256	12094	40986	86274	56308	90786	10928	96320	25634	26051
	79123	31298	74081	18590	59632	34981	63785	57132	05391	97142	21435	73654	48571	09781	39874
645	45382	96438	87094	97248	09473	76120	92867	91854	72196	08164	83497	41035	08725	74629	96401
	90176	50271	63215	06513	52618	85349	05314	20637	40835	92573	15620	28697	36914	53108	85237
660	01647	52768	81069	83214	56798	26389	03915	02863	79316	54760	28397	23751	48205	03721	54806
	35289	34019	32745	59760	14230	05471	47628	91574	84520	21389	06154	69084	13697	86594	29173
675	47521	89230	95326	08697	20185	76951	70563	19368	85093	85736	82094	54672	26459	23541	83602
	09368	45761	14087	53124	76349	38204	81294	24570	24176	02419	56137	91038	30187	70968	19547
690	07615	81049	54628	61394	28097	47302	47235	81075	90841	63914	96820	86312	86921	78610	92605
	92843	27653	93071	28507	31546	89165	86901	29346	62573	02857	34715	59704	50734	93245	47831
705	85390	02581	43582	63014	76039	92063	38045	63781	54231	85490	92183	50239	12497	97208	45312
	24167	73694	60971	92785	41528	71845	97261	05924	79806	73261	56047	61487	08563	63514	68970
720	06591	24503	06538	96532	73256	47510	30421	78439	87591	12087	15897	38795	92130	93825	63025
	74283	17986	41297	41087	40891	23869	79856	50126	02463	96534	04326	06241	67584	76401	89174
735	73502	62893	81025	93681	05681	86431	21863	73920	81930	53829	35492	78532	52134	40175	73854
	46981	05714	63479	70542	32497	50927	70495	41586	54726	16074	70816	96140	86790	32869	20691

Table XXXIII2. Random Permutations of 20 Numbers

	0	1	2	3	4	5	6	7	8	9
0	1 15 6 8 5	2 13 7 4 10	11 9 10 1 6	10 2 12 13 9	7 6 10 0 1	16 19 14 8 18	1 15 3 10 17	3 16 0 15 9	14 19 12 17 10	13 17 11 12 14
	10 17 12 11 9	18 19 9 3 15	8 5 14 2 16	0 14 7 4 8	19 8 11 17 5	0 12 13 3 4	18 7 2 8 16	17 8 6 14 7	11 13 7 5 3	10 4 8 6 9
	2 16 18 0 3	11 1 6 12 0	0 7 15 4 17	5 15 6 3 19	12 16 9 2 4	6 10 2 5 7	0 11 5 12 9	5 12 1 10 13	0 2 6 8 18	1 0 3 18 16
	14 7 4 13 19	17 5 14 8 16	3 18 13 19 12	16 1 17 11 18	13 15 18 3 14	1 17 9 15 11	14 19 13 6 4	11 4 19 18 2	1 16 9 4 15	2 7 19 5 15
1	11 19 2 15 1	3 16 13 7 12	3 15 5 0 11	12 9 16 10 1	2 16 6 18 19	11 14 10 13 5	4 14 3 10 1	6 4 19 14 17	18 17 9 19 6	0 13 18 16 2
	7 14 5 4 3	6 18 8 2 19	7 12 13 18 4	6 13 2 19 5	0 7 10 15 3	0 17 4 9 7	8 11 18 9 6	16 7 3 11 2	5 14 13 1 4	11 6 12 14 17
	16 9 18 0 12	9 1 11 10 5	8 10 17 6 16	18 4 0 8 7	8 14 5 9 13	8 12 16 6 15	16 2 15 12 13	10 8 0 9 5	7 15 12 3 2	8 15 1 19 10
	8 13 10 17 6	0 17 14 15 4	9 1 14 19 2	11 17 14 15 3	1 11 4 12 17	3 18 19 2 1	19 7 17 5 0	1 18 15 12 13	0 10 8 11 16	7 5 4 9 3
2	18 8 11 9 15	19 13 4 1 11	18 6 3 17 11	3 1 14 12 0	2 17 0 7 19	1 2 10 9 12	17 7 4 3 19	13 0 9 5 17	1 12 9 3 16	0 11 8 4 12
	4 5 14 3 0	2 0 7 16 18	4 14 9 10 16	8 2 10 13 18	14 9 12 11 8	6 11 15 0 17	10 0 13 11 14	15 11 1 12 7	5 17 0 7 13	7 2 15 3 18
	7 10 16 12 6	5 17 6 8 14	1 12 2 8 19	17 7 11 16 4	6 3 16 13 5	16 8 4 18 5	8 9 5 2 18	10 14 4 18 16	18 4 19 11 15	10 13 6 16 5
	19 2 1 13 17	9 10 12 15 3	15 5 0 13 7	5 15 9 19 6	15 10 1 4 18	13 3 14 19 7	16 12 6 15 1	3 2 19 6 8	8 6 2 10 14	9 17 19 1 14
3	4 0 1 12 7	12 9 0 5 19	15 16 2 5 19	5 14 16 11 6	11 7 5 12 2	15 11 8 5 3	18 14 8 19 15	8 16 19 0 4	13 8 5 1 14	1 11 3 17 7
	3 14 5 15 6	7 3 4 6 1	7 8 3 4 12	15 13 3 1 17	13 19 15 14 3	1 9 18 0 14	6 4 3 7 1	13 5 11 14 6	18 11 2 4 12	9 16 0 6 5
	8 17 11 19 10	8 13 2 18 16	6 1 0 17 10	19 2 7 8 18	9 10 8 6 1	4 12 16 6 17	0 13 12 2 10	1 18 17 7 12	10 9 3 15 6	4 18 8 15 2
	9 18 16 13 2	11 17 14 10 15	9 13 18 11 14	12 9 0 10 4	16 0 4 18 17	19 10 2 7 13	17 9 16 5 11	15 10 3 2 9	17 16 0 7 19	13 14 12 19 10
4	0 2 5 1 18	14 16 19 7 8	14 6 8 1 19	5 2 3 14 13	18 9 2 12 16	0 1 18 4 15	13 14 0 2 11	0 14 5 12 13	0 6 18 7 16	7 9 19 11 5
	15 6 14 12 8	1 11 9 3 4	10 7 12 13 5	18 9 0 12 16	3 10 7 8 13	17 8 13 6 16	3 19 18 1 6	9 16 7 6 4	8 11 1 10 15	18 16 2 12 9
	13 16 17 3 19	10 6 5 17 13	17 0 15 18 4	10 6 15 17 7	15 14 17 0 6	14 9 12 19 3	12 10 4 15 17	3 19 18 10 1	3 17 5 2 13	14 13 1 17 0
	10 9 7 4 11	12 15 2 18 0	3 9 11 16 2	8 19 1 11 4	11 19 4 1 5	11 5 2 10 7	5 9 16 7 8	2 17 15 11 8	14 4 9 12 19	3 15 4 10 8
5	6 2 8 4 14	7 16 6 2 1	14 8 3 19 4	7 16 11 19 17	2 11 0 1 7	10 4 12 5 19	4 13 11 9 2	1 6 19 16 11	5 3 11 9 0	3 6 7 9 1
	18 1 10 16 9	5 8 17 0 12	9 15 0 11 5	5 14 18 9 3	19 13 18 8 6	8 7 14 2 18	10 15 16 17 6	17 5 14 9 13	2 7 12 15 10	11 16 4 12 10
	7 15 5 3 12	9 18 10 3 11	1 6 2 18 7	0 10 4 8 12	9 15 3 17 16	0 3 9 15 13	19 8 14 3 7	10 18 8 2 15	8 6 19 1 14	17 8 2 13 19
	13 0 11 19 17	15 19 4 14 13	13 16 17 10 12	15 13 1 6 2	4 12 5 10 14	6 17 1 16 11	18 0 1 12 5	3 0 7 12 4	16 17 13 4 18	0 5 14 18 15
6	19 13 17 0 9	8 7 10 5 4	12 15 5 17 9	17 19 0 14 11	4 10 7 19 2	17 10 3 0 14	19 9 4 17 1	11 7 17 3 10	5 18 1 13 8	13 10 18 16 19
	3 5 1 7 14	9 1 15 6 12	6 1 18 14 4	3 2 18 8 16	0 16 3 8 6	2 9 18 19 1	10 14 15 18 3	14 1 2 19 6	11 0 6 3 19	1 11 14 8 4
	8 6 15 10 4	13 18 19 2 11	2 16 19 7 11	4 10 6 12 1	14 11 17 18 15	8 15 6 11 12	6 16 5 13 7	15 0 5 9 8	17 9 16 10 15	0 6 5 9 17
	2 11 12 18 16	14 17 16 0 3	0 8 13 3 10	7 15 5 9 13	12 9 13 5 1	4 5 13 16 7	0 8 11 12 2	12 16 13 4 18	4 14 12 2 7	15 2 12 3 7
7	3 1 17 13 19	14 1 12 6 8	16 12 6 7 1	2 18 3 8 6	11 1 6 8 9	6 1 10 11 15	2 11 10 9 8	17 7 6 15 19	10 0 14 3 19	1 3 13 9 14
	5 18 12 15 2	18 17 2 3 5	15 5 18 0 17	4 7 14 0 10	18 12 15 0 19	8 7 2 14 4	1 13 18 3 16	18 0 16 4 3	13 18 1 8 2	6 12 4 18 11
	0 9 16 4 10	15 19 13 16 4	8 10 19 3 11	19 1 9 11 13	13 2 10 14 5	13 17 0 9 19	12 14 17 6 19	10 2 9 5 12	12 11 15 9 7	16 8 2 17 7
	11 8 14 7 6	0 7 10 9 11	2 4 9 13 14	15 12 16 5 17	4 3 7 16 17	16 3 12 18 5	5 15 7 4 0	14 8 1 11 13	4 5 16 6 17	0 19 15 5 10
8	19 16 6 8 15	10 11 9 3 12	13 6 4 8 5	13 7 6 1 2	12 15 13 5 3	10 15 0 11 9	19 18 7 11 17	13 14 10 6 18	18 10 17 7 1	8 10 6 18 11
	2 10 4 14 18	14 2 4 17 0	1 2 9 14 11	8 9 0 12 11	4 9 18 16 7	3 16 13 5 2	4 5 0 16 2	2 12 4 1 3	12 13 4 8 9	9 17 13 19 1
	9 0 12 5 13	15 5 16 18 6	3 12 0 15 7	19 14 5 4 3	2 10 0 17 1	18 19 14 4 12	12 1 6 10 3	17 19 9 16 15	14 16 3 15 5	15 14 4 16 3
	11 1 7 3 17	19 7 13 8 1	16 10 17 18 19	15 17 10 18 16	6 11 19 8 14	17 6 1 7 8	13 14 15 9 8	11 8 7 0 5	2 0 19 11 6	2 5 12 0 7
9	2 7 18 0 17	5 6 13 15 16	4 15 9 3 10	13 18 19 4 3	9 17 15 11 10	13 10 11 3 14	6 13 1 19 12	9 16 2 8 6	13 2 1 9 5	3 15 10 5 1
	3 11 1 16 15	7 9 19 10 4	13 8 6 14 12	9 16 14 5 0	6 1 4 16 7	7 9 0 12 5	4 7 15 8 3	19 14 5 15 10	8 6 14 4 16	16 8 12 13 0
	13 19 12 6 8	3 1 12 18 8	0 16 7 5 2	17 1 11 15 10	19 5 14 3 0	18 19 17 8 6	2 18 14 16 5	1 4 13 18 11	0 18 11 17 12	7 9 4 2 11
	9 4 10 14 5	17 11 2 0 14	18 17 11 1 19	2 6 8 12 7	12 2 8 13 18	2 16 15 4 1	9 0 17 11 10	3 17 12 7 0	19 15 10 3 7	18 6 17 14 19

Table XXXIII2. Random Permutations of 20 Numbers—*continued*

	0	1	2	3	4	5	6	7	8	9
10	17 7 3 13 12	17 1 12 3 2	2 9 16 4 6	3 4 10 13 16	6 17 13 19 0	7 13 17 11 1	14 10 16 4 7	10 12 15 17 5	15 2 3 12 1	0 9 11 8 6
	16 15 1 10 14	13 10 14 15 16	3 5 13 10 17	14 0 18 5 9	12 15 18 9 3	0 4 14 6 19	8 11 13 19 15	2 9 4 1 8	13 4 10 18 19	7 18 15 12 1
	2 19 8 4 5	5 6 7 9 19	18 8 7 12 1	17 15 12 11 6	14 2 1 7 4	8 5 2 16 18	0 9 18 6 3	13 0 14 16 7	16 9 7 8 0	19 13 10 17 5
	0 9 11 6 18	18 0 11 8 4	15 0 11 14 19	7 1 19 2 8	10 16 5 8 11	12 10 15 3 9	12 1 5 2 17	11 6 19 18 3	14 17 11 5 6	14 16 2 4 3
11	16 1 17 15 11	6 16 1 18 17	8 16 11 3 9	4 10 15 16 12	6 13 8 12 1	1 7 2 9 4	14 3 4 1 10	10 16 0 15 3	1 19 8 6 13	0 14 18 11 5
	8 6 14 7 10	19 9 10 3 4	12 0 4 2 13	1 9 8 6 5	9 17 5 10 11	10 0 17 19 3	2 0 12 17 19	7 8 19 13 17	17 3 16 15 18	3 19 13 12 4
	9 0 3 5 4	13 0 12 11 2	6 5 19 14 7	18 7 14 13 0	18 14 7 4 0	8 16 13 5 11	15 9 16 7 18	5 1 9 2 4	4 5 14 0 12	10 7 2 15 1
	19 18 2 12 13	7 8 5 14 15	10 18 1 15 17	3 19 11 17 2	3 19 15 2 16	12 6 14 15 18	11 13 5 6 8	18 14 12 6 11	2 10 7 11 9	6 16 9 17 8
12	4 13 10 18 14	6 10 8 11 9	3 7 4 2 14	15 17 12 6 0	8 0 13 12 18	0 16 6 4 18	17 11 6 16 15	1 19 0 4 8	17 19 5 8 16	18 19 17 6 10
	2 5 0 1 11	16 13 5 0 3	1 5 13 18 15	14 4 16 19 5	2 3 19 9 5	14 5 1 2 3	13 1 18 7 19	16 6 18 14 17	3 7 18 4 0	3 1 0 11 16
	19 7 3 16 6	12 2 14 4 7	0 9 10 11 17	3 11 13 7 2	7 11 14 10 15	15 13 19 7 12	5 3 9 2 12	2 12 13 7 9	13 1 14 6 11	2 13 5 4 8
	15 12 9 8 17	17 18 1 15 19	19 16 6 12 8	10 9 8 18 1	6 16 1 17 4	11 8 17 9 10	14 4 0 8 10	15 5 3 11 10	12 15 2 10 9	7 14 9 12 15
13	11 1 5 0 14	0 7 17 19 3	16 2 18 15 1	10 0 1 17 11	19 6 18 5 16	15 11 5 14 8	11 14 15 18 5	16 18 7 17 11	7 19 12 2 13	8 13 4 0 3
	19 9 2 4 8	11 12 13 10 9	9 4 5 3 6	16 19 7 6 3	10 15 13 11 7	10 1 12 3 19	13 7 3 8 10	13 9 10 19 1	10 15 3 1 17	5 10 7 9 17
	17 10 6 12 15	5 15 6 8 1	8 13 17 14 7	15 8 5 4 2	12 0 14 1 4	4 18 7 13 0	16 4 1 0 19	0 12 8 6 14	9 16 5 4 14	15 11 14 2 16
	16 13 7 18 3	14 18 2 4 16	11 19 10 0 12	14 18 13 9 12	3 9 8 2 17	9 6 17 16 2	12 6 9 2 17	5 4 2 15 3	0 6 8 18 11	6 19 18 1 12
14	18 5 12 3 6	11 16 1 8 3	16 6 0 8 18	13 17 14 15 0	9 5 16 4 18	2 7 13 3 12	8 0 3 15 10	8 7 5 15 10	1 15 0 13 11	18 2 5 6 19
	1 2 11 15 7	18 14 19 9 13	17 5 7 19 10	8 12 2 16 11	19 1 11 17 10	10 17 5 11 8	16 6 2 14 9	1 9 4 6 0	8 17 3 10 12	8 13 9 10 4
	14 4 10 9 17	12 2 5 17 4	4 9 12 3 2	9 18 6 4 10	12 14 2 13 6	14 9 19 16 6	17 18 4 19 11	11 19 13 17 2	5 4 6 18 7	12 15 7 3 1
	16 8 0 13 19	7 0 10 6 15	14 13 1 15 11	19 3 7 1 5	0 7 8 3 15	4 15 18 1 0	1 13 12 5 7	12 16 18 14 3	19 14 9 16 2	16 0 17 14 11
15	5 1 11 7 10	2 14 4 6 18	18 4 17 16 5	15 0 1 14 4	14 13 16 7 12	19 5 15 7 9	19 8 3 1 12	14 6 8 7 18	7 17 14 13 4	18 10 15 16 0
	14 16 6 15 12	8 10 1 13 19	9 6 1 19 11	8 6 2 10 18	6 2 8 10 19	17 18 13 11 16	4 17 13 18 2	9 5 19 17 2	12 8 19 1 3	13 14 17 19 9
	17 3 4 2 0	17 16 12 3 11	7 2 13 15 8	12 9 7 16 3	3 1 11 9 18	14 3 6 10 1	10 16 9 7 0	1 4 12 11 16	0 10 5 11 2	6 2 3 12 4
	18 13 9 19 8	15 0 7 5 9	14 3 12 0 10	5 13 11 17 19	17 5 15 4 0	4 12 8 0 2	14 15 6 11 5	15 0 13 10 3	9 16 6 15 18	11 8 1 7 5
16	15 14 0 13 3	19 4 12 3 6	6 15 9 10 12	8 7 0 17 15	12 18 17 3 5	7 14 0 2 3	12 9 14 13 7	13 1 11 19 4	18 3 4 0 1	13 19 3 7 4
	4 17 2 11 16	14 11 13 18 17	4 19 14 7 5	10 13 1 2 6	15 19 14 7 16	1 11 9 5 16	2 1 11 15 0	2 16 5 9 3	8 2 14 17 7	16 10 9 14 15
	10 18 12 6 8	0 7 10 1 8	17 3 16 2 18	19 4 9 12 11	8 4 13 6 11	17 13 10 12 18	16 3 19 17 4	15 7 10 8 18	15 13 6 19 12	6 2 11 8 18
	7 9 1 19 5	9 2 16 5 15	13 8 1 0 11	3 18 16 5 14	1 2 9 0 10	4 19 8 6 15	18 5 8 6 10	0 17 6 12 14	10 16 9 11 5	1 0 5 17 12
17	4 7 18 3 8	1 15 16 18 13	2 7 12 4 0	14 12 15 17 0	16 17 6 5 11	7 2 4 1 9	16 8 14 6 19	2 16 9 6 12	4 16 2 7 18	19 3 6 2 16
	13 15 5 6 0	2 8 6 10 3	9 15 19 13 5	8 18 4 2 16	10 4 15 7 13	15 19 11 16 14	2 5 11 7 0	8 15 10 14 18	0 17 15 10 6	14 15 12 5 4
	10 2 9 17 12	17 5 19 4 0	18 16 10 8 6	13 1 10 7 3	14 1 3 2 0	0 10 18 17 13	15 17 4 3 9	5 13 19 1 4	1 12 5 14 19	9 7 11 10 8
	11 14 1 19 16	11 9 7 14 12	3 11 17 1 14	9 19 6 5 11	18 19 8 9 12	5 8 12 6 3	1 12 10 18 13	7 3 0 11 17	3 13 9 8 11	18 13 17 1 0
18	7 10 16 12 1	19 0 16 14 8	14 11 12 2 5	4 0 15 5 9	11 18 17 4 2	5 14 11 16 1	5 3 16 2 0	16 7 6 8 1	0 13 4 11 14	14 9 16 15 1
	18 8 14 3 19	2 9 18 10 4	15 1 0 6 18	7 18 13 17 1	15 0 12 5 8	18 17 0 10 15	1 13 10 19 7	4 15 14 17 10	15 9 6 8 1	13 7 8 11 17
	13 17 9 5 15	5 6 13 1 11	8 16 13 10 3	19 10 14 11 2	14 7 19 1 10	2 13 6 8 9	18 14 12 17 6	9 12 19 0 11	3 17 19 10 2	5 18 3 10 12
	2 6 4 11 0	3 12 7 17 15	19 4 7 9 17	12 8 3 16 6	13 16 6 9 3	4 12 7 19 3	11 9 8 15 4	18 2 13 3 5	5 18 12 7 16	6 4 0 2 19
19	4 1 10 18 13	17 15 9 19 4	7 4 19 12 16	6 7 9 13 3	4 2 19 3 0	12 7 6 19 0	12 14 7 19 16	8 12 10 3 19	5 6 3 16 10	1 2 6 9 5
	0 8 14 16 12	7 18 12 3 14	5 0 9 3 10	0 12 8 17 18	9 13 7 14 5	2 14 8 16 18	4 8 18 10 2	11 4 6 5 13	19 1 4 12 13	17 19 15 14 13
	9 15 7 3 17	1 8 13 2 5	2 14 8 17 11	16 1 2 4 5	15 12 17 8 6	5 13 11 17 4	13 3 0 5 6	2 16 9 18 0	17 2 0 15 9	10 18 3 12 7
	2 6 5 19 11	0 10 6 16 11	18 13 15 1 6	15 10 19 14 11	10 1 18 16 11	9 15 10 3 1	17 11 1 9 15	1 15 17 7 14	18 8 7 11 14	0 8 4 16 11

Table XXXIV. Constants, Weights and Measures, Etc.

Mathematical Constants

$\pi = 3 \cdot 14159\ 26535\ 89793\ 23846$	$e = 2 \cdot 71828\ 18284\ 59045\ 23536$	$\gamma = 0 \cdot 57721\ 56649\ 01532\ 86061$
$\log_{10}\pi = 0 \cdot 49714\ 98726\ 94133\ 85435$	$\log_{10}e = 0 \cdot 43429\ 44819\ 03251\ 82765$	$e^{\gamma} = 1 \cdot 78107\ 24179\ 90197\ 98522$
$1/\sqrt{2\pi} = 0 \cdot 39894\ 22804\ 01432\ 67794$	$\log_e 10 = 2 \cdot 30258\ 50929\ 94045\ 68402$	$e^{-\gamma} = 0 \cdot 56145\ 94835\ 66885\ 16983$

Physical Constants

Velocity of light $= 3 \cdot 00 \times 10^{10}$ cm./sec. $=$ 186,000 miles/sec. The first number is the ratio of electromagnetic to electrostatic unit of charge.

Ionic (electronic) charge $(e) = 4 \cdot 80 \times 10^{-10}$ E.S.U. $= 1 \cdot 60 \times 10^{-20}$ E.M.U. Mass of electron $(m_0) = 9 \cdot 11 \times 10^{-28}$ g.

Planck's constant $(h) = 6 \cdot 55 \times 10^{-27}$ erg secs.

Energy of gas molecule per 1° C. absolute $= 2 \cdot 07 \times 10^{-16}$ ergs.

Number of molecules in 1 gm. molecule of gas $= 6 \cdot 02 \times 10^{23}$.

Electrical charge of 1 gm. equivalent in electrolysis (Faraday) $= 9 \cdot 65 \times 10^{3}$ E.M.U.

Gas constant $(R) = 8 \cdot 315 \times 10^{7}$ c.g.s. ($pv = RT \times$mass in gram molecules).

Constant of gravitation $(G) = 6 \cdot 67 \times 10^{-8}$ c.g.s. ($f = Gmm'/r^2$).

Acceleration due to gravity (g). Teddington (N. 51° 28·1′): 981·181 cm./sec.^{2} = 32·191 ft./sec.^{2}; Washington (38° 56·5′): 980·082; whence at lat. λ, height H m.: $980 \cdot 618 - 2 \cdot 586 \cos 2\lambda + \cdot 003 \cos 4\lambda, - \cdot 0003086\, H$.

The earth. Equatorial radius = 3963·3 miles, polar radius = 3950·0 miles. 1° of Latitude = 68·70 mls. (equator), 69·41 mls. (pole). Mean density = 345 lb./ft.^{3} = 5·517 g./cm.^{3}

Distance of sun $= 93 \times 10^{6}$ miles. Distance of moon = 239,000 miles.

Solar radiation (Washington, D.C.) = ·035 cal./cm.^{2}/sec.

Velocity of sound (6° C.) $= 3 \cdot 35 \times 10^{4}$ cm./sec. = 1100 ft./sec. Audible sound: Lower limit, 30 vibr./sec.; upper limit, 24,000-41,000 vibr./sec. Middle C of piano, 261 vibr./sec. Wave-lengths of visible light: 4 to 7×10^{-5} cm.

Temperature Equivalents												
Centigrade	−40	−30	−20	−10	0	10	20	30	40	50	**100**
Fahrenheit	−40	−22	−4	14	**32**	50	68	86	104	122	**212**
Réaumur	−32	−24	−16	−8	0	8	16	24	32	40	**80**

1° F. $= \frac{5}{9}$° C. 1° R. $= \frac{5}{4}$° C. Absolute zero of temperature = −273·1° C.

Mechanical and Electrical Units

Force. g dynes = 1 gm. weight, g poundals = 1 lb. weight, where g is acceleration due to gravity measured in appropriate units (cm./sec.^{2} or ft./sec.^{2}). 1 poundal $= 1 \cdot 3825 \times 10^{4}$ dynes.

Pressure. 1 atmosphere = 760 mm. (29·921 ins.) of mercury (0° C., Lat. 45°) = 33·899 ft. of water (4° C.) = 14·696 lb./in.^{2} $= 1 \cdot 01325 \times 10^{6}$ dynes/cm.^{2} 1 bar $= 10^{6}$ dynes/cm.^{2} = ·98692 atmos. 1 lb./in.^{2} = 70·307 (1/·014223) gm./cm.^{2}

Work, Power, etc. 1 erg = 1 dyne cm. 1 joule $= 10^{7}$ ergs = 1 watt second. 1 foot pound $= 1 \cdot 356 \times 10^{7}$ ergs. 1 horse power (H.P.) = 550 ft. lb. per sec. = 746 watts (continental H.P. = 736 watts). 1 calorie = heat to raise 1 gm. of water 1° C. at a specified temperature. 1 15° C. calorie = 1 mean calorie (0° C.-100° C.) $= 4 \cdot 185 \times 10^{7}$ ergs. 1 British Thermal Unit (1 lb. water × 1° F.) = 252·0 calories $= 2 \cdot 930 \times 10^{-4}$ kilowatt hours $= 3 \cdot 929 \times 10^{-4}$ horse power hours $= 10^{-5}$ gas therms. 1 Board of Trade Unit = 1 kilowatt hour.

Electrical. Relations between engineering units and electromagnetic units. Charge: coulomb $= 10^{-1}$ E.M.U. Current: ampere $= 10^{-1}$ E.M.U. Potential: volt $= 10^{8}$ E.M.U. Resistance: ohm $= 10^{9}$ E.M.U. Capacity: farad $= 10^{-9}$ E.M.U. Inductance: henry $= 10^{9}$ E.M.U. Flux: maxwell $= 10^{0}$ E.M.U. Magnetic force: gauss $= 10^{0}$ E.M.U.

Supposed Duration of Geologic Periods (E. W. Barnes)

Era	Periods	Duration Million Years.
Tertiary or Cainozoic era; Age of Mammals	Pleistocene, Pliocene, Miocene, Oligocene, Eocene	60
Secondary or Mesozoic era; Age of reptiles	Cretaceous, Jurassic, Triassic	85
Newer Palæozoic era	Permian, Carboniferous, Devonian	175
Older Palæozoic era	Silurian, Ordovician, Cambrian	250
Archæozoic era	. . . . .	300
Eozoic era	. . . . .	500

Possible age of planet 2000-4000 million years.

The Elements and their Atomic Weights

H	Hydrogen	1·0080	Fe	Iron	55·85	Sb	Antimony	121·76	Re	Rhenium	186·31
He	Helium	4·003	Co	Cobalt	58·94	Te	Tellurium	127·61	Os	Osmium	190·2
Li	Lithium	6·940	Ni	Nickel	58·69	I	Iodine	126·91	Ir	Iridium	192·2
Be	Beryllium	9·013	Cu	Copper	63·54	Xe	Xenon	131·3	Pt	Platinum	195·23
B	Boron	10·82	Zn	Zinc	65·38	Cs	Caesium	132·91	Au	Gold	197·0
C	Carbon	12·011	Ga	Gallium	69·72	Ba	Barium	137·36	Hg	Mercury	200·61
N	Nitrogen	14·008	Ge	Germanium	72·60	La	Lanthanum	138·92	Tl	Thallium	204·39
O	Oxygen	16·000	As	Arsenic	74·91	Ce	Cerium	140·13	Pb	Lead	207·21
F	Fluorine	19·00	Se	Selenium	78·96	Pr	Praseodymium	140·92	Bi	Bismuth	209·00
Ne	Neon	20·183	Br	Bromine	79·916	Nd	Neodymium	144·27	Po	Polonium	209
Na	Sodium	22·991	Kr	Krypton	83·80	Pm	Promethium	145	At	Astatine	210
Mg	Magnesium	24·32	Rb	Rubidium	85·48	Sm	Samarium	150·43	Em	Emanation	222
Al	Aluminium	26·98	Sr	Strontium	87·63	Eu	Europium	152·0	Fr	Francium	223
Si	Silicon	28·09	Y	Yttrium	88·92	Gd	Gadolinium	156·9	Ra	Radium	226·05
P	Phosphorus	30·975	Zr	Zirconium	91·22	Tb	Terbium	158·93	Ac	Actinium	227
S	Sulphur	32·066	Nb	Niobium	92·91	Dy	Dysprosium	162·46	Th	Thoriumn	232·05
Cl	Chlorine	35·457	Mo	Molybdenum	95·95	Ho	Holmium	164·94	Pa	Protactium	231
A	Argon	39·944	Tc	Technetium	99	Er	Erbium	167·2	U	Uraniuim	238·07
K	Potassium	39·100	Ru	Ruthenium	101·1	Tm	Thulium	168·94	Np	Neptunium	237
Ca	Calcium	40·08	Rh	Rhodium	102·91	Yb	Ytterbium	173·04	Pu	Plutonium	239
Sc	Scandium	44·96	Pd	Palladium	106·7	Lu	Lutetium	174·99	Am	Americium	243
Ti	Titanium	47·90	Ag	Silver	107·880	Hf	Hafnium	178·6	Cm	Curium	245
V	Vanadium	50·95	Cd	Cadmium	112·41	Ta	Tantalum	180·95	Bk	Berkelium	245
Cr	Chromium	52·01	In	Indium	114·76	W	Tungsten	183·92	Cf	Californium	246
Mn	Manganese	54·94	Sn	Tin	118·70		(Wolfram)				

Densities

Aluminium	2·70	Iron	7·86	Platinum	21·50	Coal	1·2-1·8	White pine	·4 - ·5	Glass	2·4 -4·5
Copper	8·93	Lead	11·37	Silver	10·5	Granite	2·5-3	Beech, oak	·7 - ·9	Sea water	1·01-1·05
Diamond	3·52	Magnesium	1·74	Tin	7·29	Quartz	2·66	Ebony	1·1 -1·3	Loose snow	·12 approx.
Gold	19·32	Mercury	13·56	Zinc	7·1	Slate	2·5-2·7	Cork	·22 - ·26	Soil	1·6 approx.

Unit density = 1 g. per cu. cm. = ·036128 lb. per cu. in.

Equation of Time

Jan. 1	+ 3m. 11s.	Mar. 1	+12m. 34s.	May 1	−2m. 57s.	July 1	+3m. 32s.	Sept. 16	− 5m. 6s.	Nov. 16	−15m. 10s.
,, 16	+ 9 33	,, 16	+ 8 51	,, 14	−3 49*	,, 26	+6 18*	Oct. 1	−10 16	Dec. 1	−10 56
Feb. 1	+13 37	Apr. 1	+ 4 1	June 1	−2 27	Aug. 16	+4 11	,, 16	−14 20	,, 12	− 6 15
,, 12	+14 25*	,, 16	0 0	,, 15	0 0	Sept. 1	0 0	Nov. 3	−16 21*	,, 25	0 0

When positive add the equation of time to apparent solar time to obtain mean solar time. The values given are mean values. Maxima and minima are marked with asterisks. 1 sidereal day = 23 h. 56 m. 4·09 s.

Saturated Vapour Pressure of Water (Millibars)

°C.	−50	−40	−30	−20	−10	−5	−2
Ice	·039	·129	·381	1·035	2·600	4·017	5·173
Water	—	—	—	1·270	2·865	4·217	5·274

°C.	0	1	2	3	4	5	6	7	8	9
0	6·105	6·567	7·058	7·579	8·134	8·723	9·350	10·02	10·73	11·48
10	12·28	13·12	14·02	14·97	15·98	17·05	18·18	19·37	20·63	21·97
20	23·38	24·86	26·43	28·09	29·83	31·67	33·61	35·65	37·80	40·05
30	42·43	44·92	47·55	50·31	53·19	56·23	59·41	62·75	66·25	69·92

°C.	
40	73·76
50	123·3
60	199·2
70	311·6
80	473·4
90	701·0
100	1013·2
200	15544
300	85903

Critical temperature : 374°. Critical pressure : 220·6 bars. Interpolate logarithmically.

*Composition of Fertilizers (Commercial Grades **on** the English Market)*

	%N		$\%P_2O_5$			$\%K_2O$
Sulphate of ammonia . .	20·6	Superphosphate .	16-18 } water soluble		Sulphate of potash .	48
Nitrate of soda . . .	16	Triple superphosphate	48 } water soluble		Muriate of potash .	50-60
Nitrochalk . . .	15·5	Basic slag . .	10-18 } total		Potash manure . .	40
		Rock phosphate .	26-28 } total			

Some of these fertilizers are mixtures of two or more constituents in naturally occurring or conventional proportions. 1952 values are given.

English Weights and Measures

Length

1 foot = 12 ins. 1 yard = 3 ft.
1 mile = 1760 yds. = 5280 ft.
(60 m.p.h. = 88 ft. per sec.)

1 surveyor's chain = 22 yds.
(1 engineer's chain = 100 ft.)
1 link = $\frac{1}{100}$ chain = 7·92 ins.
(1 rod, pole, or perch = $\frac{1}{4}$ chain.)
1 furlong = 10 chains = $\frac{1}{8}$ mile.

1 fathom = 6 ft. 1 cable = 100 fath.
1 nautical mile = 6080 ft.
= 1 minute of arc of equator.
(1 knot = 1 nautical mile per hour.)

Archaic (approximate):
1 hand = 4 ins. 1 cubit = 18 ins.
1 ell = 45 ins. 1 league = 3 miles.

Area

1 acre = 10 sq. chains=4840 sq. yds.
1 sq. mile = 640 acres.
(1 rod = $\frac{1}{160}$ ac., 1 rood = $\frac{1}{4}$ ac.)

Weight

Avoirdupois (General System)

1 ounce (oz.) = 16 drams (dm.).
1 pound (lb.) = 16 oz.
1 stone = 14 lb. 1 quarter = 28 lb.
1 hundredweight (cwt.) = 112 lb.
= 4 quarters = 8 stones.
1 ton = 20 cwt. = 2240 lb.

Apothecary (Drugs)	*Troy (Precious Metals)*
1 drachm=60 grs.	1 pennyweight
(1 scruple=20 grs.)	(dwt.) = 24 grs.

1 ounce = 20 dwt. = 8 drachms.
1 pound = 12 oz. = 5760 grs.
(1 lb. avoirdupois = 7000 grs.)

Money

1 shilling (s.) = 12 pence (d.).
1 pound (£) = 20s.
Also farthing ($\frac{1}{4}$d.), florin (2s.), half-crown (2s. 6d.) and guinea (21s.).

Volume

(Liquid)

1 gallon = 4 quarts = 8 pints
= 32 gills = ·16054 (1/6·2290) cu. ft.
1 gallon water (62° F.) weighs 10 lb. av. (in air against brass weights).

1 fluid drachm (ʒ) = 60 minims (♏).
1 fluid oz. (℥) = 8 dr. = $\frac{1}{20}$ pint.
1 fluid oz. water (62° F.) weighs 1 oz. av.

(Dry)

1 peck = 2 gals. 1 bushel = 8 gals.
1 sack = 4 bush. 1 quarter = 8 bush.
(1 firkin = 9 gals. 1 barrel = 36 gals.)

1 register ton = 100 cu. ft.
1 rod = 1000 cu. ft.

Weights and Measures of the United States of America

Weights and measures are the same as the English systems, except for small differences in standards, and the following exceptions:—

1 hundredweight (short) = 100 lb., and 1 ton (short) = 20 cwt. = 2000 lb. The English (long) ton, 2240 lb., is also used.

Dry measures of volume are 3 per cent. less than the corresponding English measures, being based on the Winchester bushel instead of the Imperial bushel. Fluid measures of volume are $\frac{5}{6}$ of the corresponding English measures. (Hence in the U.S. system 1 bushel = 9·3092 fluid gallons.) Also 1 U.S. fluid ounce = $\frac{1}{16}$ U.S. pint, so that a fluid ounce of water weighs 1·0408 oz. av. (See also conversion factors.)

A township = 36 sq. miles. A billion (French *Milliard*) = 10^9, a trillion = 10^{12}. (English billion = 10^{12}, trillion = 10^{24}.)

The Metric System

The metre is (very nearly) 10^{-7} of the distance from the pole to the equator.

The gram is (very nearly) the mass of 1 cubic cm. of water at maximum density (3·98° C.). (Actual mass of 1 c.c. of water = ·999973 g.)

The litre is (very nearly) 1000 c.c. (1 litre = the volume of 1000 g. of water at 3·98° C. = 1000·027 c.c.)

Also 1 are = 100 sq. metres, so that 1 hectare = 10,000 sq. m. = $\frac{1}{100}$ sq. km.; 1 metric quintal (doppel zentner, U.S.S.R. centner) = 100 kg.; 1 metric ton = 1000 kg.; 1 Continental pound = $\frac{1}{2}$ kg.; 1 stere = 1 cu. m. 1 micron (μ) = 10^{-6} m. 1 angstrom unit (Å) = 10^{-10} m.

Prefixes: micro-, 10^{-6}; mille-, 10^{-3}; centi-, 10^{-2}; deci-, 10^{-1}; Deka-, 10; Hecto-, 10^2; Kilo-, 10^3; Myria-, 10^4: Mega-, 10^6.

Conversion Factors

Basic Factors

1 yard (G.B.) = ·914,399 metres.
1 yard (U.S.A.) = ·914,402 metres.
1 pound = 453·5924 grams.
1 gallon (G.B.) = 4·545,963 litres.
1 gallon (U.S.A.) = 3·785,332 litres.
1 bushel (U.S.A.) = 35·238,33 litres.

Length

1 inch = 2·5400 (1/·39370) cm.
1 foot = ·30480 (1/3·2808) m.
1 yard = ·91440 (1/1·0936) m.
1 mile = 1·6093 (1/·62137) km.

Weight

1 grain = ·064799 (1/15·432) g.
1 ounce = 28·350 (1/·035274) g.
1 pound = ·45359 (1/2·2046) kg.
1 cwt. = ·50802 (1/1·9684) m. quintals.
1 short cwt. = ·45359 (1/2·2046) m.q.
1 ton = 1·0160 (1/·98421) metric tons.
1 short ton = ·90718 (1/1·1023) m.t.

Capacity (Great Britain)

1 pint = ·56825 (1/1·7598) litres.
1 gallon = 4·5460 (1/·21998) litres.
1 bushel = 36·368 (1/·027497) litres.

Capacity (U.S.A.)

Fluid: 1 pint = ·47317 (1/2·1134) lit.
Fluid: 1 gall. = 3·7853 (1/·26418) lit.
Dry: 1 bush. = 35·238 (1/·028378) lit.

1 British pint, gallon, etc. = 1·20094 (1/·83268) U.S. fluid pints, gallons, etc.
1 British bushel, etc. = 1·03205 (1/·96895) U.S. bushels, etc. (dry).

Volume

1 cu. inch = 16·387 (1/·061024) c.c.
1 cu. foot = ·028317 (1/35·315) cu. m.
1 cu. yard= ·76455 (1/1·3080) cu.m.

Area, etc.

1 sq. inch = 6·4516 (1/·15500) sq. cm.
1 sq. foot = ·092903 (1/10·764) sq. m.
1 sq. yard = ·83613 (1/1·1960) sq. m.
1 sq. mile = 2·5900 (1/·38610) sq. km.

1 acre = ·40468 (1/2·4711) hectares.
1 lb. per acre = 1·1209 (1/·89218) kg. per ha.
1 cwt. per acre = 1·2554 (1/·79659) m.q. per ha.

1 ton per acre = 2·5107 (1/·39829) m. tons per ha.
1 bushel per acre = 89·867 (1/·011128) litres per ha. (U.S.A.) = 87·076 (1/·011484) litres per ha.

Bushel weights (average): wheat, 63 lb.; oats, 42 lb.; barley, 56 lb.; maize, 60 lb.; beans, peas, 64 lb.

PRINTED IN GREAT BRITAIN BY
OLIVER AND BOYD LTD.
EDINBURGH